COOPERATIVE CONTROL:
MODELS, APPLICATIONS AND ALGORITHMS

T0191683

Cooperative Systems

Volume 1

Cooperative Control: Models, Applications and Algorithms

Edited by

Sergiy Butenko

University of Florida,
Gainesville, Florida, U.S.A.

Robert Murphey

Air Force Research Laboratory,
Eglin AFB, Florida, U.S.A.

and

Panos M. Pardalos

University of Florida,
Gainesville, Florida, U.S.A.

KLUWER ACADEMIC PUBLISHERS
DORDRECHT / BOSTON / LONDON

A C.I.P. Catalogue record for this book is available from the Library of Congress.

ISBN 978-1-4419-5241-7

Published by Kluwer Academic Publishers,
P.O. Box 17, 3300 AA Dordrecht, The Netherlands.

Sold and distributed in North, Central and South America
by Kluwer Academic Publishers,
101 Philip Drive, Norwell, MA 02061, U.S.A.

In all other countries, sold and distributed
by Kluwer Academic Publishers,
P.O. Box 322, 3300 AH Dordrecht, The Netherlands.

Printed on acid-free paper

"But the bravest are surely those who have the clearest vision of what is before them, glory and danger alike, and yet notwithstanding go out to meet it."

-Thucydides,
The History of the Peloponnesian War [431-413 B. C.]

Contents

Preface

A cooperative system is a collection of dynamical objects which communicate and cooperate in order to achieve a common or shared objective. Examples of cooperative systems are found in many human activities and other biological systems where the processes of cooperation are often taken for granted. Indeed, our knowledge of how cooperation manifests itself is sparse. Consequently, it is no surprise that the active control of a collection of machines is not well understood. Nonetheless, the potential for cooperating machines appears huge, with a wide range of applications in robotics, medicine, commercial transportation, emergency search-and-rescue operations, and national defense.

The cooperation of entities in a system is achieved through communication; either explicitly by message passing, or implicitly via observation of another entities' state. As in natural systems, cooperation may assume a hierarchical form and the control processes may be distributed or decentralized. Due to the dynamic nature of individuals and the interaction between them, the problems associated with cooperative systems typically involve many uncertainties. Moreover, in many cases cooperative systems are required to operate in a noisy or hazardous environment, which creates special challenges for designing the control process.

During the last decades, considerable progress has been observed in all aspects regarding the study of cooperative systems including modeling of cooperative systems, resource allocation, discrete event driven dynamical control, continuous and hybrid dynamical control, and theory on the interaction of information, control, and hierarchy. Solution methods have been proposed using control and optimization approaches, emergent rule based techniques, game theoretic and team theoretic approaches. Measures of performance have been suggested that include the effects of hierarchies and information structures on solutions, performance bounds, concepts of convergence and stability, and problem complexity.

These and other topics were discussed at the Second Annual Conference on Cooperative Control and Optimization, November 2001 in Gainesville, Florida.

Refereed papers written by selected conference participants of the conference are gathered in this volume, which present problem models, theoretical results, and algorithms for various aspects of cooperative control.

We would like to thank the authors of the papers, the Air Force Research Laboratory and the University of Florida College of Engineering for financial support, the anonymous referees, and Kluwer Academic Publishers for making the conference successful and the publication of this volume possible.

EDITORS

Chapter 1

N-OCULAR VOLUME HOLOGRAPHIC IMAGING

George Barbastathis
Mechanical Engineering Department
Massachusetts Institute of Technology
gbarb@mit.edu

Arnab Sinha
Mechanical Engineering Department
Massachusetts Institute of Technology
arnab@mit.edu

Abstract   We describe a novel class of distributed imaging systems based on several volume holographic imaging elements collaborating to produce high resolution images of surrounding targets. The principles and theory of imaging based on volume diffraction are presented along with results from proof-of-principle experimental system.

Keywords: Volume holography, three-dimensional imaging, triangulation

Introduction

Cooperative systems are becoming increasingly significant in industrial, biomedical, and military applications. Some advantages of cooperative systems are cheaper cost for individual components, reconfigurability, and graceful degradation if some of these components fail. In this chapter, we are interested in cooperative imaging systems, where image information is gathered by a number of Sensing Agents (SA's). In particular, we will describe cooperative vision systems where each SA carries a novel class of imaging element: a volume hologram.

1

S. Butenko et al. (eds.), Cooperative Control: Models, Applications and Algorithms, 1-21.
© 2003 *Kluwer Academic Publishers.*

Volume Holographic Imaging (VHI) was invented in 1999 [1, 2]. It refers to a class of imaging techniques which use a volume hologram in at least one location within the optical path. The volume hologram acts as a "smart lens," which processes the optical field to extract spatial information in three dimensions (lateral as well as longitudinal) and spectral information. Image formation in VHI systems is based on the matched filtering properties of volume holograms: the hologram can "see" a specific subset of the object only and reject the remainder. This corresponds to a "virtual slicing" operation on the object. The slice shape depends on the way the hologram was recorded. Since many objects of interest are four dimensional (three spatial plus one spectral dimension) and detectors can capture at most a two-dimensional grid of image data, an array of similar holograms (each one of them tuned to a different 2D slice) can span the entire 4D object space simultaneously. [3]

In this chapter we will primarily be describing telescopic VHI systems. These are designed to acquire 4D information about moderately remote objects, at distances of $\sim$ 1km away from the imaging system. The cooperative aspect of the imaging system turns out to be beneficial in improving the system resolution: whereas each VHI sensor alone would perform poorly in the longitudinal direction (*i.e.* ranging along the direction of the optical axis), the collective system resolves the increased longitudinal ambiguity by virtue of the different orientations of the cooperating sensors.

The structure of the chapter is as follows: in Section 1 we make some general remarks about hybrid imaging systems that contain analog (optical) and digital (electronic or software) processing elements. Section 2 describes the basic physics of VHI using a single SA qualitatively. A more quantitative description is given in Section 3. Collective VHI is described in Section 4. Conclusions and ongoing work are described in Section 5.

1.   Hybrid Imaging

Until the development of CCD cameras in the early to mid-70's, the almost universal purpose of imaging systems had been the production of geometrically exact projections of the 3D world onto 2D sensor arrays, permanent (such as photographic film) or instantaneous (such as an observer's retina). Remarkable exceptions such as computed tomography and radio astronomy, known earlier, had limited or no use in the domain of optical frequencies. However, as digital cameras become increasingly available, ample digital computing power to process the images, and digital networks to distribute them produced a revolutionary shift in optical imaging design. The necessity to produce "images" on the detector plane that are physically analogous to the imaged objects became secondary, as long as the detector captured sufficient information to allow the recovery of the objects after appropriate processing operations. This paradigm

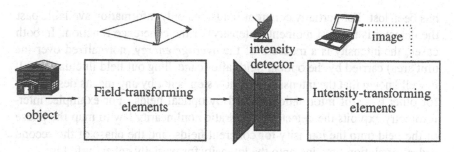

Figure 1.1. General classification of elements composing an imaging system.

shift had two apparent additional benefits. Firstly, it simplified the design of optical elements, since several geometric imaging imperfections (*e.g.*, defocus) could be corrected digitally, to some extent. Secondly, and most importantly, digital processing allows the user to recover *more* object information from the sensor data. For example, by processing several 2D intensity patterns one may recover surface topography data about opaque objects (2 + 1/2D images) or volumetric structure data (3D images) and even spectral data in addition (4D images) about semi-transparent objects. Examples of this principle in action are triangulation-based imaging systems, the confocal microscope, and coherence imaging systems.

A generic imaging system is shown in Figure 1.1. Field-transforming elements are the first to receive the radiation from the object. The field produced by this stage is converted to electrical signal by the intensity detector. The electrical signal is then fed to the intensity-transforming stages, which produce the final image.

As the respective names suggest, there is a fundamental difference in the nature of transformations that can be effected by field- and intensity-transforming imaging elements. Field-transforming elements operate directly on the electromagnetic field that composes the optical wave. For coherent fields, this property allows field-transforming elements to modify the amplitude as well as the phase of the (deterministic) input wave. For example, a spherical lens imposes a quadratic phase modulation in the paraxial approximation. In the case of partially coherent fields, field-transforming elements operate rather on the random process which represents one realization of the optical field (different at every experiment); as a result, these elements are capable of modifying all the moments of the random field process.

At optical frequencies, the detection step (*i.e.*, the conversion of the electromagnetic field to an electrical signal, such as a voltage) is fundamentally limited to return the intensity of the field. For coherent fields, the intensity is proportional to the modulus of the complex amplitude; *i.e.*, phase information has been lost. For partially coherent fields, the only information available past the detector is the first moment (intensity) of the coherence function. In both cases, the intensity is a measure of the *average energy* (normalized over the unit area) carried by the optical field, after cancelling out field fluctuations. It is well known that the intensity detection step is not by any means destructive for other types of information carried by optical fields. For example, interferometry exploits the detectors' quadratic nonlinearity law to map the phase of the field onto the intensity for coherent fields, and the phase of the second order correlation function onto the intensity for partially coherent fields.

Therefore, the designer has considerable freedom in selecting the location of the detection step in an optical imaging system and, subsequently, the division of labor between the field- and intensity-transforming parts of the system. In practice, some configurations work better than others, depending on the application. However, the analog part of the system sets an upper bound on the amount of information that can be retrieved from the system. We have shown [4] that volume holographic elements, when properly designed, can increase that upper bound.

2.   Principles of Volume Holographic Imaging

Volume holography was introduced in a seminal paper by van Heerden [5] as extension of well-known results about Bragg diffraction to the domain of optical wavelengths. Since then, volume holograms have been used in data storage [5, 6, 7, 8], optical interconnects and artificial neural networks [9, 10], and communications [11, 12, 13, 14, 15]. To date commercial applications of volume holograms are for spectral filtering [16] and 3D storage devices. In this section we describe the function of volume holograms as imaging elements.

Figure 1.2 is a generic VHI system. The object is either illuminated by a light source (*e.g.*, sunlight or a pump laser) as shown in the figure, or it may be self-luminous. Light scattered or emitted by the object is first transformed by an objective lens and then illuminates the volume hologram. The role of the objective is to form an intermediate image which serves as input to the volume hologram. The volume hologram itself is modeled as a three-dimensional (3D) modulation $\Delta\epsilon(\mathbf{r})$ of the dielectric index within a finite region of space. The light entering the hologram is diffracted by $\Delta\epsilon(\mathbf{r})$ with efficiency η, defined as

$$\eta = \frac{\text{Power diffracted by the volume hologram}}{\text{Power incident to the volume hologram}}. \tag{1}$$

We assume that diffraction occurs in the Bragg regime. The diffracted field is Fourier-transformed by the collector lens, and the result is sampled and measured by an intensity detector array (such as a CCD or CMOS camera).

Intuitively, we expect that a fraction of the illumination incident upon the volume hologram is Bragg-matched and is diffracted towards the Fourier-trans-

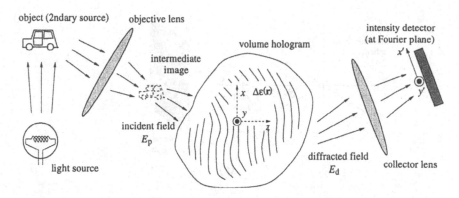

Figure 1.2. Volume Holographic Imaging (VHI) system.

forming lens. The remainder of the incident illumination is Bragg-mismatched, and as result is transmitted through the volume hologram undiffracted. Therefore, the volume hologram acts as a filter which admits the Bragg-matched portion of the object and rejects the rest. When appropriately designed, this "Bragg imaging filter" can exhibit very rich behavior, spanning the three spatial dimensions and the spectral dimension of the object.

To keep the discussion simple, we consider the specific case of a transmission geometry volume hologram, described in Figure 1.3. The volume hologram is created by interfering a spherical wave and a plane wave, as shown in Figure 1.3(a). The spherical wave originates at the coordinate origin. The plane wave is off-axis and its wave-vector lies on the xz plane. As in most common holographic systems, the two beams are assumed to be at the same wavelength λ, and mutually coherent. The volume hologram results from exposure of a photosensitive material to the interference of these two beams.

First, assume that the object is a simple point source. The intermediate image is also approximately a point source, that we refer to as "probe," located somewhere in the vicinity of the reference point source. Assuming the wavelength of the probe source is the same as that of the reference and signal beams, volume diffraction theory shows that:

1. If the probe point source is displaced in the y direction relative to the reference point source, the image formed by the volume hologram is also displaced by a proportional amount. Most common imaging systems would be expected to operate this way.
2. If the probe point source is displaced in the x direction relative to the reference point source, the image disappears (*i.e.*, the detector plane remains dark).

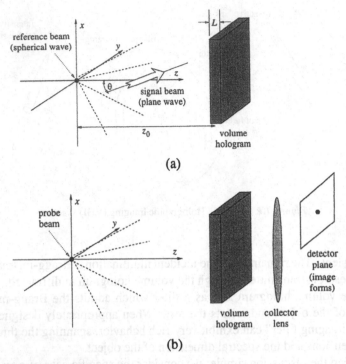

Figure 1.3. (a) Recoding of a transmission-geometry volume hologram with a spherical wave and an off-axis plane wave with its wave-vector on the xz plane. (b) Imaging of a probe point source that replicates the location and wavelength of the reference point source using the volume hologram recorded in part (a).

3. If the probe point source is displaced in the z direction relative to the reference point source, a defocused and faint image is formed on the detector plane. "Faint" here means that the fraction of energy of the defocused probe transmitted to the detector plane is much smaller than the fraction that would have been transmitted if the probe had been at the origin.

Now consider an extended, monochromatic, spatially incoherent object and intermediate image, as in Figure 1.3(f). According to the above description, the volume hologram acts as a "Bragg slit" in this case: because of Bragg selectivity, the volume hologram transmits light originating from the vicinity of the y axis, and rejects light originating anywhere else. For the same reason, the volume hologram affords depth selectivity (like the pinhole of a confocal microscope). The width of the slit is determined by the recording geometry, and the thickness of the volume hologram. For example, in the transmission recording geometry of Figure 1.3(a), where the reference beam originates a

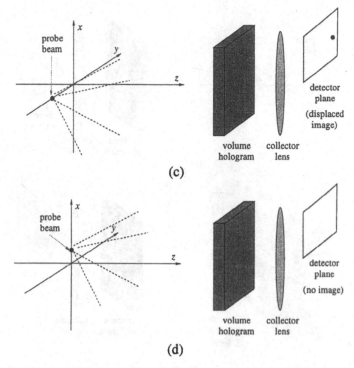

Figure 1.3. [continued] (c) Imaging of a probe point source at the same wavelength but displaced in the y direction relative to the reference point source. (d) Imaging of a probe point source at the same wavelength but displaced in the x direction relative to the reference point source.

distance z_0 away from the hologram, the plane wave propagates at angle θ with respect to the optical axis z (assuming $\theta \ll 1$ radian), and the hologram thickness is L (assuming $L \ll z_0$), the width of the slit is found to be

$$\Delta x \approx \frac{\lambda z_0}{L\theta}. \tag{2}$$

The imaging function becomes richer if the object is polychromatic. In addition to its Bragg slit function, the volume hologram exhibits then dispersive behavior, like all diffractive elements. In this particular case, dispersion causes the hologram to image simultaneously multiple Bragg slits, each at different color and parallel to the original slit at wavelength λ, but displaced along the z axis. Light from all these slits finds itself in focus at the detector plane, thus forming a "rainbow image" of an entire slice through the object, as shown in Figure 1.3(g).

To further exploit the capabilities of volume holograms, we recall that in general it is possible to "multiplex" (super-impose) several volume gratings

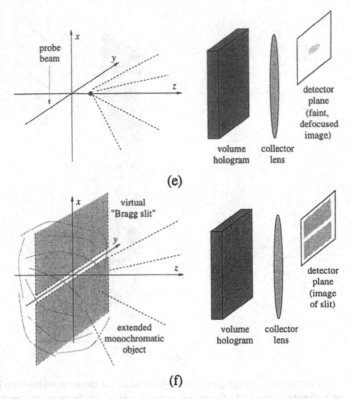

Figure 1.3. [continued] (e) Imaging of a probe point source at the same wavelength but displaced in the z direction relative to the reference point source. (f) "Bragg slitting:" Imaging of an extended monochromatic, spatially incoherent object using a volume hologram recorded as in (a).

within the same volume by successive exposures. In the imaging context, suppose that we multiplex several gratings similar to the grating described in Figure 1.3(a) but with spherical reference waves originating at different locations and plane signal waves at different orientations. When the multiplexed volume hologram is illuminated by an extended polychromatic source, each grating forms a separate image of a rainbow slice, as described earlier. By spacing appropriately the angles of propagation of the plane signal waves, we can ensure that the rainbow images are formed on non-overlapping areas on the detector plane, as shown in Figure 1.3(h). This device is now performing true four-dimensional (4D) imaging: it is separating the spatial and spectral components of the object illumination so that they can be measured independently by the detector array. Assuming the photon count is sufficiently high and the number of detector pixels is sufficient, this "spatio-spectral slicing" operation can be performed in real time, without need for mechanical scanning.

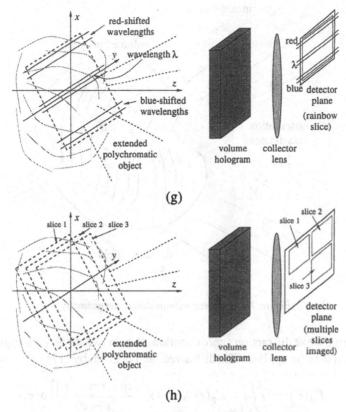

Figure 1.3. [continued] (g) Joining Bragg slits from different colors to form rainbow slices: Imaging of an extended polychromatic object using a volume hologram recorded as in (a). (h) Multiplex imaging of several slices using a volume hologram formed by multiple exposures.

3. Volume Diffraction Theory

A volume hologram is a spatial modulation of the refractive index of a dielectric material, expressed as a function $\Delta\varepsilon(\mathbf{r})$. The modulation is effective for values of the space coordinate $\mathbf{r}$ satisfying $\mathbf{r} \in V_{\mathcal{H}}$, where $V_{\mathcal{H}}$ is the volume occupied by the holographic material. When the hologram is illuminated by a probe field $E_p(\mathbf{r})$, as in Figure 1.4, the diffracted field $E_d(\mathbf{r}')$ is found as the solution to Maxwell's equations in an inhomogeneous medium, with modulated refractive index. The solution is simplified if we assume that the magnitude of the modulation is much smaller than the unmodulated refractive index ε_0:

$$|\Delta\varepsilon(\mathbf{r})| \ll \varepsilon_0, \qquad \mathbf{r} \in V_{\mathcal{H}}, \qquad (3)$$

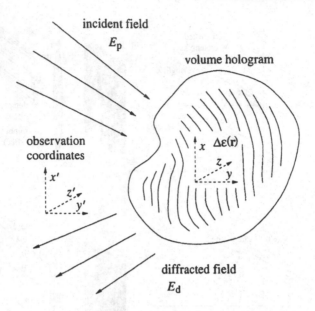

Figure 1.4. General volume diffraction geometry.

because the weak diffraction approximation (also known as "Born's approximation") can then be applied. The diffracted field is given by [17]

$$E_d(\mathbf{r}') = \iiint_{V_{\mathcal{H}}} E_p(\mathbf{r}) \Delta\varepsilon(\mathbf{r}) \times \frac{\exp\{ik\,|\mathbf{r} - \mathbf{r}'|\}}{|\mathbf{r} - \mathbf{r}'|} d^3\mathbf{r}, \qquad (4)$$

where $k = 2\pi/\lambda$ is the wavenumber, and the last term in the integrand is recognized as the scalar Green's function for free space. The derivation of (4) from Maxwell's equations is beyond the scope of this chapter.

Equation (4) has a simple interpretation: Assume that the volume grating is composed of infinitesimal scatterers, the strength of the scatterer located at $\mathbf{r} \in V_{\mathcal{H}}$ being $\Delta\varepsilon(\mathbf{r})$. Then the diffracted field is the coherent summation of the fields emitted by all the scatterers when they are excited by the incident field $E_p(\mathbf{r})$. Naturally, this picture omits higher-order scattering, *i.e.* fields generated when the field scattered from one infinitesimal scatterer reaches other infinitesimal scatterers. This omission, though, is consistent with the weak scattering approximation, which says that these higher order effects are even weaker, and, therefore, negligible.

Expression (4) is computationally efficient when spherical waves are involved in the recording of the hologram. For other types of fields, a representation of the diffracted field and the grating in wave-vector space works better [17], but is beyond the scope of this chapter. We will here limit our discussion to the "transmission-geometry" (Figure 1.5), where the reference

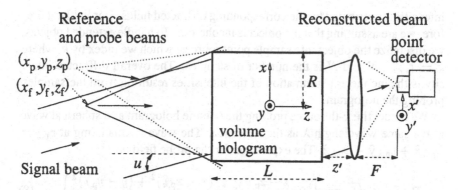

Figure 1.5. Schematic and notation for the transmission holographic recording geometry.

and signal beam are both incident on the front face of a cube-like recording medium. The probe beam is incident from the same direction as the reference, and the reconstruction appears as a continuation of the signal.

We assume that the holographic medium is disk-shaped with thickness L and radius R. The hologram is recorded by the interference of a plane wave signal beam and a spherical reference beam. The reference emanates from a point source at $\mathbf{r}_f = x_f \hat{\mathbf{x}} + y_f \hat{\mathbf{y}} + z_f \hat{\mathbf{z}}$. We express this wave in the paraxial approximation as

$$E_f(\mathbf{r}) = \exp\left\{ i2\pi \frac{z - z_f}{\lambda} + i\pi \frac{(x - x_f)^2 + (y - y_f)^2}{\lambda(z - z_f)} \right\}. \tag{5}$$

Note that here and in the sequel we neglect a term of the form $1/\lambda(z - z_f)$ because it varies with z much slower than the exponential term. The signal beam is a plane wave propagating at angle $u \ll 1$ with respect to the $\hat{\mathbf{z}}$-axis. In the paraxial approximation, it is expressed as

$$E_s(\mathbf{r}) = \exp\left\{ i2\pi \left(1 - \frac{u^2}{2} \right) \frac{z}{\lambda} + i2\pi u \frac{x}{\lambda} \right\}. \tag{6}$$

After recording is complete, the index modulation recorded in the hologram is

$$\Delta\epsilon(\mathbf{r}) \propto E_f^*(\mathbf{r}) E_s(\mathbf{r}), \tag{7}$$

where the "*" denotes complex conjugate. Note that the actual interference pattern is given by $|E_f + E_s|^2$, but out of the four resulting product terms only the one in (7) results in significant diffraction (the remaining three terms are Bragg-mismatched.)

We will calculate the intensity response of the volume hologram to individual point sources at arbitrary spatial coordinates. Then the intensity collected at any individual point in the detector plane is given simply by the sum of the intensities contributed by the corresponding diffracted fields. Implicitly, therefore, we are assuming that the object is incoherent. To handle extended objects, we discretize the object into sample point sources, which we index by k, where $k = 1, \ldots, N$ and N is the number of samples. The overall diffracted intensity is the incoherent summation of the intensities resulting from the samples probing the hologram.

We model the k-th source probing the volume hologram as a spherical wave at the same wavelength λ as the reference. The wave is emanating at $\mathbf{r}_{p,k} = x_{p,k}\hat{\mathbf{x}} + y_{p,k}\hat{\mathbf{y}} + z_{p,k}\hat{\mathbf{z}}$. The expression for the probe field is

$$E_p(\mathbf{r}) = \sqrt{I_k} \exp\left\{ i2\pi \frac{z - z_{p,k}}{\lambda_p} + i\pi \frac{(x - x_{p,k})^2 + (y - y_{p,k})^2}{\lambda_p(z - z_{p,k})} \right\}. \qquad (8)$$

where I_k represents the intensity emitted by the k-th source. The diffracted field at detector coordinates $\mathbf{r}'$ on the focal plane of the Fourier-transforming lens (see Figure 1.5) we use (4), substitute expressions (5), (6), (7) and (8) and perform a Fourier transformation. The details of the calculation are given in [17, pp. 38-42]. The result is

$$\tilde{E}_d(\mathbf{r}') = 2\pi R^2 \sqrt{\eta I_k} \int_{-L/2}^{L/2} \exp\left\{ i\pi C(z) \right\} \mathcal{L}\left(2\pi A(z)R^2, 2\pi B(z)R \right) dz,$$

$$(9)$$

where the coefficients $A(z)$, $B_x(z)$, $B_y(z)$, $C(z)$ are given by

$$A(z) = \frac{1}{\lambda(z - z_f)} - \frac{1}{\lambda(z - z_{p,k})}, \qquad (10)$$

$$B_x(z) = -\frac{x_{p,k}}{\lambda(z - z_{p,k})} + \frac{x_f}{\lambda(z - z_f)} - \frac{x'}{\lambda F} + \frac{u}{\lambda}, \qquad (11)$$

$$B_y(z) = -\frac{y_{p,k}}{\lambda(z - z_{p,k})} + \frac{y_f}{\lambda(z - z_f)} - \frac{y'}{\lambda F}, \qquad (12)$$

$$C(z) = \frac{x^2_{p,k} + y^2_{p,k}}{\lambda(z - z_{p,k})} - \frac{x_f^2 + y_f^2}{\lambda(z - z_f)} + \left(\frac{x'^2 + y'^2}{\lambda F^2} - \frac{u^2}{\lambda} \right) z. \qquad (13)$$

The function

$$\mathcal{L}(u, v) = \int_0^1 \exp\left\{ -\frac{i}{2}u\rho^2 \right\} J_0(v\rho)\rho d\rho \qquad (14)$$

is the same integral which occurs in the calculation of the three dimensional light distribution near the focus of a lens [18].

The intensity contributed by the probe source to the detector plane is

$$\tilde{I}_k(\mathbf{r}') = \left| \tilde{E}_d(\mathbf{r}') \right|^2. \qquad (15)$$

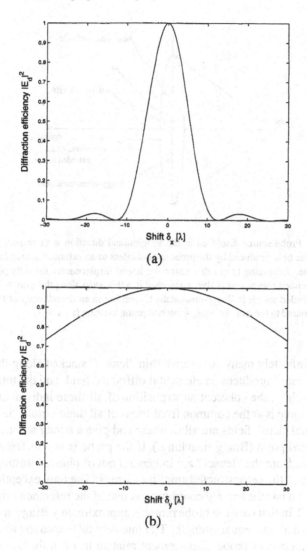

Figure 1.6. Dependence of the intensity diffracted towards the focal point for displacements along *(a)* x and *(b)* y directions. The calculation used $\mathbf{r}_f = (0, 0, -10^4 \lambda)$, $u = 0.2$, $R = 500\lambda$, $L = 4 \times 10^3 \lambda$.

The total intensity is the sum of all the contributing point sources, *i.e.*

$$\tilde{I}(\mathbf{r}') = \sum_{k=1}^{N} \tilde{I}_k(\mathbf{r}'). \tag{16}$$

The result (9) for the diffracted field has an interesting interpretation based on the nature of the function $\mathcal{L}(.,.)$. The volume hologram can be decom-

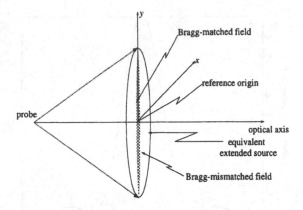

Figure 1.7. Probe source displaced in the longitudinal direction with respect to the reference plane. The field produced by the probe is equivalent to an extended source located at the reference plane. According to our discussion for lateral displacement, the only portion of the equivalent extended source that is Bragg-matched is a thin strip along the y-axis. The remainder of the extended source is Bragg-mismatched, resulting in an overall drop of the diffracted intensity compared to the exactly Bragg-matched probe location ($\mathbf{r}_{p,k} = \mathbf{r}_f$.)

posed into infinitely many successive thin "lenses" stacked along the $\hat{z}$ direction. Each "lens" produces an elemental diffracted field, and the total volume-diffracted light is the coherent superposition of all these individual fields. If the probe source is at the common front focus of all these virtual "lenses," then the individual "lens" fields are all in phase and give a strong reconstruction in the back focal point (Bragg matching.) If the probe is at a different location, contributions from the "lenses" are in general out of phase, resulting in Bragg mismatch, and the reconstructed amplitude drops. The only exception is if the probe location has the same y coordinate as that of the reference origin $\mathbf{r}_f$ (*i.e.*, if $y_{p,k} = y_f$.) In that case the probe remains approximately Bragg-matched and diffracts with almost equal strength. The intensity diffracted towards the focal spot as a function of probe displacement relative to $\mathbf{r}_f$ in the two orthogonal directions x, y is compared in Figure 1.6.

Even richer behavior is obtained if the wavelength of the probe source is different than λ [2, 19]. These properties are of interest for spectrally resolved volumetric imaging, *i.e.* for imaging in four dimensions (three spatial and one spectral.) However, a detailed description of the coupling between probe wavelength, location, and Bragg matching is beyond the scope of this chapter.

According to volume diffraction theory, the behavior of the hologram in response to probe displacement is asymmetric with respect to two lateral directions. The hologram behaves similarly to a lens in the y direction, but in the x direction it rather acts as a pinhole: if the probe is displaced in that direction,

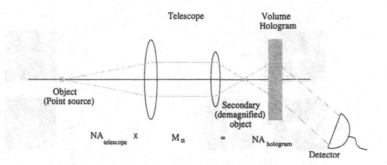

Figure 1.8. Schematic of the volume holographic telescope. The angular magnification $M\alpha$ increases the NA of the volume hologram and enables better resolution.

the diffracted beam is "vignetted out" by Bragg mismatch. Therefore, the volume hologram is actually shift-invariant in the y direction and shift-variant in the x direction. This property is well known [20] and has formed the basis of many optical pattern recognition systems in the past [21].

4. Telescopic Imaging with Cooperative Volume Holographic Sensors

Figure 1.8 is a schematic of the volume holographic telescope. The telescope produces a laterally demagnified real image of the object in front of the volume hologram. This intermediate image serves to reduce the physical aperture requirement on the volume hologram. After data acquisition is completed on this demagnified intermediate image, the image results are re-scaled to the size of the actual object using the magnification of the telescope (which is known separately or pre-calibrated with a well-characterized target).

The resolution of the volume holographic telescope is measured by the full width at half maximum (FWHM) of the point spread function (PSF). The longitudinal resolution of the volume holographic telescope is found by using (9) and fitting numerically. The result is

$$\Delta z_{\text{FWHM}} = \frac{1.7\lambda}{(\text{NA})_{\text{tel}}^2}, \qquad (17)$$

where NA is the numerical aperture of the telescope. The lateral resolution of the volume holographic telescope is similarly found to be

$$\Delta x_{\text{FWHM}} = \frac{0.5\lambda}{\theta(\text{NA})_{\text{tel}}}, \qquad (18)$$

where θ is the angle of the signal beam with the optical axis.

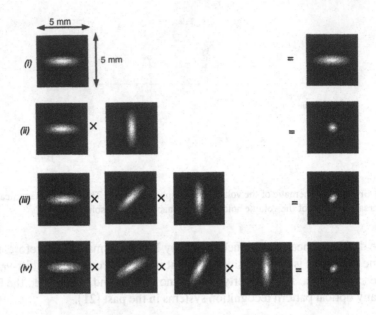

Figure 1.9. Theoretical performance of imaging with multiple SAs. *(i)* PSF of single SA. *(ii)* PSF of two SAs at $\theta = 90°$. *(iii)* PSF of three SAs at 45°. *(iv)* PSF of four SAs at 30°.

Based on equations (17) and (18), the value of the longitudinal and lateral resolution for objects that are a kilometer away are 41.11 m and 5.08 mm for a VH-telescope with $NA_{tel} = 0.1/1000$. The depth resolution of most imaging systems degrades as the square of the distance of the object from the sensor. Equivalently, the resolution varies inversely as the square of the numerical aperture of the system (17). As a result of that observation, the performance of the volume holographic telescope rapidly deteriorates at distances longer than a few meters. The ability to use multiple sensing agents (SAs), each equipped with a volume holographic sensor leads to a solution of this problem with dramatic performance improvement, as we show later. This approach is inspired by triangulation, found in many animals as well as artificial machine vision systems.

There are some important differences between triangulation and the VHI sensor with multiple SAs. Triangulating systems need to establish correspondence between the location of a particular object feature on the two acquired images. (Many active triangulation schemes do not need correspondence, but they require structured light which is difficult to implement at long ranges.) The cooperative VHI sensor does not require correspondence. Moreover, since multiple SAs might be configured to be looking at the same object, it is possible to achieve N-ocular imaging. This improves resolution only marginally,

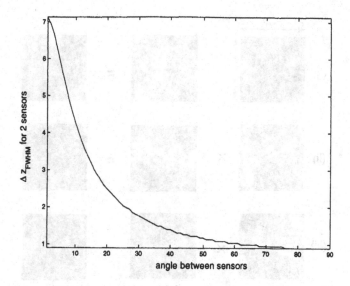

Figure 1.10. Depth resolution *vs.* angle θ between two SAs. Each individual sensor has $\Delta z_{FWHM} = 10$mm; $\Delta x_{FWHM} = 0.9$mm.

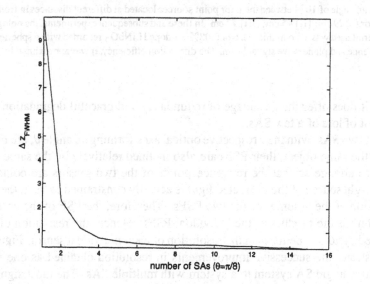

Figure 1.11. Graceful degradation of depth resolution as number of sensors are lost. The sensors are assumed to be oriented at an angle of $\theta = \pi/8$ with respect to each other. Each individual sensor has $\Delta z_{FWHM} = 10$mm; $\Delta x_{FWHM} = 0.9$mm.

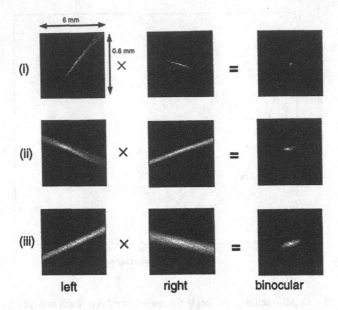

Figure 1.12. Experimental PSFs for two individual SAs oriented at $\theta = 10°$ are for two SAs with an angle of $10°$ between them for point sources located at different distances in front of the sensor. (i) 29 cm; (ii) 45 cm; (iii) 72 cm. In these and subsequent experiments, the holographic material used was a 2 mm thick slab of 0.03% Fe-doped $LiNbO_3$ recorded with a spherical wave reference and plane wave signal beam. The diffraction efficiency, η was approximately 10%.

but it does offer the advantage of redundancy and graceful degradation in the event of loss of a few SAs.

If two SAs, with their respective optical axes forming an angle θ, are observing the same object, their PSFs are also inclined relatively by the same angle. If we arrange so that the reference points of the two sensors are coincident, the light source of the diffracted light is actually constrained to lie at the intersection of the volumes of the two PSFs. Therefore, the PSF of the combined system is the product of the individual PSFs. Hence, the resolution of combined system is better than the resolution of each individual sensor. Figure 1.9 illustrates the successive improvements in resolution obtained as one moves from a single SA system to a system with multiple SAs. The most significant improvement in resolution is obtained when comparing a single SA to two SAs mutually perpendicular to each other (*i.e.*, at $\theta = 90°$). Beyond this, adding sensors improves the resolution only marginally, as noted earlier.

In Figure 1.10, depth resolution is plotted against the angle θ between two sensors. According to this calculation, even an angle of $\theta = 45°$ between the two sensors is sufficient to significantly improve the depth resolution over

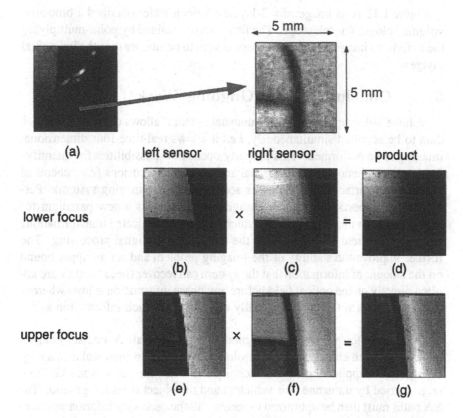

Figure 1.13. Image of a three-layered Silicon structure using a H–LADAR with two SA's oriented at 10° with each other. The Silicon structure was located 50 cm away from the entrance pupil of the telescopes of $M\alpha=4$. (a) is a picture of the object taken with a digital camera. (b) depicts the region that we imaged experimentally using the binocular VHI system. Each of the three layers are at different reference depths. (c) and (d) are the VH-images with the middle layer of the object in focus. e is the image obtained by multiplying images (c) and (d). It is seen that the features are sharper for (e). (f) and (g) are images of the object with the focus raised (1 mm) from the middle layer. As a result of translating the sensor with respect to the object, the uppermost layer is now Bragg-matched and appears bright. The two lower layers appear progressively darker. (h) is the image obtained by multiplying the two individual images. It is sen that this image has sharper features than both (f) and (g).

the single SA configuration. Figure 1.11 illustrates the graceful degradation of depth resolution as number of SAs decrease.

Figure 1.12 is an experimental demonstration of the improvement in range resolution by the use of two SAs at angle $\theta = 10°$. Despite the relatively small angle, still the resolution of the binocular system is considerably better than the corresponding resolution for individual SAs. Optimum θ is 90° as was pointed out earlier.

Figure 1.13 is an image of a 3-layered Silicon wafer obtained a binocular volume holographic telescope. The final image obtained by point-multiplying the individual images from each sensor is seen to be sharper than the individual images.

5.   Conclusions and Ongoing Work

Volume holographic imaging is unusual in that it allows depth and spectral data to be acquired simultaneously, *i.e.* it allows real-time four dimensional imaging to be performed. This property opens new possibilities for quantifying complex phenomena with spatial and spectral dimensions (*e.g.* chemical reactions on surfaces or in volumes accessible to the imaging system). Perhaps more importantly, volume holographic imaging is a new paradigm for next-generation imaging systems, which rely on sophisticated transformations of the optical field directly, before the application of digital processing. The former improve the stability of the imaging problem and set an upper bound on the amount of information that the system can recover (because they are applied directly on the optical field before any phase information is lost) whereas the latter perform the task of actually extracting as much information as the bound allows.

In this chapter, we described a binocular (in general, N-ocular) imaging technique which enhances depth resolution compared to monocular imaging. One possible application of this technique is when the sensors are airborne (*e.g.* carried by unmanned air vehicles) and the object is on the ground. The SA paths must then be optimized to recover all the necessary information from the object in minimum time.

References

[1] G. Barbastathis, M. Balberg, and D. J. Brady, "Confocal microscopy with a volume holographic filter", *Opt. Lett.*, 24(12):811–813, 1999.

[2] G. Barbastathis and D. J. Brady, "Multidimensional tomographic imaging using volume holography", *Proc. IEEE*, 87(12):2098–2120, 1999.

[3] W. Liu, D. Psaltis, and G. Barbastathis, "Real time spectral imaging in three spatial dimensions", to appear in *Opt. Lett.*

[4] G. Barbastathis and A. Sinha, "Information content of volume holographic images", *Trends in Biotechnology*, 19(10):383–392, 2001.

[5] P. J. van Heerden, "Theory of optical information storage in solids", *Appl. Opt.*, 2(4):393–400, 1963.

[6] D. Psaltis, "Parallel optical memories", *Byte*, 17(9):179, 1992.

[7] J. F. Heanue, M. C. Bashaw, and L. Hesselink, "Volume holographic storage and retrieval of digital data", *Science*, 265(5173):749–752, 1994.

[8] D. Psaltis and F. Mok, "Holographic memories", *Sci. Am.*, 273(5):70–76, 1995.

[9] Y. S. Abu-Mostafa and D. Psaltis, "Optical neural computers", *Sci. Am.*, 256(3):66–73, 1987.

[10] J. Hong, "Applications of photorefractive crystals for optical neural networks", *Opt. Quant. Electr.*, 25(9):S551–S568, 1993.

[11] D. J. Brady, A. G.-S. Chen, and G. Rodriguez, "Volume holographic pulse shaping", *Opt. Lett.*, 17(8):610–612, 1992.

[12] P.-C. Sun, Y. Fainman, Y. T. Mazurenko, and D. J. Brady, "Space-time processing with photorefractive volume holography", *SPIE Proceedings*, 2529:157–170, 1995.

[13] P.-C. Sun, Y. T. Mazurenko, W. S. C. Chang, P. K. L. Yu, and Y. Fainman, "All-optical parallel-to-serial conversion by holographic spatial-to-temporal frequency encoding", *Opt. Lett.*, 20(16):1728–1730, 1995.

[14] K. Purchase, D. Brady, G. Smith, S. Roh, M. Osowski, and J. J. Coleman, "Integrated optical pulse shapers for high-bandwidth packet encoding", *SPIE Proceedings*, 2613:43–51, 1996.

[15] D. M. Marom, P.-C. Sun, and Y. Fainman, "Analysis of spatial-temporal converters for all-optical communication links", *Appl. Opt.*, 37(14):2858–2868, 1998.

[16] G. A. Rakuljic and V. Levya, "Volume holographic narrow-band optical filter", *Opt. Lett.*, 18(6):459–461, 1993.

[17] H. Coufal, D. Psaltis, and G. Sincerbox, editors, *Holographic data storage*, Springer, 2000.

[18] M. Born and E. Wolf, *Principles of optics*, Pergamon Presss, 6th edition, 1980.

[19] G. Barbastathis and D. Psaltis, "Shift-multiplexed holographic memory using the two-lambda method", *Opt. Lett.*, 21(6):429–431, 1996.

[20] M. Levene, G. J. Steckman, and D. Psaltis, "Method for controlling the shift invariance of optical correlators", *Appl. Opt.*, 38(2):394–398, 1999.

[21] H.-Y. S. Li, Y. Qiao, and D. Psaltis, "Optical network for real-time face recognition", *Appl. Opt.*, 32(26):5026–5035, 1993.

Chapter 2

MULTI-TASK ALLOCATION AND PATH PLANNING FOR COOPERATING UAVS

John Bellingham
Department of Aeronautics and Astronautics
Massachusetts Institute of Technology
john_b@mit.edu

Michael Tillerson
Department of Aeronautics and Astronautics
Massachusetts Institute of Technology
mike_t@mit.edu

Arthur Richards
Department of Aeronautics and Astronautics
Massachusetts Institute of Technology
arthurr@mit.edu

Jonathan P. How
Department of Aeronautics and Astronautics
Massachusetts Institute of Technology
jhow@mit.edu

Abstract   This paper presents results on the guidance and control of fleets of cooperating Unmanned Aerial Vehicles (UAVs). A key challenge for these systems is to develop an overall control system architecture that can perform optimal coordination of the fleet, evaluate the overall fleet performance in real-time, and quickly reconfigure to account for changes in the environment or the fleet. The optimal fleet coordination problem includes team composition and goal assignment, resource allocation, and trajectory optimization. These are complicated optimization problems for scenarios with many vehicles, obstacles, and targets. Fur-

23

S. Butenko et al. (eds.), Cooperative Control: Models, Applications and Algorithms, 23-41.
© 2003 *Kluwer Academic Publishers*.

thermore, these problems are strongly coupled, and optimal coordination plans cannot be achieved if this coupling is ignored. This paper presents an approach to the combined resource allocation and trajectory optimization aspects of the fleet coordination problem which calculates and communicates the key information that couples the two. Also, this approach permits some steps to be distributed between parallel processing platforms for faster solution. This algorithm estimates the cost of various trajectory options using the distributed platforms and then solves a centralized assignment problem to minimize the mission completion time. The detailed trajectory planning for this assignment can then be distributed back to the platforms. During execution, the coordination and control system reacts to changes in the fleet or the environment. The overall approach is demonstrated on several example scenarios to show multi-task allocation and cooperative path planning.

Keywords:　Distributed coordination, unmanned aerial vehicles, task allocation, trajectory design, mixed-integer linear programming

1.　Introduction

The capabilities and roles of Unmanned Aerial Vehicles (UAVs) are evolving, and require new concepts for their control. Today's UAVs typically require several operators for control, but future UAVs will be designed to make their own tactical decisions autonomously and will be integrated into teams that coordinate to achieve high-level goals, thereby allowing one operator to control a fleet of UAVs [1]. This level of autonomy will require new methods in planning and execution to coordinate the achievement of goals between the UAVs in the fleet.

The simplest form of a mission for a fleet of UAVs (*e.g.* a Suppression of Enemy Air Defenses (SEAD) mission) can be generalized as visiting a set of N_W waypoints, while avoiding the "No Fly Zones". Further constraints can be added to this problem, including waypoint types that only a subset of the fleet is capable of visiting; simultaneous, delayed or ordered arrival at waypoints; and collision avoidance between UAVs. Numerous changes can also occur during the mission execution, such as movement, addition, or removal of waypoints and No Fly Zones, and the addition or loss of members of the fleet. An example of a fleet coordination scenario including capability constraints is shown in Figure 2.1.

Designing a coordinated mission plan that satisfies these constraints can be viewed as three coupled decisions: (i) Teams are formed and group goals are assigned to each team; (ii) Tasks that achieve the group goals are assigned to each team member; and (iii) A path is designed for each team member that achieves their tasks while adhering to spatial constraints, timing constraints, and the dynamic capabilities of the aircraft. The coordination plan is designed to minimize some cost, such as the completion time or the probability of mission failure. The overall control system then monitors the execution of the coordinated plan, and reacts to changes in the fleet, the environment, or the goals. This work assumes that lower-level controllers are present on each vehicle that are capable of following the planned path and performing the activities required at each waypoint.

Even if each of the three fleet coordination decisions is considered in isolation, it is clear that they are computationally demanding tasks. For even moderately sized problems, the number of combinations of possible teams, task allocations, and waypoint orderings that must be considered for the team formation and task assignment decisions is very large, growing at a non-polynomial rate that is at least the number of permutations of N_W elements taken N_W at a time. The problem of planning kinematically and dynamically constrained optimal paths, even for one aircraft, is also a very high dimension nonlinear optimization problem [2]. The optimal path planning problem requires a trajectory of control inputs to be designed that guide the UAV to its destination in minimum time, subject to control input limits, the differential equations representing the aircraft dynamics, and kinematic constraints presented by the No Fly Zones.

As difficult as each of these three decisions is to make in isolation, they are in fact strongly coupled because the optimality of the overall coordination plan is strongly limited by the team partitioning and task allocation. However, it is not clear how to make these decisions optimally until detailed trajectories have been planned, because the cost to be minimized by these decisions is a function of the resulting detailed trajectories. This coupling has been handled in one approach [3] by forming a large optimization problem that simultaneously assigns the tasks to vehicles and plans corresponding detailed trajectories. This method is computationally intensive, but it is guaranteed to find the globally-optimal solution to the problem and thus can be used as a benchmark against which the techniques presented in this paper can be compared [3].

Another approach to this problem is to decouple the decisions to some degree in order to make the problem computationally tractable, while maintaining the essential aspects of the coupling in order to approach optimality. This paper presents such a partially-decoupled approach to the task allocation and trajectory optimization problems which calculates and communicates the key information that couples the two problems, and distributes the computational effort of some steps between parallel processing platforms to improve the solution time. This partially-decoupled approach is shown to yield coordinated mission plans that are very close to the optimal solution [3].

Several previous studies have investigated methods of trajectory planning for coordination and control. Trajectory generation methods include the use of Voronoi diagrams [4], adaptive random search algorithms [5], model predictive control [6], and mixed-integer linear programming [3, 7]. However most of these methods are too computationally expensive for the multi-vehicle, multi-

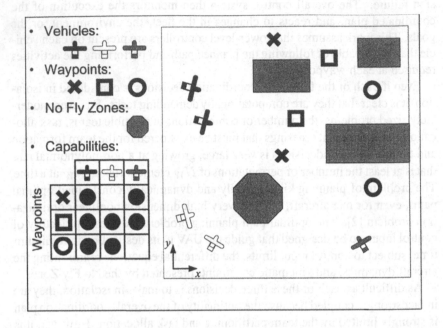

Figure 2.1. Schematic of a typical mission scenario for a UAV fleet with numerous waypoint: No Fly Zones and capabilities

waypoint scenarios considered in this work. Thus, this paper investigates al approximate method that yields a fast estimate of the finishing times for th UAV trajectories which can then be used in the task allocation problem. It per forms this estimation by using straight line path approximations. This method takes advantage of the fact that, for typical missions, the shortest paths for th UAVs tend to resemble straight lines that connect the UAVs' starting position the vertices of obstacle polygons, and the waypoints. A more detailed trajec tory generator is used to plan the UAV paths once the waypoints have beer assigned.

The task allocation problem has also been considered in various applica tions. One method of determining the task allocation is through a networl flow analogy [8]. This leads to a linear assignment problem that can be solve using integer linear programming. The task allocation problem has also beer studied using a market based approach [9, 10]. In this approach, vehicles bi on each possible task presented. A central auction receives the bids and send out the current price quote for each task. A decision is reached when no nev bids arrive. The approaches discussed above handle some aspects of the tasl assignment, but they cannot easily include more detailed constraints such a selecting the sequence of the tasks and the timing of when tasks must be completed.

This paper presents a mathematical approach to solve the task allocation problem with the flexibility to include more detailed constraints through the use of mixed-integer linear programming (MILP). The resulting MILP problems can be readily solved using commercially available software such as CPLEX [11]. The combination of the approximate cost algorithm and task allocation formulation presented in this paper provides a flexible and efficient means of assigning multiple objectives to multiple vehicles under various detailed constraints.

2.   Problem Formulation

The algorithms described here assume that the team partitioning has already been performed, and that a set of tasks has been identified which must be performed by the team. This paper presents algorithms that assign the tasks to team members and design a detailed trajectory for each member to achieve its tasks. The team is made up of N_V UAVs with known starting states and maximum velocities. The starting state of UAV p is given by the pth row $[\ x_{0p}\ \ y_{0p}\ \ \dot{x}_{0p}\ \ \dot{y}_{0p}\ ]$ of the matrix S_0, and the maximum velocity of UAV p is given $v_{\text{max},p}$. The waypoint locations are assumed to be known, and the position of waypoint i is given by the ith row $[\ W_{ix}\ \ W_{iy}\ ]$ of the matrix $\mathbf{W}$. The application of the algorithms presented in this paper to No Fly Zones that are bounded by polygons is straightforward, but the case where the polygons are rectangles will be presented here for simplicity. The location of the lower-left corner is given by (Z_{j1}, Z_{j2}), and the upper-right corner by (Z_{j3}, Z_{j4}). Together, these two pairs make up the jth row of the matrix $\mathbf{Z}$. Finally, the UAV capabilities are represented by a binary capability matrix $\mathbf{K}$. The entry K_{pi} is 1 if vehicle p is capable of performing the tasks associated with waypoint i, and 0 if not.

This algorithm produces a trajectory for each vehicle, represented for the pth vehicle by a series of states $s_{tp} = [\ x_{tp}\ \ y_{tp}\ \ \dot{x}_{tp}\ \ \dot{y}_{tp}\ ]$, $t \in \{1, \ldots, t_p\}$, where t_p is the time at which aircraft p reaches its final waypoint. The finishing times of all vehicles make up the vector $\mathbf{t}$.

This work is concerned with coordination and control problems in which the cost is a function of the resulting trajectories. This is a broad category of coordination and control problems, and includes costs that involve completion time or radar exposure, and constraints on coordinated arrival or maximum range. While a cost function has been chosen that penalizes both the maximum completion time and the average completion times over all UAVs, the approach presented here can be generalized to costs that involve other properties of the trajectories. The cost used in this paper can be written as

$$\bar{t} = \max_p t_p, \tag{1}$$

$$J_1(\bar{t}, t) = \bar{t} + \frac{\alpha}{N_V} \sum_{p=1}^{N_V} t_p, \tag{2}$$

where $\alpha \ll 1$ weights the average completion time compared to the maximum completion time. If the penalty on average completion time were omitted (*i.e.*, $\alpha = 0$), the solution could assign unnecessarily long trajectories to all UAVs except for the last to complete its mission. Note that, because this cost is a function of the completion time for the entire fleet, it cannot be evaluated exactly until detailed trajectories have been planned that visit all the waypoints and satisfy all the constraints. The minimum cost coordination problem could be solved by planning detailed trajectories for all possible assignments of waypoints to UAVs and all possible orderings of those waypoints, then choosing the detailed trajectories that minimize cost function $J_1(\bar{t}, t)$. However, the computational effort required to plan one detailed trajectory is large, and given all possible assignments and orderings, there exist a very large number of potential detailed trajectories that would have to be designed. For the relatively small coordination problem shown in Figure 2.1, there are 1296 feasible allocations, and even more possible ordered arrival permutations.

3.   Algorithm Overview

Clearly, planning detailed trajectories for all possible task allocations is not computationally feasible. Instead, the algorithm presented in this paper constructs estimates of the finishing times for a subset of the feasible allocations, then performs the allocation to minimize the cost function evaluated using the estimates. Next, detailed UAV trajectories are designed, and checked for collisions between vehicles. The main steps in the algorithm are shown in Figure 2.2 and are described in the following.

First, a list of all unordered feasible task combinations is enumerated for every UAV, given its capabilities. Next, the length of the shortest path made up of straight line segments between the waypoints and around obstacles is calculated for all possible order-of-arrival permutations of each combination. The construction of these paths can be performed extremely rapidly using graph search techniques. The minimum finishing time for each combination is estimated by dividing the length of the shortest path by the UAV's maximum speed. Some of the task allocations and orderings have completion times that are so high that they can confidently be removed from the list.

With these estimated finishing times available, the task allocation problem can be performed to find the minimum of the estimated costs. MILP is well

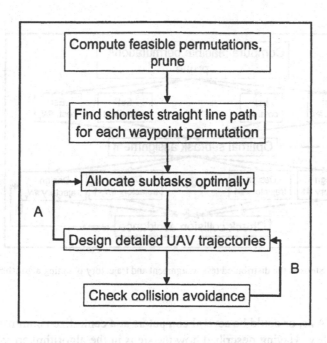

Figure 2.2. Steps in the task assignment and trajectory planning algorithm

suited to solving this optimization problem, because it allows Boolean logic to be incorporated into constraints [12, 13]. These constraints can naturally express concepts such as "exactly one aircraft must visit every waypoint", and extend well to more complex concepts such as ordered arrival and obstacle avoidance [14]. Once the optimal task allocation is found using the estimated completion times, detailed kinematically and dynamically feasible trajectories that visit the assigned waypoints can be planned and checked for collision avoidance between UAVs [15]. If the minimum separation between UAVs is violated, the trajectories can be redesigned to enforce a larger separation distance (shown by loop B in Figure 2.2). If desired, or if the completion time of the detailed trajectory plan is sufficiently different from the estimate, detailed trajectories can be planned for several of the task allocations with the lowest estimated completion times. The task allocation can then be performed using these actual completion times (shown by loop A in Figure 2.2).

This strategy also casts the task allocation and detailed trajectory planning problems in a form that allows parts of the computation to be distributed to parallel platforms, as shown in Figure 2.3. By making the processes of estimating the costs and designing detailed trajectories independent for each vehicle, they can be performed separately. The parallel platforms could be processors on-

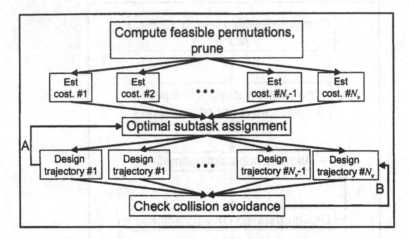

Figure 2.3.   Steps in the distributed task assignment and trajectory planning algorithm

board the UAVs, or could be several computers at a centralized command and control facility. Having described how the steps in the algorithm are related, methods for performing them will be described next.

4.   Finding Feasible Permutations and Associated Costs

This section presents a detailed analysis of the process for developing a list of feasible task assignments, finding approximate finishing times for each task assignment, and pruning the list. This step accepts the aircraft starting states S_0, capabilities K, obstacle vertex position Z, and waypoint positions W. The algorithm also accepts two upper boundaries: n_{max} specifies the maximum number of waypoints that a UAV can visit on its mission, and t_{max} specifies the maximum time that any UAV can fly on its mission. From these inputs this algorithm finds, for each UAV and each combination of n_{max} or fewer waypoints, the order in which to visit the waypoints that gives the shortest finishing time.

The steps in this algorithm are depicted in Figs. 2.4–2.6, in which a fleet of UAVs (shown with o) must visit a set of waypoints (shown with ×). The first step is to find the visibility graph between the UAV starting positions, way-points, and obstacle vertices. The visibility graph is shown in Figure 2.4 with grey lines. Next, UAV 6 is considered, and the visibility graph is searched to find the shortest paths between its starting point and all waypoints, as shown in Figure 2.4 with black lines. In Figure 2.5, a combination of three waypoints

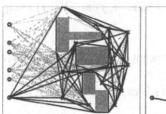 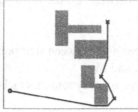 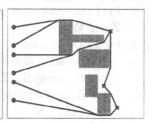

Figure 2.4. Visibility graph and shortest paths between UAV 6 and all waypoints.

Figure 2.5. Shortest path for UAV 6 over one combination of waypoints.

Figure 2.6. Shortest paths for all UAVs over same combination of waypoints.

has been chosen, and the fastest path from UAV 6's starting position through them is shown. The order-of-arrival for this path is found by forming all possible order-of-arrival permutations of the unordered combination of waypoints, then summing the distance over the path associated with each order-of-arrival from UAV 6's starting point. The UAV is assumed to fly this distance at maximum speed, and the order-of-arrival with the shorted associated finishing time is chosen. In Figure 2.6, the fastest path to visit the same combination of waypoints is shown for each vehicle. Note that the best order-of-arrival at these waypoints is not the same for all vehicles.

The algorithm produces five matrices whose j^{th} columns, taken together, fully describe one permutation of waypoints. These are the row vector $\mathbf{u}$, whose u_j entry identifies which UAV is involved in the jth permutation; $\mathbf{P}$, whose P_{ij} entry identifies the ith waypoint visited by permutation j; $\mathbf{V}$, whose V_{ij} entry is 1 if waypoint i is visited by permutation j, and 0 if not; $\mathbf{T}$, whose T_{ij} entry is the time at which waypoint i is visited by permutation j, and 0 if waypoint i is not visited; and $\mathbf{c}$, whose c_j entry is the completion time for the jth permutation. All of the permutations produced by this algorithm are guaranteed to be feasible given the associated UAV's capabilities.

All steps in this approach are shown in Algorithm 1. In this algorithm, finding the shortest distance between a set of waypoints and starting points is performed by finding the visibility graph between the points and vertices of obstacles, then applying an appropriate shortest path algorithm, such as the Floyd-Warshall All-Pairs Shortest Path algorithm [16]. Note that the iterations through the "**for** loop" between lines 2 and 23 of Algorithm 1 are independent, and can be distributed to parallel processors. The corresponding matrices from each processor can then be combined and passed onto the next stage in the algorithm, the task allocation problem.

1: Find shortest distances between all waypoint pairs (i, j) as $D(i, j)$ using $\mathbf{Z}$ and $\mathbf{W}$.

2: **for all** UAVs p **do**

3:     Find shortest distances $d(i)$ between start point of UAV p, and all waypoints i using $\mathbf{S}_0$, $\mathbf{Z}$, and $\mathbf{W}$.

4:     **for all** combinations of n_C waypoints that p is capable of visiting, $n_C = 1 \ldots n_{max}$ **do**

5:         **for** $j = 1 \ldots n_C P n_C$ **do**

6:             Make next unique permutation $P'_{1j} \ldots P'_{n_C j}$ of waypoints in the combination

7:             $c'_j = \frac{d(P'_{1j})}{v_{max,p}}$

8:             $T'_{P'_{1j}j} = c'_j$

9:             **for** $i = 2 \ldots n_C$ **do**

10:                 **if** $c'_j > t_{max}$ **then**

11:                     go to next j

12:                 **end if**

13:                 $c'_j \leftarrow c'_j + \frac{D(P'_{(i-1)j}, P'_{ij})}{v_{max,p}}$

14:                 $T'_{P'_{ij}j} = c'_j$

15:             **end for**

16:         **end for**

17:         Append p to $\mathbf{u}$

18:         Append a column to $\mathbf{V}$, whose i^{th} element is 1 if waypoint i is visited, 0 if not.

19:         $j_{min} = \text{minarg}_j \, c'_j$

20:         Append $c_{j_{min}}$ to $\mathbf{c}$

21:         Append column j_{min} of $\mathbf{T}'$ to $\mathbf{T}$ and column j_{min} of $\mathbf{P}'$ to $\mathbf{P}$

22:     **end for**

23: **end for**

Algorithm 1: Algorithm for finding shortest paths between waypoints

5.   Task Allocation

The previous section outlined a method of rapidly estimating completion times for individual vehicles for the possible waypoint allocations. This section presents a mathematical method of selecting which of these assignments to use for each vehicle in the fleet, subject to fleet-wide task completion and arrival timing constraints.

The basic task allocation problem is formulated as a *Multi-dimensional Multiple-Choice Knapsack Problem* (MMKP) [17]. In this classical problem, one element must be chosen from each of multiple sets. Each chosen element uses an amount of each resource dimension, but yields a benefit. The choice from each set is made to maximize the benefit subject to multi-dimensional resource constraints.

In the UAV task allocation problem, the choice of one element from each set corresponds to the choice of one of the N_M permutations for each vehicle. Each resource dimension corresponds to a waypoint, and a permutation uses 1 unit of resource dimension i if it visits the ith waypoint. The arrival constraints are then transformed into constraints on each resource dimension. The negative of the completion times in this problem is equivalent to the benefit. Thus the overall objective is to assign one permutation (element) to each vehicle (set) that is combined into the mission plan (knapsack), such that the cost of the mission (knapsack) is minimized and the waypoints visited (resources used) meet the constraint for each dimension. The problem can be written as

$$\min \ J_2 = \sum_{j=1}^{N_M} c_j x_j$$

$$\text{subject to} \quad \sum_{j=1}^{N_M} V_{ij} x_j \geq w_i : \forall i \in \{1, \ldots, N_W\} \tag{3}$$

$$\sum_{j=N_p}^{N_{p+1}-1} x_j = 1 : \forall p \in \{1, \ldots, N_V\},$$

where the permutations of vehicle p are numbered N_p to $N_{p+1} - 1$, with $N_1 = 1$ and $N_{N_V+1} = N_M + 1$ and the indices have the ranges $i \in \{1, \ldots, N_W\}$, $j \in \{1, \ldots, N_M\}$, $p \in \{1, \ldots, N_V\}$. c_j is a vector of costs (mission times) for each permutation. x_j is a binary decision variable equal to one if permutation j is selected, and 0 otherwise. The cost in this problem formulation minimizes the sum of the times to perform each selected permutation. The first constraint enforces that waypoint i is visited at least w_i times (typically $w_i = 1$). The second constraint prevents more than one permutation being assigned to each vehicle. The MMKP formulation is the basic task allocation algorithm. However, modifications are made to the basic problem statement to include additional cost considerations and constraints.

Modified Cost: Total Mission Time. The first modification for the UAV allocation problem is to change the cost. The cost in Eq. 2 is a weighted combination of the sum of the individual mission times (as in the MMKP problem) and the total mission time. The new cost is as follows,

$$J_3 = \bar{t} + \frac{\alpha}{N_V} \sum_{i=1}^{N_M} c_i x_i. \tag{4}$$

The solution to the task allocation problem is a set of ordered sequences of waypoints for each vehicle which ensure that each waypoint is visited the correct number of times while minimizing the desired cost (mission completion time).

Timing Constraints. Solving the task allocation as a centralized problem allows the inclusion of timing constraints on when a waypoint is visited. For example, a typical constraint might be for one UAV to eliminate a radar site at waypoint A before another vehicle can proceed to a waypoint B. The constraint would then be that waypoint A must be visited t_D time units before waypoint B, which can be included in the task allocation problem. The timing constraint is met by either altering the order in which waypoints are visited, delaying when a vehicle begins a mission, or assigning a loitering time to each waypoint. The constraint formulation in which vehicle p starts at VT_p and then executes its mission without delay is presented below. To construct the constraint, an intermediate variable, WT_i, is used to determine the departure time of the vehicle that visits waypoint i. The constraint can be written as

$$\sum_{j=1}^{N_M}(T_{Bj}x_j - T_{Aj}x_j) + WT_B - WT_A \geq t_D, \tag{5}$$

$$WT_A \leq VT_p + R(1 - \sum_{j=N_p}^{N_{p+1}-1} V_{Aj}x_j) \ \forall \, p \in \{1, \ldots, N_V\}, \tag{6}$$

$$WT_A \geq VT_p - R(1 - \sum_{j=N_p}^{N_{p+1}-1} V_{Aj}x_j) \ \forall \, p \in \{1, \ldots, N_V\}, \tag{7}$$

$$WT_B \leq VT_p + R(1 - \sum_{j=N_p}^{N_{p+1}-1} V_{Bj}x_j) \ \forall \, p \in \{1, \ldots, N_V\}, \tag{8}$$

$$WT_B \geq VT_p - R(1 - \sum_{j=N_p}^{N_{p+1}-1} V_{Bj}x_j) \ \forall \, p \in \{1, \ldots, N_V\}. \tag{9}$$

Constraint Eq. 5 enforces waypoint A to be visited t_D time units before B. Constraint Eqs. 6 – 9 are used to determine the start time for the vehicles that are assigned the waypoints in the timing constraint. R is a large number that relaxes the constraint if vehicle p does not visit the waypoint in question. If vehicle p does visit waypoint A, R multiplies 0, and Equations 6 and 7 combine to enforce an equality relationship $WT_A = VT_p$. Note that using this formulation does not allow the same vehicle to visit both waypoints unless one of the original permutations met the timing constraint. This is because the start times WT_A and WT_B would be the same and cancel in Eq. 5. Again these constraints are added to the original problem in Eq. 3 to form a task allocation problem including timing constraints. The cost must also be altered to include the UAV start times as follows

$$J_4 = \bar{t} + \frac{\alpha}{N_V} \sum_{i=1}^{N_M} c_i x_i + \frac{\alpha}{N_V} \sum_{p=1}^{N_V} VT_p. \tag{10}$$

The constraints presented here for delaying individual start times can be generalized to form other solutions to the timing constraint, such as allowing a UAV to loiter at a waypoint before going to the next objective.

6. Reaction to Dynamic Environment

The task allocation problem is used to assign a sub-team of vehicles to visit a set of waypoints based on the information (vehicle states, waypoint locations, and obstacles) known at the beginning of the mission. However, throughout the execution of the mission the environment can (and most likely will) change. As a result the optimal task allocation could be dramatically altered. Note that if the problem size is sufficiently small, it would be practical to perform a complete re-allocation using a new set of costs based on the updated environment. However for larger problems it would be beneficial to re-solve smaller parts of the allocation problem. Two smaller problems are presented in this section. One is a *local repair* where only one vehicle assignment is altered to meet the new environment. Another is a *sub-team allocation problem* where only those "directly influenced" by the change in environment are re-assigned.

Addition of Waypoint. As the mission is executed, it is possible that further reconnaissance will identify a new goal or waypoint. The set of waypoints could be re-allocated amongst the entire fleet, but several alternatives exist that can result in much smaller optimization problems.

The *local repair* method estimates the cost of adding the new waypoint to each vehicle's list of objectives using Alg. 1. The cost for each UAV is determined using the UAV's current state, the remaining waypoints assigned to the UAV from the original problem, and the new waypoint. The assignment of this waypoint that results in the smallest increase in the cost function is then chosen. The local repair can be solved very quickly, but it is a sub-optimal solution because it does not allow the vehicles to trade previously assigned waypoints.

A *sub-team* problem can be formulated which only considers those vehicles capable of visiting the new waypoint. These vehicles, their previously assigned waypoints, and the new waypoint are then considered as a smaller task allocation problem. This allows any waypoints within this group to be traded amongst the vehicles (subject to each vehicle's capabilities), but it is still sub-optimal because it does not consider the possibility that some waypoints of a type different than the new one could be traded to UAVs that are not in the sub-team considered. The *full re-allocation* problem would, of course, consider this coupling in the problem, but this would typically take longer to compute.

Addition/Removal of Obstacle. The addition or removal of an obstacle is considered by estimating the new cost for each vehicle given their current assigned waypoints with (and without) the obstacle in question. If the vehicle's cost estimate changes, then that vehicle is considered to be influenced by the obstacle.

If the vehicle is influenced by the obstacle, the *local repair* method does not change its assignment of waypoints, but redesigns its detailed trajectory to account for the change. The *sub-team* problem considers all vehicles that are influenced by the obstacle. The vehicles and their previously assigned waypoints are grouped into a new allocation problem and the re-assignment is performed for this subset of the fleet. The *full re-allocation* problem could also be solved for this situation.

The methods presented in this paper only consider unexpected changes in the environment, such as a waypoint suddenly appearing, or an obstacle appearing/disappearing. Another dynamic problem to consider is when knowledge of the environment includes a change that will occur at some future time or due to the actions of the vehicles, such as removal of an obstacle by a vehicle. Formulations of the cost estimation and task allocation steps that take advantage of expected changes are areas of current research.

7.   Simulations

The problem formulation presented in this paper leads to cooperative path planning in the sense that the task assignments to each UAV are made in a centralized way. In particular, the tasks are divided amongst the vehicles in the fleet to minimize the overall cost. The problem formulation also allows multi-task assignments, possibly assigning a sequence of several tasks to a single UAV. The complete approach is illustrated in this section using several examples.

A small problem is first considered to show how the assignment changes when constraints are added. The basic scenario includes three UAVs and four waypoints to be visited. There are also two obstacles in the environment. The objective is to visit each waypoint once and only once in minimum time. Each vehicle also must take at least one waypoint. The costs for each vehicle to visit a sequence of waypoints is determined using the approximate cost algorithm presented earlier. The task allocation is then determined through the modified MMKP problem. The basic problem is formed using Eqs. 1–3.

The first solution, shown in Figure 2.7, is the basic allocation problem. The UAVs are homogeneous so any vehicle can visit any waypoint. The solution for this problem is relatively straightforward, with each vehicle visiting waypoints

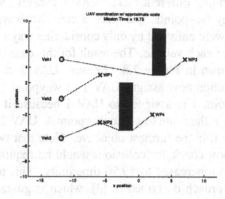

Figure 2.7. Scenario has 3 homogeneous vehicles, which is the basic task allocation problem.

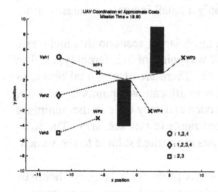

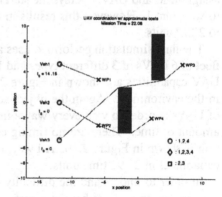

Figure 2.8. Same problem as Figure 2.7, plus heterogenous capabilities shown in the legend.

Figure 2.9. Same problem as Figure 2.8, plus waypoint 4 must be visited 5 time units before waypoint 1.

that are close. The mission time in this case is 19.75. The second scenario is the same basic problem except now the vehicles are heterogenous. Each vehicle is capable of performing different tasks. UAV 1 can visit waypoints 1, 2, 4, UAV 2 can visit any waypoints, and UAV 3 can visit waypoints 2, 3. The vehicle capabilities were enforced by only considering waypoint permutations that were feasible for each vehicle. The result for this case is the slightly less obvious solution shown in Figure 2.8. Because UAV 3 can no longer visit waypoint 4, the solution now assigns UAV 1 to waypoint 1 before going to waypoint 4. Waypoint 1 is assigned to UAV 1 because it can be achieved with little deviation in the route to get to waypoint 4. UAV 2 is assigned only waypoint 3 because it is the furthest objective. Note that two of the vehicle paths cross and a post check for collisions would be required. The mission time for this scenario increased to 19.90 time units. This problem was also solved using the approach described in [3], which is guaranteed to find the globally optimal coordinated mission plan. The allocation found in Figure 2.8 was verified to be the global optimum, and was found much more rapidly by the approach presented here. The third scenario is the same as the second problem with the addition of the timing constraint that waypoint 4 must be visited 5 time units before waypoint 1. The timing constraint is included using Eq. 5. The solution to this problem (see Figure 2.9) results in a very different set of assignments and UAV 1 delays the start time by 14.15 units before proceeding to waypoint 1. However, this results in only a small increase in mission time to 22.08 units.

The final simulation performed uses a much larger scenario that includes a fleet of 6 UAVs of 3 different types and 12 waypoints of 3 different types. The UAV capabilities are shown in Figure 2.10. There are also several obstacles in the environment. Again the objective is to allocate waypoints to the team of UAVs in order to visit every waypoint once and only once in the minimum amount of time. There are no timing constraints in this scenario. The solution is shown in Figure 2.10. All waypoints are visited subject to the vehicle capabilities in 23.91 time units.

In order to understand the difficulty of this problem, a "greedy" heuristic was applied to it. This heuristic performs allocation decisions one waypoint at a time. It calculates the increase in the cost function in Eq. 2 associated with allocating each waypoint to each capable vehicle. The vehicle-waypoint allocation with the smallest associated increase in the cost is selected. The allocated waypoint is removed from consideration, and the receiving vehicle is considered to be at the waypoint's location for the next allocation. This procedure is repeated until all waypoints are allocated. The greedy heuristic solved the large scenario shown in Figure 2.10 with a maximum completion time of 28.20, an increase of 17.9%. In this coordination plan, the last waypoint to be allocated caused a large increase in the completion time. This clearly shows

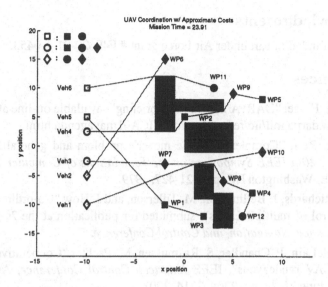

Figure 2.10.   Scenario has three pairs of heterogenous vehicles with 12 waypoints (3 different types). Figure legend shows which vehicles can visit each waypoint. The scenario demonstrates task allocation for a large problem with heterogenous vehicles using the approximate cost method.

that locally justified decisions do not provide globally optimal fleet coordination plans, and that the MILP-based method presented here provides significantly better results than heuristics with local scope.

8.   Conclusions

This paper presents an approach to the task allocation and detailed trajectory design components of the optimal fleet coordination problem. The approach presented here partially decouples these problems. It efficiently estimates finishing times associated with the different allocation options, and provides this information to the allocation optimization. The allocation is an extension of the MMKP problem, and is solved as a MILP problem. A small number of detailed trajectories are then designed to perform the allocated tasks. The results from several examples clearly illustrate the impact of including various constraints in the fleet assignment problem. They also clearly illustrate that the partially-decoupled method presented here was capable of solving a large problem involving many obstacles and waypoints. Future work will compare these approximate results with the globally-optimal solution available from the full MILP solution.

Acknowledgments

Research funded in part under Air Force grant # F49620-01-1-0453.

References

[1] S. A. Heise, "DARPA industry day briefing", available on-line at www.darpa.mil/ito/research/mica/MICA01mayagenda.html

[2] J. H. Reif, "Complexity of the mover's problem and generalizations", *Proc. 20th IEEE Symposium on the Foundations of Computer Science,* IEEE, Washington DC, pp. 421-427, 1979.

[3] A. Richards, J. Bellingham, M. Tillerson, and J. How "Co-ordination and control of multiple UAVs", submitted for publication at the 2002 *AIAA Guidance, Navigation, and Control Conference.*

[4] T. McLain, P. Chandler, S. Rasmussen, M. Pachter,"Cooperative control of UAV rendezvous", IEEE *American Control Conference,* Arlington, VA, June 25-27, pp. 2309–2314, 2001.

[5] R. Kumar, D. Hyland, "Control law design using repeated trials", IEEE *American Control Conference,* Arlington, VA, June 25-27, pp. 837–842, 2001.

[6] L. Singh and J. Fuller, "Trajectory generation for a UAV in urban terrain, using nonlinear MPC", IEEE *American Control Conference,* Arlington, VA, June 25-27, pp. 2301-2308, 2001.

[7] J. Bellingham, A. Richards, J. How, "Receding horizon control of autonomous aerial vehicles", to appear in the IEEE *American Control Conference,* May 2002.

[8] C. Schumacher, P. Chandler, and S. Rasmussen, "Task allocation for wide area search munitions via network flow optimization", AIAA *Guidance, Navigation, and Control Conference,* Montreal, Canada, Aug. 6-9, 2001.

[9] P. Chandler, M. Pachter, "Hierarchical control for autonomous teams", AIAA *Guidance, Navigation, and Control Conference,* Montreal, Canada, Aug. 6-9, 2001.

[10] J. Tierno, "Distributed autonomous control of concurrent combat tasks", IEEE *American Control Conference,* Arlington, VA, June 25-27, pp. 37-42, 2001.

[11] *ILOG CPLEX User's guide,* ILOG, 1999.

[12] C. A. Floudas, *Nonlinear and Mixed-Integer Programming – Fundamentals and Applications,* Oxford University Press, 1995.

[13] H. P. Williams and S. C. Brailsford, "Computational logic and integer programming", in *Advances in Linear and Integer Programming,* Editor J. E. Beasley, pp. 249–281, Clarendon Press, Oxford, 1996.

[14] A. Richards, and J. P. How, "Aircraft trajectory planning with collision avoidance using mixed integer linear programming", to appear at the *IEEE American Controls Conference*, 2002.

[15] T. Schouwenaars, B. D. Moor, E. Feron, and J. P. How, "Mixed integer programming for multi-vehicle path planning" presented at the *European Controls Conference*, 2001.

[16] T. H. Cormen, C. E. Leiserson, R. L. Rivest. *Introduction to Algorithms*, MIT Press, 1990.

[17] M. Moser, D. Jokanovic, N. Shiratori, "An algorithm for the multidimensional multiple-choice knapsack problem" *IEICE Trans. Fundamentals*, Vol. E80-A, No.3 March 1997.

Chapter 3

ON THE CONSTRUCTION OF VIRTUAL BACKBONE FOR AD HOC WIRELESS NETWORK

Sergiy Butenko
Department of Industrial and Systems Engineering,
University of Florida, Gainesville, FL 32611, USA.
butenko@ufl.edu

Xiuzhen Cheng
Department of Computer Science and Engineering,
University of Minnesota, Minneapolis, MN 55455, USA.
cheng@cs.umn.edu

Ding-Zhu Du
Department of Computer Science and Engineering,
University of Minnesota, Minneapolis, MN 55455, USA.
dzd@cs.umn.edu

Panos M. Pardalos
Department of Industrial and Systems Engineering,
University of Florida, Gainesville, FL 32611, USA.
pardalos@ufl.edu

Abstract   Ad hoc wireless network is featured by a dynamic topology. There is no fixed infrastructure as compared with wired network. Every host can move to any direction at any speed. This characteristic puts special challenges in routing protocol design. Most existing well-known routing protocols use flooding for route construction. But, flooding suffers from the notorious *broadcast storm problem* which causes excessive *redundancy*, *contention* and *collision* in the network. One solution to overcome this problem is to compute a virtual backbone based on the physical topology, and run any existing routing protocol over the virtual backbone. In our study, the virtual backbone is approximated by a *minimum*

S. Butenko et al. (eds.), Cooperative Control: Models, Applications and Algorithms, 43-54.

connected dominating set (MCDS). We propose a distributed algorithm which computes a small CDS. The performance of our algorithm is witnessed by simulation results and theoretical analysis.

Keywords: Multihop ad hoc wireless network, connected dominating set, virtual backbone routing

1.   Introduction

Ad hoc wireless network has applications in emergency search-and-rescue operations, decision making in the battlefield, data acquisition operations in inhospitable terrain, etc. It is featured by dynamic topology (infrastructureless), multihop communication, limited resources (bandwidth, CPU, battery, etc) and limited security. These characteristics put special challenges in routing protocol design.

Existing routing protocols (see [10, 12, 14, 15]) can be classified into three categories: *proactive, reactive* and the combination of the two. Proactive routing protocols ask each host to maintain global topology information, thus a route can be provided immediately when requested. Protocols in this category suffer from lower scalability and high protocol overhead. Reactive routing protocols have the feature *on-demand*. Each host computes route for a specific destination only when necessary. Topology changes that do not influence active routes do not trigger any route maintenance function, thus communication overhead is lower compared to proactive routing protocol. The third category maintains partial topology information in some hosts. Routing decisions are made either proactively or reactively. One important observation on these protocols is that none of them can avoid the involvement of *flooding*. For example, proactive protocols rely on flooding for the dissemination of topology update packets, and reactive protocols rely on flooding for route discovery.

Flooding suffers from the notorious *broadcast storm problem* [13]. Broadcast storm problem refers to the fact that flooding may result in excessive *redundancy, contention*, and *collision*. This causes high protocol overhead and interference to ongoing traffic. On the other hand, flooding is very *unreliable* [11], which means that *not all* hosts get all the broadcast messages when free from collisions. Sinha et. al. in [16] claimed that " in moderately sparse graphs the expected number of nodes in the network that will receive a broadcast message was shown to be as low as 80%." In reactive protocols, the unreliability of flooding may obstruct the detection of the shortest path, or simply can't detect any path at all, even though there exists a path. In proactive protocols, the unreliability of flooding may cause the global topology information to become obsolete, thus causing the newly-computed path obsolescence.

Recently an approach based on overlaying a virtual infrastructure on an ad hoc network was proposed in [16]. Routing protocols are operated over this infrastructure, which is termed *core*. All core hosts form a dominating set. The key feature in this approach is the new *core broadcast mechanism* which uses unicast to replace the flooding mechanism used by most on-demand routing protocols. The unicast of route requests packets to be restricted to core nodes and a (small) subset of non-core nodes. Simulation results when running DSR (Dynamic Source Routing [12]) and AODV (Ad hoc On-demand Distance Vector routing [15]) over the core indicate that the core structure is *effective* in enhancing the performance of the routing protocols. Prior to this work, inspired by the *physical backbone* in a wired network, many researchers proposed the concept of *virtual backbone* for unicast, multicast/broadcast in ad hoc wireless networks (see [2, 8, 17]).

In this chapter, we will study the problem of efficiently constructing virtual backbone for ad hoc wireless networks. The number of hosts forming the virtual backbone must be as small as possible to decrease protocol overhead. The algorithm must be time/message efficient due to resource scarcity. We use a connected dominating set (CDS) to approximate the virtual backbone. We assume a given ad hoc network instance contains n hosts. Each host is in the ground and is mounted by an omni-directional antenna. Thus the transmission range of a host is a disk. We further assume that each transceiver has the same communication range R. Thus the footprint of an ad hoc wireless network is a unit-disk graph. In graph-theoretic terminology, the network topology we are interested in is a graph $G = (V, E)$ where V contains all hosts and E is the set of links. A link between u and v exists if they are separated by the distance of at most R. In a real world ad hoc wireless network, sometimes even when v is located in u's transmission range, v is not reachable from u due to *hidden/exposed terminal problems*. But in this chapter we only consider bidirectional links. From now on, we use host and node interchangeably to represent a wireless mobile.

There exist several distributed algorithms ([1, 8, 17]) for MCDS computation in the context of ad hoc wireless networking. The one in [1] first builds a rooted tree distributedly. Then the status (inside or outside of the CDS) is assigned for each host based on the level of the host in the tree. Das and Bharghavan [8] provide the distributed implementation of the two centralized algorithms given by Guha and Khuller [9]. Both implementations suffer from high message complexities. The one given by Wu and Li in [17] has no performance analysis, but it needs at least two-hop neighborhood information. The status of each host is assigned based on the connectivity of its neighbors. We will compare our algorithm with the other approaches [1, 8, 17] in Sections 3 and 4.

The remainder of this chapter is organized as follows. Section 2 provides basic concepts related to this topic. Our algorithm and its theoretic performance analysis are presented in Section 3. Simulation results are demonstrated in Section 4. Section 5 concludes the chapter.

2.   Preliminaries

Given graph $G = (V, E)$, two vertices are *independent* if they are not neighbors. For any vertex v, the set of *independent neighbors* of v is a subset of v's neighbors such that any two vertices in this subset are independent. An *independent set (IS)* S of G is a subset of V such that $\forall u, v \in S, (u, v) \notin E$. S is *maximal* if any vertex not in S has a neighbor in S (denoted by MIS).

A *dominating set (DS)* D of G is a subset of V such that any node not in D has at least one neighbor in D. If the induced subgraph of D is connected, then D is a *connected dominating set (CDS)*. Among all CDSs of graph G, the one with minimum cardinality is called a *minimum connected dominating set (MCDS)*. Computing an MCDS in a unit graph is NP-hard [7]. Note that the problem of finding an MCDS in a graph is equivalent to the problem of finding a spanning tree (ST) with maximum number of leaves. All non-leaf nodes in the spanning tree form the MCDS. An MIS is also a DS.

For a graph G, if $e = (u, v) \in E$ iff $length(e) \leq 1$, then G is called a *unit-disk graph*. We will only consider unit-disk graphs in this chapter. From now on, when we say a "graph G", we mean a "unit-disk graph G". The following lemma was proved in [1]. This lemma relates the size of any MIS of unit-disk graph G to the size of its optimal CDS.

Lemma 2.1. *[1] The size of any MIS of G is at most $4 \times opt + 1$, where opt is the size of any MCDS of G.*

For a minimization problem $\mathcal{P}$, the *performance ratio* of an approximation algorithm A is defined to be $\rho_A = sup_{i \in I} \frac{A_i}{opt_i}$, where I is the set of instances of $\mathcal{P}$, A_i is the output from A for instance i and opt_i is the optimal solution for instance i. In other words, ρ is the supreme of $\frac{A}{opt}$ among all instances of $\mathcal{P}$.

3.   An 8-Approximate Algorithm to Compute CDS

In this section, we propose a distributed algorithm to compute CDS. This algorithm contains two phases. First, we compute a maximal independent set (MIS); then we use a tree to connect all vertices in the MIS. We will show that our algorithm has performance ratio at most 8 and is message and time efficient.

3.1. Algorithm Description

Initially each host is colored *white*. A dominator is colored *black*, while a dominatee is colored *gray*. The *effective degree* of a vertex is the total number of white neighbors. We assume that each vertex knows its distance-one neighbors and their effective degrees d^*. This information can be collected by periodic or event-driven hello messages.

We also designate one host as the *leader*. This is a realistic assumption. For example, the leader can be the commander's mobile for a platoon of soldiers in a mission. If it is impossible to designate any leader, a distributed leader-election algorithm can be applied to find out a leader. This adds message and time complexity. The best leader-election algorithm (see [4]) takes time $O(n)$ and message $O(n \ log \ n)$ and these are the best-achievable results. Assume host s is the leader.

Phase 1. Host s first colors itself black and broadcasts message DOMINATOR. Any white host u receiving DOMINATOR message the first time from v colors itself gray and broadcasts message DOMINATEE. u selects v as its dominator. A white host receiving at least one DOMINATEE message becomes active. An active white host with highest (d^*, id) among all of its active white neighbors will color itself black and broadcast message DOMINATOR. A white host decreases its effective degree by 1 and broadcasts message DEGREE whenever it receives a DOMINATEE message. Message DEGREE contains the sender's current effective degree. A white vertex receiving a DEGREE message will update its neighborhood information accordingly. Each gray vertex will broadcast message NUMOFBLACKNEIGHBORS when it detects that none of its neighbors is white. Phase 1 terminates when no white vertex left.

Phase 2. When s receives message NUMOFBLACKNEIGHBORS from all of its gray neighbors, it starts phase 2 by broadcasting message M. A host is "ready" to be explored if it has no white neighbors. We will use a tree to connect all black hosts generated in Phase 1. The idea is to pick those gray vertices which connect to many black neighbors. We will modify the classical distributed depth first search spanning tree algorithm given in [3] to compute the tree.

A black vertex without any dominator is *active*. Initially no black vertex has a dominator and all hosts are *unexplored*. Message M contains a field *next* which specifies the next host to be explored. A gray vertex with at least one active black neighbor is *effective*. If M is built by a black vertex, its *next* field contains the *id* of the unexplored gray neighbor which connects to maximum number of active black hosts. If M is built by a gray vertex, its *next* field contains the *id* of any unexplored black neighbor. Any black host u receiving an M message the first time from a gray host v sets its dominator to v by broadcasting message PARENT. When a gray host u receives message M from v that specifies u to be explored next, u then colors itself black, sets its dominator to v and broadcasts its own M message. Any gray vertex receiving message PARENT from a black neighbor will broadcast message NUMOFBLACKNEIGHBORS, which contains the number of active black neighbors. A black vertex becomes *inactive* after its dominator is set. A gray vertex becomes *ineffective* if none of its black neighbors is active. A gray vertex without active black neighbor, or a black vertex without effective gray neighbor, will send message DONE to the host which activates its exploration or to its dominator. When s gets message DONE and it has no effective gray neighbors, the algorithm terminates.

Note that phase 1 sets the dominators for all gray vertices. Phase 2 may modify the dominator of some gray vertex. The main job for phase 2 is to set a dominator for each black vertex. All black vertices form a CDS.

In Phase 1, each host broadcasts each of the messages DOMINATOR and DOMINATEE at most once. The message complexity is dominated by message DEGREE, since it may be broadcasted Δ times by a host, where Δ is the maximum degree. Thus the message complexity of Phase 1 is $O(n \cdot \Delta)$. The time complexity of Phase 1 is $O(n)$.

In phase 2, vertices are explored one by one. The total number of vertices explored is the size of the output CDS. Thus the time complexity is at most $O(n)$. The message complexity is dominated by message NUMOFBLACKNEIGHBORS, which is broadcasted at most 5 times by each gray vertex because a gray vertex has at most 5 black neighbors in a unit-disk graph. Thus the message complexity is also $O(n)$.

From the above analysis, we have

Theorem 3.1. *The distributed algorithm has time complexity $O(n)$ and message complexity $O(n \cdot \Delta)$.*

Note that in phase 1 if we use (id) instead of (d^*, id) as the parameter to select a white vertex to color it black, the message complexity will be $O(n)$ because no DEGREE messages will be broadcasted. $O(n \cdot \Delta)$ is the best result we can achieve if effective degree is taken into consideration.

3.2. Performance Analysis

In this subsection, we study the performance of our algorithm.

Lemma 3.2. *Phase 1 computes an MIS which contains all black nodes.*

Proof. A node is colored black only from white. No two white neighbors can be colored black at the same time since they must have different (d^*, id).

When a node is colored black, all of its neighbors are colored gray. Once a node is colored gray, it remains in color gray during Phase 1. ∎

From the proof of Lemma 3.2, it is clear that if (id) instead of (d^*, id) is used, we still get an MIS. Intuitively, this would yield an MIS of a larger size.

Lemma 3.3. *In phase 2, at least one gray vertex which connects to maximum number of black vertices will be selected.*

Proof. Let u be a gray vertex with maximum number of black neighbors. At some step in phase 2, one of u's black neighbors v will be explored. In the following step, u will be explored. This exploration is triggered by v. ∎

Lemma 3.4. *If there are c black hosts after phase 1, then at most $c - 1$ gray hosts will be colored black in phase 2.*

Proof. In phase 2, the first gray vertex selected will connect to at least 2 black vertices. In the following steps, any newly selected gray vertex will connect to at least one new black vertex. ∎

Lemma 3.5. *If there exists a gray vertex which connects to at least 3 black vertices, then the number of gray vertices which are colored black in phase 2 will be at most $c - 2$, where c is the number of black vertices after phase 1.*

Proof. From Lemma 3.3, at least one gray vertex with maximum black neighbors will be colored black in phase 2. Denote this vertex by u. If u is colored black, then all of its black neighbors will choose u as its dominator. Thus, the selection of u causes more than 1 black hosts to be connected. ∎

Theorem 3.6. *Our algorithm has performance ratio at most 8.*

Proof. From Lemma 3.2, phase 1 computes an MIS. We will consider two cases here.

If there exists a gray vertex which has at least 3 black neighbors after phase 1, from Lemma 2.1, the size of the MIS is at most $4 \cdot opt + 1$. From lemma 3.5, we know the total number of black vertices after phase 2 is at most $4 \cdot opt + 1 + ((4 \cdot opt + 1) - 2) = 8 \cdot opt$.

If the maximum number of black neighbors a gray vertex has is 2, then the size of the MIS computed in phase 1 is at most $2 \cdot opt$ since any vertex in MCDS connects to at most 2 vertices in the MIS. Thus from Lemma 3.4, the total number of black hosts will be $2 \cdot opt + 2 \cdot opt - 1 < 4 \cdot opt$. ∎

Note that from the proof of Theorem 3.6, if (id) instead of (d^*, id) is used in phase 1, our algorithm still has performance ratio at most 8.

We compare the theoretical performance of our algorithm with the algorithms proposed in [1, 8, 17] in Table 3.1. The parameters used for comparison include the (upper bound of the) cardinality of the generated CDS (CDSC), the message and time complexities (MC and TC, respectively), the message length (ML) and neighborhood information (NI).

Table 3.1. Performance comparison of the algorithms in [8, 17, 1] and the one proposed in this chapter. Here *opt* is the size of the MCDS; Δ is the maximum degree; $|C|$ is the size of the generated connected dominating set; m is the number of edges; n is the number of hosts.

	[8]-I	[8]-II	[17]	[1]	A
CDSC	$\leq (2ln\Delta + 3)opt$	$\leq (2ln\Delta + 2)opt$	N/A	$\leq 8\,opt + 1$	$< 8\,opt$
MC	$O(n|C| + m + nlog\,n)$	$O(n|C|)$	$O(n\Delta)$	$O(nlog\,n)$	$O(n)$
TC	$O((n + |C|)\Delta)$	$O(|C|(\Delta + |C|))$	$O(\Delta^2)$	$O(n\Delta)$	$O(n\Delta)$
ML	$O(\Delta)$	$O(\Delta)$	$O(\Delta)$	$O(1)$	$O(1)$
NI	2-hop	2-hop	2-hop	1-hop	1-hop

Note that the last column (labeled by A) in Table 3.1 corresponds to our algorithm. We see our algorithm is superior over the two algorithms in [8] for all parameters. Algorithm in [17] takes less time than our algorithm but it has much higher message complexity and it uses more complicated message information. The algorithm in [1] is comparable with our algorithms in many parameters. But the simulation results in Section 4 show that our algorithm computes smaller in average connected dominating sets for both random and uniform graphs.

4.   Simulation

Table 3.1 in the previous section compares our algorithms with others in [1, 8, 17] theoretically. In this section, we will compare the size of the CDSs computed by different algorithms. As mentioned earlier, the virtual backbone is mainly used to disseminate control packets. Thus the most important parameter is the number of hosts in the virtual backbone after it is constructed. The larger the size of a virtual backbone, the larger number of transmissions to broadcast a message to the whole network is needed. Note that the message complexities of the algorithms in [8] and [17] are too high compared to other algorithms and they need 2-hop neighborhood information. Thus we will not consider them in the simulation study. We will compare our algorithm with the one given by [1].

We will consider two kinds of topologies: random and uniform. We assume there are N hosts distributed randomly or uniformly in a 100×100 square

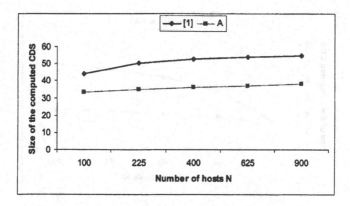

Figure 3.1.   Averaged results for R=15 in random graphs.

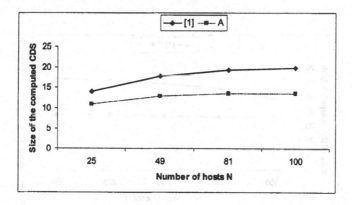

Figure 3.2.   Averaged results for R=25 in random graphs.

units. Transmission range R is chosen to be 15, 25 or 50 units. For each value of R, we run our algorithms 100 times for different values of N. The averaged results are reported in Figures 3.1, 3.2 and 3.3 for random graphs, and in Figures 3.4, 3.5 and 3.6 for uniform graphs. From these figures it is clear that in all of our simulation scenarios, our algorithm performs better than the algorithm in [1].

5.   Conclusion

In this chapter we provide a distributed algorithm which computes a connected dominating set of a small size. Our algorithm has performance ratio at most 8 which is the best to our knowledge. This algorithm takes time $O(n)$ and message $O(n \cdot \Delta)$. Our future work is to investigate the performance of vir-

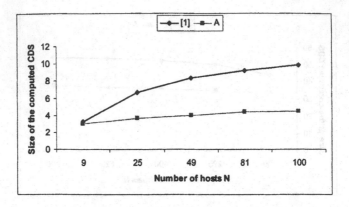

Figure 3.3.   Averaged results for R=50 in random graphs.

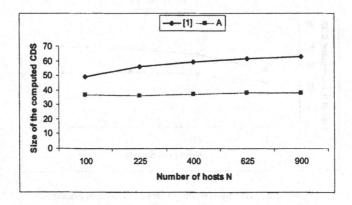

Figure 3.4.   Averaged results for R=15 in uniform graphs.

tual backbone routing and to study the problem of maintaining the connected
dominating set in a mobility environment.

References

[1] K.M. Alzoubi, P.-J. Wan and O. Frieder, "Message-efficient distributed
    algorithms for connected dominating set in wireless ad hoc networks",
    manuscript, 2001.

[2] A.D. Amis and R. Prakash, "Load-balancing clusters in wireless ad hoc
    networks", *Proc. of Conf. on Application-Specific Systems and Software
    Engineering Technology (ASSET 2000), Richardson, Texas*, pp. 25-32,

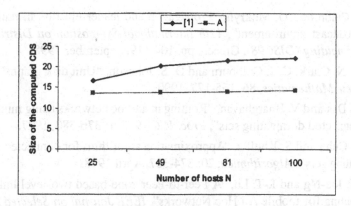

Figure 3.5.   Averaged results for R=25 in uniform graphs.

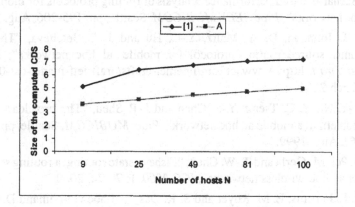

Figure 3.6.   Averaged results for R=50 in uniform graphs.

2000.

[3] H. Attiya and J. Welch, *Distributed computing: fundamentals, simulations and advanced topics*, McGraw-Hill Publishing Company, 1998.

[4] B. Awerbuch, "Optimal distributed algorithm for minimum weight spanning tree, counting, leader election and related problems", *Proceedings of the 19th ACM Symposium on Theory of Computing, ACM*, pp. 230-240, 1987.

[5] V. Bharghavan and B. Das, "Routing in ad hoc networks using minimum connected dominating sets", *Proc. ICC 1997*, Montreal, Canada, June 1997.

[6] I. Cidon and O. Mokryn, "Propagation and leader election in multihop broadcast environment", *12th International Symposium on Distributed Computing (DISC98)*, Greece, pp. 104-119, September 1998.

[7] B. N. Clark, C. J. Colbourn and D. S. Johnson, "Unit disk graphs", *Discrete Mathematics*, 86: 165-177, 1990.

[8] B. Das and V. Bharghavan, "Routing in ad-hoc networks using minimum connected dominating sets", *Proc. ICC 1997*, 1: 376-380, 1997.

[9] S. Guha and S. Khuller, "Approximation algorithms for connected dominating sets", *Algorithmica*, 20: 374-387, April 1998.

[10] M. Joa-Ng and I.-T. Lu, "A Peer-to-Peer zone-based two-level link state routing for mobile Ad Hoc Networks", *IEEE Journal on Selected Areas in Communications*, 17: 1415-1425, 1999.

[11] P. Johansson, T. Larsson, N. Hedman, B. Mielczarek and M. Degermark, "Scenario-based performance analysis of routing protocols for mobile ad hoc networks", *Proc. IEEE MOBICOM*, Seattle, pp. 195-206, Aug. 1999.

[12] D.B. Johnson, D. A. Maltz, Y.-C. Hu and J. G. Jetcheva, "The dynamic source routing protocol for mobile ad hoc networks", *Internet Draft* http://www.ietf.org/internet-drafts/draft-ietf-manet-dsr-07.txt, March 2001.

[13] S.-Y. Ni, Y.-C. Tseng, Y.-S. Chen and J.-P. Sheu, "The broadcast storm problem in a mobile ad hoc network", *Proc. MOBICOM*, Seattle, pp. 151-162, Aug. 1999.

[14] G. Pei, M. Gerla and T.-W. Chen, "Fisheye state routing: a routing scheme for ad hoc wireless networks", *ICC 2000*, 1: 70-74, 2000.

[15] C. E. Perkins, E. M. Royer and S. R. Das, "Ad hoc On-demand Distance Vector (AODV) Routing", *Internet Draft* http://www.ietf.org/internet-drafts/draft-ietf-manet-aodv-08.txt, March 2001.

[16] P. Sinha, R. Sivakumar and V. Bharghavan, "Enhancing ad hoc routing with dynamic virtual infrastructures", *Proc. INFOCOM 2001*, 3: 1763-1772, 2001.

[17] J. Wu and H. Li, "On calculating connected dominating set for efficient routing in ad hoc wireless networks", *Proc. 3rd International Workshop on Discrete Algothrithms and Methods for MOBILE Computing and Communications*, Seattle, WA USA, pp.7-14, 1999.

Chapter 4

CONTROL GRAPHS FOR ROBOT NETWORKS

Aveek Das
GRASP Laboratory,
University of Pennsylvania, Philadelphia PA
aveek@grasp.cis.upenn.edu

Rafael Fierro
School of Electrical and Computer Engineering,
Oklahoma State University
rfierro@okstate.edu

Vijay Kumar
GRASP Laboratory,
University of Pennsylvania, Philadelphia PA
kumar@grasp.cis.upenn.edu

Abstract   In this paper we address the problem of stabilizing a group of mobile robots in formation. The group is required to follow a prescribed trajectory, while achieving and maintaining a desired formation. We describe algorithms for assigning control policies to different robots, based on sensor and actuator constraints. This assignment is described by a *control graph*. We relate the structure of the control graph to the stability of the dynamics of the formation. We examine both holonomic and nonholonomic mobile robot formations, and present analytical results and numerical simulations illustrating our approach.

Keywords: Formation control, cooperative control, nonholonomic mobile robots, stability, control graphs

S. Butenko et al. (eds.), Cooperative Control: Models, Applications and Algorithms, 55-73.
© 2003 *Kluwer Academic Publishers.*

1.　　Introduction

In real world situations multi-agent robotic systems are subject to sensor, actuator and communication constraints, and have to operate within uncertain or unstructured environments. We are interested in tasks that include exploration [3], surveillance [11], search and rescue [14] operations, mapping of unknown or partially known environments [28], and distributed manipulation [21, 19] and transportation of large objects [24, 25]. In all these applications, there is a need to have the robots estimate their relative positions and orientations with respect to their neighbors and maintain a desired formation.

Many formation control approaches use the *leader-following* framework [29, 8, 6]. In this case, a *follower* uses the state of its *leader(s)* to compute its control signals such that predefined separations and orientations are maintained. The behavior of such systems can be analyzed using the theory of interconnected systems [16]. For example, the analysis of convoy-like formations [4] leads to important considerations like string stability [26] and mesh stability. Asymptotic stabilization of nonholonomic mobile robot formations [30], one-dimensional swarm stability [18], and stability with changes in controllers [12] have also been investigated. The stability of coupled linear decentralized control systems [22, 11] has been shown to depend on the eigenvalues of the formation graph Laplacian [10].

In [17], leader-following and virtual structures [27] are combined. Thus, two different error functions are defined, namely a formation error and an individual tracking error. In a stable maneuver, both errors should converge to zero simultaneously. In [15] authors use the concept of *action reference* to develop a higher level controller that coordinates a group of autonomous vehicles. Individual controllers can be designed separately using any well-known technique. A similar approach has been proposed in [9] where a *formation constrain* function is defined and used as a coordination controller.

Figure 4.1.　Our Clodbuster wheeled mobile robot platform with omnidirectional cameras and wireless networking (left). Typical view from omnidirectional camera (right).

In this work, we are interested in the leader-follower assignment paradigm where each robot follows one or two leaders [8]. The goal is to achieve and maintain a desired *shape*. Here shape refers to the distribution of the robots without considering the position and orientation of the group. The choice of leader-follower controllers leads to the description of a *control graph*, and the stability of the system depends on this graph. We are motivated by our experimental platform of wheeled mobile robots with omnidirectional cameras and IEEE 802.11b wireless networking (see Figure 4.1). Our previous work in modeling this system and developing vision based controllers is described in a series of papers [6, 23, 12, 1]. In this paper, we address the issue of selecting control policies for each robot based on sensing constraints, and deriving measures of performance for the group. We start with some stability results for holonomic and nonholonomic robots in Section 3 and illustrate them with examples. Next, in Section 4 we see how these and other measures are incorporated in our control graph assignment algorithm. We illustrate through examples how different assignments for similar trajectories of the leader affect the performance of the formation. Finally some future work ideas and conclusions are presented in Sections 5 and 6.

2.   Formation Model

A team of N robots is built on three different networks: a *physical network* that captures the physical constraints on the dynamics, control and sensing of each robot; a *communication network* that describes the information flow between the robots; and a *computational network* that describes the computational resources available to each robot.

We model each network by a graph with N nodes, one node for each agent. R is a finite set of nodes, $R_1, R_2, \ldots, R_N$. The physical network is a directed graph, $G_p = (R, E_p)$, where E_p consists of edges each of which represent the flow of sensory information (relative state). Thus the edge $(R_i, R_j) \in E_p$ whenever robot R_i can see robot R_j. $G_c = (R, E_c)$ is an undirected graph representing the communication network. The edge set E_c consists of pairs of robots that can communicate with each other (assuming omnidirectional transmitters and receivers on each robot) G_p and G_c are determined by constraints of the hardware, the physical distribution of the robots, and the characteristics of the environment.

The key goal is to design the computational network. This network is modeled by a directed acyclic graph $H = (R, E)$. In our work, E will consist of edges that belong to $E_p \cap E_c$. In other words, we will be interested in robots that can see each other and talk to each other [1]. The design of the graph H is based on the task. For example, if we want a team of robots to transport a payload through a field of obstacles to a destination, a control-centric point of view leads to the assignment of edges with the goal of maximizing performance goals such as stability, robustness to performance, and time-optimality. On the other hand, if the task is to explore the environment and build a three-dimensional reconstruction of the environment, H must be designed with different perception-centric performance measures. In tasks involving target detection, pursuit and evasion of threats, the performance measures are more intertwined.

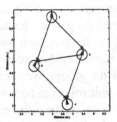

Figure 4.2.   Control graph for a 4 robot formation

In this paper we will primarily be interested in controlling the formation of the groups of mobile robots. Hence the use of the term *control graph* for the graph H. Edges $(R_i, R_j) \in H$ are associated with the agent R_j following R_i. In addition, for the physical network, we are interested in: (a) the position and orientation of the team in space denoted by g; and (b) the shape of the formation denoted by r. g is an element of the related motion group (the special Euclidean group in two dimensions $SE(2)$, in our case), while $r \in \mathbb{R}^m$, the shape vector, describes the distribution of the robots around g. Thus we represent the group as a tuple $\mathcal{F} = (g, r, H)$ (as defined in [8]). Since each robot is assumed to have a unique label, it is enough to describe the relative positions with respect to a designated robot. It is possible to add *virtual robots* to the group (and corresponding nodes in the graph) to represent either moving targets, or trajectories that are along (or avoiding) obstacle boundaries.

Any control graph can be represented by its *adjacency matrix* (see [13] for definition). Consider a possible control graph shown in Figure 4.2, where R_2 follows R_1, R_3 follows R_1 and R_2, while R_4 follows R_2 and R_3. The adjacency matrix is given by

$$H = \begin{bmatrix} 0 & 1 & 1 & 0 \\ 0 & 0 & 1 & 1 \\ 0 & 0 & 0 & 1 \\ 0 & 0 & 0 & 0 \end{bmatrix}. \tag{1}$$

Note that this is a directed acyclic graph (DAG) with the control flow from leader i to follower j. If a column j has a non zero entry in row i, then robot R_j is following R_i. A robot can have up to m leaders, as this is limited by the maximum number of outputs we can regulate using a nonholonomic robot controlled by m inputs. For planar robots operating in $SE(2)$, $m = 2$. The column with all zeros corresponds to the lead robot. A row with all zeros corresponds to a *terminal follower*.

The design of the computational network involves the assignment of control policies for each robot. The ability of a robot R_j to sense another robot R_i allows R_j to use a state feedback controller that regulates relative position and/or orientation of R_j with respect to R_i. The ability of R_j to listen to R_i allows R_i to broadcast feed-forward information and for R_j to use feed-forward control. We are interested in the role of feedback and feed-forward control in the performance of the group. This is addressed in the next section.

3.   Stability

In this section we analyze the effect that the structure of the control graph has on the stability of the formation. We first look at the simpler model of a point robot without nonholonomic constraints, and consider linear control laws with and without feed-forward information. We then analyze formation stability of a group of nonholonomic robots with controllers designed using input-output feedback linearization. The table below summarizes our results.

Table 4.1.   Summary of stability results

	Feedback	Feed-forward
Holonomic	Theorem 1	Theorem 2
Nonholonomic	No theoretical results, experimental results in [6]	Bounds on leader's inputs guarantee stability of zero dynamics

3.1.   Linear Holonomic System

We first consider a linear system model to better understanding the connection between stabilizability and the structure of control graphs. Let us consider a group of N holonomic point robots with kinematics of the ith robot R_i given by $\dot{x}_i = u_i$, where the states $x_i \in \mathbb{R}^2$ and velocity control inputs $u_i \in U \subset \mathbb{R}^2$. Assuming each subsystem is stabilized by feedback control laws of the form $u_i = -\sum_{j=1}^{N} K_{ij}(x_i - x_j)$, we can rewrite the closed loop dynamics as:

$$\dot{x}_i = -k_{ii}x_i + \sum_{j \in J_i} k_{ij}x_j, \qquad (2)$$

where $i \in [1, N]$, $J_i = [1, N] \backslash \{i\}$, and all $k_{ij} > 0$. Notice that each robot measures its position relative to a leader (which could be a virtual leader for the lead robot). The above interconnected system has an associated directed control graph $H = (V, E)$ represented by its adjacency matrix as described in Section 2. The graph has to be connected for the full state to be represented with respect to a single reference robot. This is a reachability criterion [22] and is a necessary condition for further stability analysis.

We now state a well known result from graph theory and show its implications on group stability.

Definition 1. *Given a system (2), the following statements are equivalent:*

1) The vertices of H can be reordered so that the resulting dynamical system is in lower (upper) triangular form.

2) The graph H is a directed acyclic graph (DAG).

If we collect the individual subsystem states into $x = [x_1 \ldots x_N]^T$, we can rewrite the Eqn (2) as

$$\dot{x} = -Kx. \tag{3}$$

Theorem 3.1. *If a linear interconnected system (2) satisfies definition 1, then the closed loop system is stable.*

Proof: Given an acyclic control graph (see definition 1) this becomes -

$$\dot{x} = \begin{pmatrix} -k_{11} & 0 & \cdots & 0 \\ k_{21} & -k_{22} & \cdots & 0 \\ \vdots & \cdots & \ddots & \vdots \\ k_{N1} & \cdots & \cdots & -k_{NN} \end{pmatrix} x. \tag{4}$$

Since the control gains $k_{ii} > 0$, the eigenvalues $\lambda_i = -k_{ii} < 0$. Thus the system is stabilized.

$$\square$$

We look at the case where we have feed-forward terms available to us for design of our local control laws. This means that R_i knows $\dot{x}_j$ in addition to x_j. For each leader follower pair the dynamics can be written as

$$\dot{x}_i - \dot{x}_j = -k_{ij}[x_i - x_j]. \tag{5}$$

Thus, through a coordinate transform $z_j = x_i - x_j$ and by stacking the $z_j(s)$ into z we can convert the interconnected system to the form

$$\dot{z} = -Kz. \tag{6}$$

Theorem 3.2. *A linear interconnected system with closed loop relative dynamics specified by Eqn (6) is stable.*

Proof: If the control graph is acyclic, the matrix K is diagonal in the relative coordinates and hence for all positive control gains the eigenvalues are negative and real.

$\square$

The next obvious question is - if the control graph has cycles, is the interconnected system stabilizable? Since K is no longer lower-triangular, the choice of gains will influence stability. It is possible to derive a set of positive gains that will drive the system unstable. Presently we have no result relating cycles to stability.

3.2. Feedback Linearized Nonholonomic System

In this section we consider a *nonlinear interconnected* formation of robots. The robots are velocity controlled nonholonomic car-like platforms and have two independent inputs. This means we are able to regulate two outputs. The kinematics of the ith robot can be abstracted by a unicycle model given by

$$\dot{x}_i = v_i \cos\theta_i, \quad \dot{y}_i = v_i \sin\theta_i, \quad \dot{\theta}_i = \omega_i, \quad (7)$$

where $x_i \equiv (x_i, y_i, \theta_i) \in SE(2)$, and inputs $u_i = [v_i \ \omega_i]^T$ with v_i and ω_i as the linear and angular control velocities, respectively.

Basic leader-following controller. First, consider a *basic leader–following* controller (denoted by *SBC* here) as shown in Figure 4.3. The follower robot R_j is to follow its leader R_i with a desired separation and relative bearing l_{ij}^d and ψ_{ij}^d, respectively. The state-vector $x_j = [x_j \ \ y_j \ \ \theta_j]^T$ is then

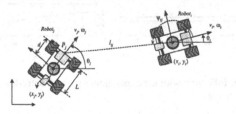

Figure 4.3. A 2-robot interconnected system (Basic leader–follower control SBC_{ij}).

transformed into a new state-vector $r_j = [l_{ij} \ \ \psi_{ij} \ \ \beta_{ij}]^T$ through a transformation $r_j = T_1(x_i, x_j)$ given by

$$
\begin{aligned}
l_{ij} &= \sqrt{(x_i - x_j - d\cos\theta_j)^2 + (y_i - y_j - d\sin\theta_j)^2}, \\
\psi_{ij} &= \pi - \arctan 2(y_j + d\sin\theta_j - y_i, x_i - x_j - d\cos\theta_j) - \theta_i, \quad (8) \\
\beta_{ij} &= \theta_i - \theta_j,
\end{aligned}
$$

d is the offset on the follower shown in Figure 4.3. As in [8], the kinematic model in the new r-coordinates is found by taking the derivative of (8)

$$\dot{l}_{ij} = v_j \cos(\beta_{ij} + \psi_{ij}) - v_i \cos \psi_{ij} + d\omega_j \sin(\beta_{ij} + \psi_{ij}),$$

$$\dot{\psi}_{ij} = \frac{1}{l_{ij}}(v_i \sin \psi_{ij} - v_j \sin(\beta_{ij} + \psi_{ij}) + d\omega_j \cos(\beta_{ij} + \psi_{ij})) - \omega_i,$$

$$\dot{\beta}_{ij} = \omega_i - \omega_j. \tag{9}$$

The two outputs of interest are $z_j = [l_{ij} \quad \psi_{ij}]^T$. The input-output feedback

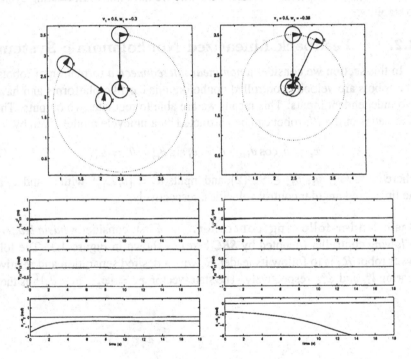

Figure 4.4. Leader following with the Separation-Bearing Controller: $|\omega_1| < W_{1,max}$ (left) and $|\omega_1| > W_{1,max}$ (right).

linearization control problem [16] is to derive a feedback law such that $\tilde{z}_j \equiv (z_j^d - z_j) \to 0$ as $t \to \infty$, and the internal dynamics of the system are stable i.e., $|\beta_{ij}| < B$ for some $B < \infty$. In [12] we show that when

$$v_i > 0,$$
$$|\omega_i| \leq W_{i,max} = v_i/(d + max\{l_{ij}^d, l_{ij}(0)\}), \tag{10}$$

and the initial relative orientation $|\beta_{ij}(0)| < \pi$, such a feedback law can be derived for the interconnected system (8). The closed-loop linearized system

becomes

$$\dot{l}_{ij} = k_1(l_{ij}^d - l_{ij}), \quad \dot{\psi}_{ij} = k_2(\psi_{ij}^d - \psi_{ij}), \quad \dot{\beta}_{ij} = \omega_i - \omega_j, \quad (11)$$

where k_1, $k_2 > 0$ are the control gains. Notice that the linearized error dynamics can be written as $\dot{\tilde{z}}_j = -K\tilde{z}_j$ which is similar to Eqn.(6).

We illustrate these bounds with the help of the example in Figure 4.4. The leader's velocity is fixed at $v_1 = 0.5 \; m/s$ and the offset for the follower is $d = 0.2 \; m$. The initial and desired conditions for (l_{12} in m, ψ_{12} in rad) are $(1.15, \frac{3\pi}{2})$ and $(1.2, \frac{3\pi}{2})$ respectively. In the case on the left, $\omega_1 = -0.3 rad/s$. The case on the right corresponds to $\omega_1 = -0.38$. In the former (left) the zero dynamics are stable; however the zero dynamics become unstable when the magnitude of ω_1 is increased (right).

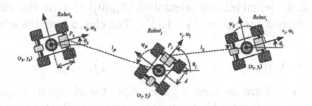

Figure 4.5. An in-line convoy of 3 robots.

Formations of 3 and more robots. Next, we address the performance of formations with more than two robots. Let us assume that a third robot joins the 2-robot formation. One possibility is that R_k follows R_j using a separation-bearing controller. Thus, the formation becomes a convoy-like interconnected system as shown in Figure 4.5. As before, to guarantee stability of the internal dynamics of R_k we need $v_j > 0$ and $|\omega_j| < W_{j,max}$, where $W_{j,max}$ is given by (10) with j, k replacing i, j. This in turn means that v_i and ω_i will have to be appropriately constrained [2], *i.e.* $v_i > V_{i,min}$ and $|\omega_i| < W_{i,max}$. An analysis of the dynamics (7)-(8) with some algebraic manipulation shows that

$$V_{i,min} > -min\{-k_1 max\{\tilde{l}_{ij}\}, \quad (12)$$
$$-k_2 max\{\tilde{\psi}_{ij}\} max\{l_{ij}\} - W_{i,max} max\{l_{ij}\}\}.$$

Notice, it is not enough that $v_i > 0$, but $v_i > V_{i,min}$ where $V_{i,min}$ will depend on the initial formation error, controller gains, and $W_{i,max}$.

A second possibility for the three-robot formation is for R_k is to use the *Separation-Separation Controller* introduced in [8]. In this case the desired formation is a rigid triangle where R_k follows R_i at a predetermined separation l_{ik}^d and R_j at l_{jk}^d. This controller is referred to as SSC_{ijk} as shown in Figure

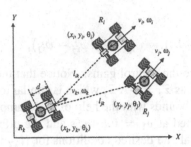

Figure 4.6. A three robot triangular formation using SSC_{ijk}.

4.6. Assuming R_j is following R_i using SBC_{ij}, in [12] we showed that the internal dynamics of the follower robot R_k are stable as long as $v_i > 0$, $|\omega_i| < W_{i,max}$, and the initial relative orientation $|\beta_{ik}(0)| < \pi$. In this case, the two outputs of interest are $z_k = [l_{ik} \quad l_{jk}]^T$. The closed-loop linearized system becomes

$$\dot{l}_{ik} = k_3(l_{ik}^d - l_{ik}), \quad \dot{l}_{jk} = k_4(l_{jk}^d - l_{jk}), \quad \dot{\beta}_{ik} = \omega_i - \omega_k, \quad (13)$$

where k_3, $k_4 > 0$ are the control gains. Also, the triangular inequality $l_{ik} + l_{jk} - l_{ij} \geq \epsilon_{ijk} > 0$ should hold otherwise the controller SSC is not defined. Under these assumptions, the velocity bounds on R_i can be computed to ensure that $v_k > 0$ (assuming R_j follows R_i with $v_j > 0$). We require $v_k > 0$ and $|\omega_k| < W_{k,max}$ so that R_k can be chosen as a leader by an additional follower R_{k+1}. If the leader satisfies these bounds, the terminal follower and hence the whole chain is stable.

These ideas can be applied to arbitrary chained combinations of basic leader–follower (SBC) and separation–separation controllers (SSC). Thus given any leader-follower graph we can recursively calculate bounds starting form the last follower to the leader. The bounds are given by Lyapunov theory and are conservative but they provide a good way of deciding whether a trajectory assigned to the leader maintains the stability of the formation.

4. Choice of Control Graphs

It is clear that the number of possible control graphs increases dramatically with the number of robots. If we constrain the control graph to be acyclic, we can always rearrange the labels on the nodes so as to get it in an upper triangular form [13] which, as we saw in Section 3, is important for formation stability. The number of possible directed acyclic graphs for even 10 robots exceeds a billion. Identifying the appropriate control graph for a given situation is an important and challenging problem.

In this work, we look at the following problem — given a distribution of N robots, find a formation control graph H assigning a controller and leader(s) for each robot subject to the following two constraints: (a) *kinematic constraints* that must be satisfied by the relative position and orientation between neighboring robots; and (b) *sensor and communication constraints* based on the limits on range and field of view of sensor and communication device(s) that prevent a robot from obtaining complete information about its neighbors. Among the feasible control graphs that satisfy the constraints, we select those control graphs that locally maximize a safety measure and a stability measure. We assume the desired formation shape is specified, and the lead robot is elected or designated [3].

4.1. Constraints

Kinematic Constraints. The nonlinear kinematic controllers in Section 3 have singularities (see [12]) which constrain the choice of controllers based on the configuration of the group. The Separation-Bearing Controller is not defined for a pair of robots (R_i, R_j) having initial relative orientation satisfying $|\theta_i - \theta_j| = \pi$, which corresponds to them facing towards or away from each other. Further, the Separation-Separation Controller is not applicable for robot R_k to follow R_i, R_j when the inter-robot separations are such that $\epsilon_{ijk} = l_{ik} + l_{jk} - l_{ij}$ is zero (refer to Figure 4.6). Thus, the assignment of edges in the control graph is constrained by these fundamental limitations of the Separation-Bearing and Separation-Separation Controllers.

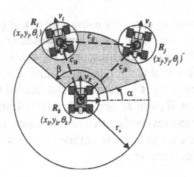

Figure 4.7. Schematic of sensor constraints for a group of three robots.

Sensing Constraints. The sensor and communication constraints limit the observations that are possible. We must restrict the choice of control graphs to those that are compatible with the sensor constraints (*e.g.*, a robot cannot follow a robot that it cannot see even if it is within communication range). We adopt a generic sensor model (omni-directional sensing [2] being a special case). The sensor[4] has a limited range r_s and a limited angular field of view parameterized by the angles α and β as shown in Figure 4.7. Notice that R_k detect R_i but not R_j.

4.2.   Measures of Performance

Measure of safety.   In order to prevent collisions, we want to ensure that the separation c_{ij} between robots R_i and R_j is above a threshold (see Figure 4.7). In addition, we will consider the rate of change of this separation and ensure that relative motion between the robots do not cause this separation to decrease below the threshold rapidly. Consider first the dynamics of the formation, where the group configuration is written as $x = [x_1\ x_2\ \ldots x_N]^T$ with x_i as defined in Section 3, $i \in [1, N]$. Then each robot R_j with control inputs u_j has dynamics given by Eqn. (7) which can be written as

$$\dot{x}_j = f(x_j)u_j, \tag{14}$$

Suppose R_j has to maintain the separation constraint $c_{ij} = c(x_i, x_j) \leq 0$ with a neighboring robot R_i.

$$\dot{c}_{ij} = \frac{dc_{ij}}{dt} = \mathcal{L}_{f_i}c_{ij} + \mathcal{L}_{f_j}c_{ij}, \tag{15}$$

where $\mathcal{L}_{f_i}c_{ij}$ denotes the Lie derivative of c_{ij} along f_i. R_j can estimate the *time to collision* with R_i as:

$$\delta t_{ij} = \frac{c_{ij}}{\dot{c}_{ij}}. \tag{16}$$

If robot R_j can estimate u_i either by using a estimator [6] or by explicit communication [23], it can compute $\dot{c}_{ij}$ and thus estimate δt_{ij}. Both the magnitude and sign of δt_{ij} can be used to identify pairs of robots (R_i, R_j), that are on a collision course. If $\delta t_{ij} = 0$ then collision is imminent. A smaller magnitude and negative δt_{ij} means R_j is closer to R_i and headed towards it. A positive time to collision means R_j and R_i are headed away from each other or are far apart. This concept captures the fact that robots which are close but are facing away from each other are less likely to collide than ones which are farther apart but are headed towards each other.

Measure of stability.   Although acyclicity of the control graph guarantees stability, the performance associated with a control graph depends on the the *maximum depth*, which we define to be the maximum length of the shortest directed path (assuming all control links have same weights) from the leader to any follower. As this depth becomes greater, the formation shape errors have a tendency to grow. Two examples of this are presented in Figures 4.8 and 4.9. We use a simple heuristic based on this idea. When deciding between two control graphs that are otherwise similar we prefer the one with smaller maximum depth.

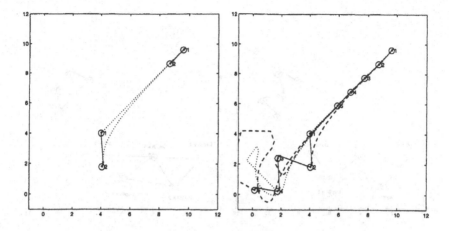

Figure 4.8. The formation on left has one SBC controller, while the one on the right represents an interconnection of 5 SBC controllers. The formation errors on the right become unacceptably large even for a simple straight line constant velocity trajectory for the leader, as the control inputs for followers R_3 and R_4 no longer satisfy the stability bounds in Eqn. (10).

The control graph assignment algorithm has to select the appropriate controller for R_k so as to maintain connectivity of the graph with the maximum allowable in-degree of two (we can regulate two outputs for our nonholonomic platforms) and to maintain stability for all followers in the formation. We extend these ideas in our algorithm in the next section for assigning a controller for each individual robot (node) to build the control graph. We will assume that each robot knows its own state. The only channel of inter-robot communication is via broadcasts on a 802.11b like wireless network.

4.3. Graph Assignment Algorithm

The problem of assigning an optimally connected acyclic graph with degree constraints is NP-hard [5]. However, we can employ heuristics for for choosing controllers (edges) that satisfy the above constraints while optimizing performance for the group. The control graph assignment algorithm in Table 4.2, ensures directed acyclic graphs while maximizing locally two measures of optimality subject to the two sets of constraints (as described in the previous subsections). For a specific choice of a lead robot and a specified desired formation shape, our algorithm assigns control links for each follower minimizing the maximum depth (as defined in Section 4) and hence the shortest path between the leader and the follower.

We divide the assignment procedure into the three sub-procedures: (a) find a spanning tree using standard algorithms [5]; (b) assign initial acyclic leader

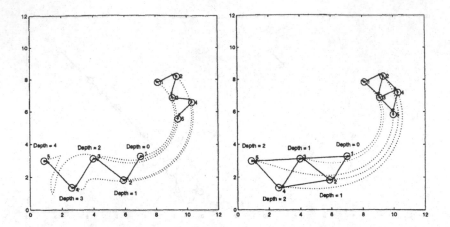

Figure 4.9. The formation on left has all SBC chains, while the one on the right has 1 SBC and 4 SSC links. The maximum depth for a follower is 4 on the left and 2 on the right. For the same leader trajectory, notice how the control graph with the larger maximum depth (left) has higher transient formation shape errors.

follower graph with single leader based control links; and (c) refinement (addition/deletion of edges) of control graph based on local optimality measures. Step (a) is a breadth-first search that is restricted to robots that are connected by edges in G_p. Note if at any point we arrive at a robot that does not have unassigned robots that it can see, there is a break in the control graph. A new subgraph has to be initiated. We use Figure 4.10 to illustrate a typical scenario showing available choices of control graphs.

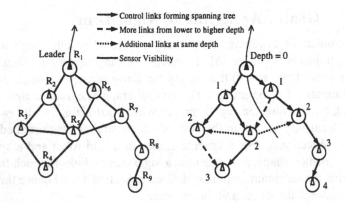

Figure 4.10. Illustration of control graph assignment procedure. Sensor visibility graph (left) and resulting tree structure shown with possible controller choices (right).

Table 4.2. Control graph assignment algorithm.

control_graph_assignment algorithm {
    initialize adjacency matrix $H(i, j) := 0 \; \forall i, j \in \{1, 2, \dots, n\}$;
    create physical network G_p from local sensed information;
    initialize communication network G_c starting from *leader*;
    find depths with $d_{leader} := 0$;
    for each robot $k \in \{1, 2, \dots, n\}$, $k \neq leader$,
        $d_k :=$ depth of node k in G_c;
        find set P_k of robots visible to k with depths $d_k, d_k - 1$;
        if $P_k = \emptyset$ (disconnected)
            report failure at k, break;
        $S_k :=$ P_k sorted by ascending timeToCollision with k;
        if numOfElements(S_k) ≥ 2
            pick last two elements $i, j \in P_k$
            assign $H(i, k) := 1$;
            if $\epsilon_{ijk} \neq 0$
                assign $H(j, k) := 1$ for SSC_{ijk};
            else
                repeat above check for remaining $j \in S_k$ in order;
        else
            assign $H(i, k) := 1$ for SBC_{ij} ($P_k = \{i\}$);
}

The algorithm could run centrally at a supervisory level with the current configurations of all the robots as inputs. This is especially useful in a client(robot)-server(host) implementation of large formations in complex scenarios to keep track of individual choice of controllers and switching between them. The output is the adjacency matrix corresponding to the assigned control graph H.

The distributed version of the control graph assignment algorithm in Table 4.2 involves implementing dynamic distributed spanning tree algorithms and to then add/delete edges locally at each node. Presently we assume that the robots communicate in broadcast mode, and that the robot labels are identifiable by sensing or communication. Optimizing the number of broadcast messages and optimal bandwidth utilization based on a given cooperative control task are important issues that are currently being addressed.

Our approach for choosing a formation control graph also works in the presence of obstacles which can be treated as virtual leaders [12] if they satisfy the sensor constraints. We add an extra row before the present first row of

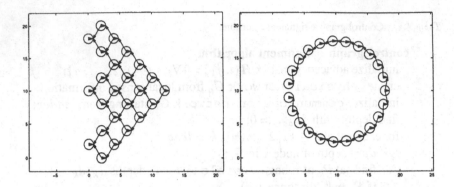

Figure 4.11. Control graph assignment 24 robot formations with various shapes

the adjacency matrix for each obstacle and use the algorithm. Some formation assignment examples are shown in Figure 4.11.

5.   Discussion

Our selection of a control graph is based only on an instantaneous description of the shape and trajectory. It is necessary to investigate the effects of changes in controller assignments on the stability and performance of the group. Our previous work [12] provides some insight into this problem for three-robot formations. It is far from clear how our assignment algorithm will affect group performance for a bigger group of robots. It may be possible to show stability results with respect to a wide class of switching rules. In fact, there are partial results for holonomic robots with random switches between nearest neighbor controllers [20]. However, it is much harder to predict group performance. This is clearly one direction for future work.

6.   Conclusions

A group of robots can be modeled by a tuple (g, r, H) where g represents the team's gross position and orientation, r the shape of the formation, and H, the control graph, represents the assignment of controllers to the robots in the team. In this paper we focus on the problem of determining the control graph for a given shape r and for a given specified trajectory $g(t)$. The assignment of controllers is constrained by constraints on visibility, communication, and relative position and orientation. In addition, the choice of controllers affect the stability of the group and its performance. We derive results for group stability for holonomic and nonholonomic robots. For nonholonomic robots, we show how the performance of the system degrades as the *maximum depth* of the control graph increases. Finally, we present an algorithm for selecting controllers and assigning edges on a control graph, which guarantees group stability for the group and the satisfaction of all constraints. The basic results in the paper are illustrated with several numerical examples.

Acknowledgments

This work was supported by the AFOSR grant no. F49620-01-1-0382 and NSF grant no. CDS-97-03220. We thank Jim Ostrowski, Jaydev Desai, Peng Song, and Joel Esposito for discussions on multi-robot formation control.

Notes

1. One can imagine variations on this theme. For example, robots are guided simply via communication or only via line of sight sensing.

2. This procedure can be applied to an N robot in-line formation. In summary, the smaller the initial formation errors and the smoother the leader's trajectory, the easier it is to maintain a formation shape (*i.e.* the smaller the bounds on the shape errors).

3. A forthcoming paper [7] discusses the leader election protocol for dynamic assignment of leaders.

4. One can imagine the availability of communication links to influence the range and field of view within which estimates of robot states can be measured.

References

[1] R. Alur, A. Das, J. Esposito, R. Fierro, Y. Hur, G. Grudic, V. Kumar, I. Lee, J. P. Ostrowski, G. Pappas, J. Southall, J. Spletzer, and C. J. Taylor, "A framework and architecture for multirobot coordination", In D. Rus and S. Singh, editors, *Experimental Robotics VII*, pages 303–312. Springer Verlag, 2001.

[2] S. Baker and S. Nayar, "A theory of catadoptric image formation", In *International Conference on Computer Vision*, pages 35–42, Bombay, India, Jan. 1998.

[3] W. Burgard, M. Moors, D. Fox, R. Simmons, and S. Thrun, "Collaborative multi-robot exploration", In *Proc. IEEE Int. Conf. Robot. Automat.*, pages 476–481, San Francisco, CA, April 2000.

[4] C. Canudas-de-Wit and A. D. NDoudi-Likoho, "Nonlinear control for a convoy-like vehicle", *Automatica*, 36:457–462, 2000.

[5] T. H. Cormen, C. E. Leiserson, and R. L. Rivest, *Introduction to Algorithms*, The MIT Press, Cambridge, Massachusetts, 1997.

[6] A. Das, R. Fierro, V. Kumar, J. Southall, J. Spletzer, and C. J. Taylor, "Real-time vision based control of a nonholonomic mobile robot", In *IEEE Int. Conf. Robot. Automat., ICRA2001*, pages 157–162, Seoul, Korea, May 2001.

[7] A. Das, V. Kumar, J. Spletzer, and C. J. Taylor, "Ad hoc networks for localization and control of mobile robots", In *IEEE Conf. on Decision and Control* (invited paper), 2002.

[8] J. Desai, J. P. Ostrowski, and V. Kumar, "Controlling formations of multiple mobile robots", In *Proc. IEEE Int. Conf. Robot. Automat.*, pages 2864–2869, Leuven, Belgium, May 1998.

[9] M. Egerstedt and X. Hu, "Formation constrained multi-agent control", In *Proc. IEEE Int. Conf. Robot. Automat.*, pages 3961–3966, Seoul, Korea, May 2001.

[10] J. A. Fax and R. M. Murray, "Graph laplacians and vehicle formation stabilization", In *IFAC 15th World Congress on Automatic Control*, July 2002.

[11] J. Feddema and D. Schoenwald, "Decentralized control of cooperative robotic vehicles", In *Proc. SPIE Vol. 4364, Aerosense*, Orlando, Florida, April 2001.

[12] R. Fierro, A. Das, V. Kumar, and J. P. Ostrowski, "Hybrid control of formations of robots", *Proc. IEEE Int. Conf. Robot. Automat., ICRA01*, pages 157–162, May 2001.

[13] F. Harary, *Graph Theory*, Addison-Wesley, Reading, Massachusetts, 1969.

[14] J. S. Jennings, G. Whelan, and W. F. Evans, "Cooperative search and rescue with a team of mobile robots", *Proc. IEEE Int. Conf. on Advanced Robotics*, 1997.

[15] W. Kang, N. Xi, and A. Sparks, "Theory and applications of formation control in a perceptive referenced frame", In *Proc. IEEE Conference on Decision and Control*, pages 352–357, Sydney, Australia, December 2000.

[16] H. Khalil, *Nonlinear Systems*, Prentice Hall, Upper Sadle River, NJ, 2nd edition, 1996.

[17] J. Lawton, B. Young, and R. Beard, "A decentralized approach to elementary formation maneuvers", In *Proc. IEEE Int. Conf. Robot. Automat.*, pages 2728–2733, San Francisco, CA, April 2000.

[18] Y. Liu, K. M. Passino, and M. Polycarpou, "Stability analysis of one-dimensional asynchronous mobile swarms", In *Proc. IEEE Conference on Decision and Control*, pages 1077–1082, Orlando, FL, 2001.

[19] M. Mataric, M. Nilsson, and K. Simsarian, "Cooperative multi-robot box pushing", In *IEEE/RSJ Int. Conf. on Intelligent Robots and Systems*, pages 556–561, Pittsburgh, PA, Aug. 1995.

[20] A. S. Morse, "Stability for switching control systems" Personal Communication, 2002.

[21] D. Rus, B. Donald, and J. Jennings, "Moving furniture with teams of autonomous robots", In *IEEE/RSJ Int. Conf. on Intelligent Robots and Systems*, pages 235–242, Pittsburgh, PA, Aug. 1995.

[22] D. D. Siljak, *Decentralized Control of Complex Systems*, Academic Press, 1991.

[23] J. Spletzer, A. Das, R. Fierro, C. J. Taylor, V. Kumar, and J. P. Ostrowski, "Cooperative localization and control for multi-robot manipulation", In *IEEE/RSJ Int. Conf. Intell. Robots and Syst., IROS2001*, Oct. 2001.

[24] D. Stilwell and J. Bay, "Toward the development of a material transport system using swarms of ant-like robots", In *IEEE International Conf. on Robotics and Automation*, pages 766–771, Atlanta, GA, May 1993.

[25] T. Sugar and V. Kumar, "Control and coordination of multiple mobile robots in manipulation and material handling tasks", In P. Corke and J. Trevelyan, editors, *Experimental Robotics VI: Lecture Notes in Control and Information Sciences*, volume 250, pages 15–24. Springer-Verlag, 2000.

[26] D. Swaroop and J. K. Hedrick, "String stability of interconnected systems", *IEEE Trans. Automat. Contr.*, 41(3):349–357, March 1996.

[27] K. H. Tan and M. A. Lewis, "Virtual structures for high precision cooperative mobile robot control", *Autonomous Robots*, 4:387–403, October 1997.

[28] C. J. Taylor, "Videoplus: A method for capturing the structure and appearance of immersive environment", *Second Workshop on 3D Structure from Multiple Images of Large-scale Environments*, 2000.

[29] P. Wang, "Navigation strategies for multiple autonomous mobile robots moving in formation", *Journal of Robotic Systems*, 8(2):177–195, 1991.

[30] H. Yamaguchi and J. W. Burdick, "Asymptotic stabilization of multiple nonholonomic mobile robots forming groups formations", In *Proc. IEEE Int. Conf. Robot. Automat.*, pages 3573–3580, Leuven, Belgium, May 1998.

Chapter 5

SEARCH, CLASSIFICATION AND ATTACK DECISIONS FOR COOPERATIVE WIDE AREA SEARCH MUNITIONS *

David R. Jacques

Asst. Prof., Dept. of Aeronautics and Astronautics

Air Force Institute of Technology, Wright-Patterson AFB OH, USA

david.jacques@afit.edu

Abstract   There are currently several wide area search munitions in the research and development phase within the Department of Defense. While the work on the air-
frames, sensors, target recognition algorithms and navigation schemes is promising, there are insufficient analytical tools for evaluating the effectiveness of these
concept munitions. Simulation can be used effectively for this purpose, but analytical results are necessary for validating the simulations and facilitating the design trades early in the development process. Recent research into cooperative
behavior for autonomous munitions has further highlighted the importance of
fundamental analysis to steer the direction of this new research venture. This paper presents extensions to some classic work in the area of search and detection.
The unique aspect of the munition problem is that a search agent is lost whenever an attack is executed. This significantly impacts the overall effectiveness
in a multi-target/false target environment. While the analytic development here
will concentrate on the single munition case, extensions to the multi-munition
will be discussed to include the potential benefit from cooperative classification
and engagement.

1.   Introduction

Several types of wide area search munitions are currently being investigated
within the U.S. Department of Defense research labs. These munitions are being designed to autonomously search, detect, recognize and attack mobile and

---

*The views expressed in this article are those of the authors and do not reflect the official policy of the U.S.
Air Force, Department of Defense, or the U.S. Government.

S. Butenko et al. (eds.), Cooperative Control: Models, Applications and Algorithms, 75-93.
© 2003 *Kluwer Academic Publishers.*

relocatable targets. Additional work at the basic research level is investigating the possibility of having these autonomous munitions share information and act in a cooperative fashion[1][2][3]. While some of the research is promising, most of it is relying heavily on simulation to evaluate the performance of the multi-munition system. Analysis appears to be lacking with regards to the fundamental nature of the wide area search munition problem, to include identification of the critical munition and target environment parameters that must be adequately modeled for a valid simulation. Some classic work has been done in the area of optimal search [4][5][6], but this work does not address the multi-target/false target scenario where an engagement comes at the expense of a search agent. Further, this work needs to be extended for application in cooperative behavior algorithms. This paper presents extensions to some of this classic work in the area of search and detection. Section 2 will present the basic method of analysis for the single munition/single target case. A uniform target distribution in a Poisson field of false targets will be considered. Section 3 will essentially repeat this analysis for the multiple target case, where a Poisson distribution will be assumed for both real and false targets. Section 4 will provide some analytic extensions for the multiple munition case and establish a basis for comparison with cooperative behavior approaches. Section 5 will use the analytical approaches to suggest methods and performance limitations for both cooperative engagement and classification. While the scenarios being considered are somewhat simplistic, the goal is to obtain closed form analytic results that can provide insight as to the fundamental nature of the wide area search munition problem.

2. The Single Munition/Single Target Case

A formula describing the probability of mission success for the single munition/single target scenario is as follows:

$$P_{MS} = P_K \cdot P_{TR} \cdot P_{LOS} \cdot P_E, \tag{1}$$

where

P_K $\equiv$ probability of target kill given Target Report (TR)
P_{TR} $\equiv$ prob. of Target Report given clear Line of Sight to target
P_{LOS} $\equiv$ prob. of clear LOS given target in Field of Regard (FOR)
P_E $\equiv$ prob. the target will appear in the FOR.

The expression in (1) is not the most general, but is easily shown to be equivalent to the more general equations. For example, P_K represents the product of guidance, hit, and kill probabilities. P_{TR} represents the product of detection and confirmation probabilities, where confirmation could be either classification or identification depending upon the level of discrimination being employed by the munition being considered. P_{LOS} could also be included in P_{TR}, and that is the convention that will be followed for the remainder of the development.

With the exception of P_E, the other probabilities are expressed as either single numerical values, or, in the case of P_{TR}, a table of values sometimes referred to as a confusion matrix. The term confusion matrix stems from the fact that it represents the probability of both correct *and* incorrect target reports. P_E is a function of the area to be searched, the density function describing the probable target location, and the ordering of the search process. Consider an autonomous munition looking for a single target (see Figure 5.1). For now we shall assume a single target is uniformly distributed amongst a Poisson field of false targets in the area A_S. A false target is considered to be something that has the potential for fooling the autonomous target recognition (ATR) algorithm (e.g., similar size, shape). Because we are considering single shot munitions, the probability of successfully engaging a target in the incremental area ΔA is conditioned on not engaging a false target prior to arriving at ΔA. The incremental probability of encountering a target in ΔA can be expressed as:

$$\Delta P_E = P_{\overline{FTA}}(A) \cdot \frac{\Delta A}{A_S}, \tag{2}$$

where $P_{\overline{FTA}}(A)$ is the probability of having no false target attacks while searching the area A leading up to ΔA. A closed form expression $P_E(A_S)$ can be obtained as follows. Let

η	$\equiv$	false target probability density
$P_{FTA\mid FT}$	$\equiv$	probability of false target attack given encounter
α	$\equiv$	False Target Attack Rate (FTAR), $\alpha = \eta\, P_{FTA\mid FT}$
$P_{FT_{j,A}}$	$\equiv$	false target attack probability distribution.

$P_{FT_{j,A}}$ represents the distribution of j, the expected number of false target attacks which would be reported by the seeker in a non-commit mode, as a function of the area searched, A. It is a Poisson distribution with parameter $\lambda_{false} = \alpha A$,

$$P_{FT_{j,A}} = \frac{(\alpha A)^j e^{-\alpha A}}{j!}. \tag{3}$$

The probability of searching A without executing a false target attack is

$$P_{\overline{FTA}}(A) = P_{FT_{0,A}} = e^{-\alpha A}. \tag{4}$$

We can now formulate and solve an expression for the probability of encountering a target within A_S,

$$P_E(A_S) = \int_0^{A_S} \frac{e^{-\alpha A}}{A_S} dA = \frac{1 - e^{-\alpha A_S}}{\alpha A_S}. \tag{5}$$

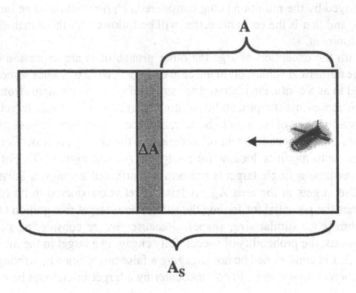

Figure 5.1.   Single Target Search

Note that the expression above assumes that the target is contained within A_S with probability one. For the case of uniform target/Poisson false target distribution, $P_E(A_S)$ can simply be multiplied by the probability that the target is contained within A_S. For cases of non-uniform target distributions a simple multiplication factor is no longer sufficient because the order of the search affects the probability of encountering the target.

Figure 5.2 shows the sensitivity of mission success to the FTAR (α) and probability of correct target report (P_{TR}) for $P_k = 0.8$ and $A_S = 50$ km^2. As shown, the probability of success begins to drop off rapidly for $\alpha > .01/km^2$. The problem is more sensitive to P_{TR} for low values of α than it is for higher values. While the probability of success may seem low, it can be improved by assigning multiple munitions to the same search area as will be discussed later.

3.   The Single Munition/Multiple Target Case

There are several ways of looking at the multiple target case. If the objective is to find a specific target within a field of other targets, this could be treated in the same manner as the single target case; the other targets merely serve to increase the density of false targets. If any of the targets is considered valid then we need to be able to evaluate the probability of a successful encounter with any one of the targets. The single target case allowed us to determine the

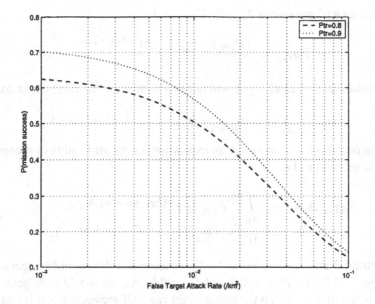

Figure 5.2. Single Target - Uniform Distribution

probability of finding and recognizing a target within a searchable area as

$$P_{RT}(A_S) = P_{TR} P_E(A_S). \qquad (6)$$

For that case P_{TR} did not appear in the formulation for $P_E(A_S)$. For the multiple target case we will formulate it in a slightly different fashion. Referring back to Figure 5.1, the ability to find and recognize a target in the element of area ΔA is now conditioned on no false target attacks *and* no real target declarations/attacks prior to getting to ΔA. Assuming a Poisson distribution for both real and false targets (with $\lambda_{real} \neq \lambda_{false}$), our new formulation for the elemental probability of recognizing the target is

$$\Delta P_{RT}(A) = P_{TR} P_{\overline{FTA}}(A) P_{\overline{RT}}(A) \eta_T \Delta A, \qquad (7)$$

where η_T is the uniform target probability density. Implicit in this formulation is the assumption that $\eta_T \Delta A$, loosely interpreted as the probability of finding a target in the elemental area ΔA, is sufficiently less than one. This assumption is typically met for munitions with relatively small instantaneous sensor footprints relative to the average target density in the area. $P_{\overline{RT}}(A)$, the probability of not having recognized a real target after searching A, is obtained in the same manner as $P_{\overline{FTA}}(A)$. Specifically, $P_{RT_{k,A}}$ represents the distribution of k, the number of target recognitions that would be reported by the seeker in a non-commit mode, as a function of the area searched, A. It is a Poisson distribution with parameter $\lambda_{real} = P_{TR} \, \eta_T \, A$,

$$P_{RT_{k,A}} = \frac{(P_{TR} \, \eta_T \, P_{LOS} \, A)^k e^{-P_{TR} \, \eta_T \, A}}{k!}. \tag{8}$$

The probability of searching A without executing a real *or* false target attack is

$$P_{\overline{RT,FTA}}(A) = P_{RT_{0,A}} \cdot P_{FT_{0,A}} = e^{-(P_{TR} \, \eta_T \, + \, \alpha) \, A}. \tag{9}$$

We can now formulate and solve an expression for the probability of recognizing a target within A_S,

$$
\begin{aligned}
P_{RT_m}(A_S) &= \int_0^{A_S} P_{TR} \, \eta_T \, e^{-(P_{TR} \, \eta_T \, + \, \alpha) \, A} \, dA \\
&= \frac{P_{TR} \, \eta_T}{(P_{TR} \, \eta_T + \alpha)} (1 - e^{-(P_{TR} \, \eta_T \, + \, \alpha) \, A_S}).
\end{aligned} \tag{10}
$$

Figure 5.3 shows P_{MS} vs. α for the Poisson distributed multi-target case, with $\eta_T = .1/km^2$, $P_{TR} = 0.8$ and $P_k = 0.8$. As one would anticipate, it is far less sensitive to α than the single target case. Of greater interest is that the sensitivity to P_{TR} is greater for low values of α than it is for higher values; the opposite of the trend for the single target case. The reason for this is that a missed target is no longer a failed mission because there are other targets to be found. Further, the probability that these other targets will be encountered is high if the FTAR is sufficiently low.

4.   Analytic Multi-Munition Extensions

The single target scenario can be extended to the multi-munition case in several ways. The easiest way is to divide the total search area by the number of munitions, and determine the P_{MS} for the munition searching the subarea that the target appears in. All other munitions find nothing for the single target case. P_{MS} will increase because A_S will decrease for all munitions, including the munition searching the subarea where the target happens to be. However, because this method assumes zero overlap in the subareas being searched, the probability of mission success is ultimately limited by the P_K for the single munition. If the warhead is not lethal enough to provide the desired probability of mission success from a single munition-target engagement, then overlapping search areas and multi-munition engagements must be considered. It should also be noted that unnecessarily limiting the area over which the munitions can search is not an efficient use of valuable search assets.

Extending expressions (1) and (5) above for multiple munitions becomes quickly complicated by path considerations and the degree of correlation assumed for the behavior of the munitions as they encounter either real or false targets. Considering only the terminal engagement for the time being, we can

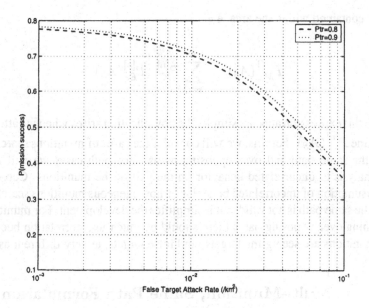

Figure 5.3. Multi Target - Uniform Distribution

assume independent events for each warhead shot yielding an expression for $P_K^{[N]}$, the probability of kill for the case of N munitions engaging the target,

$$P_K^{[N]} = 1 - (1 - P_K)^N. \tag{11}$$

One could (incorrectly) assume a similar roll-up of P_{MS} for the case of N munitions all searching the same area for a single target,

$$P_{MS}^{[N]} = P_C \left(1 - (1 - \frac{P_{MS}}{P_C})^N\right), \tag{12}$$

where P_C refers to the probability that the target is located within the area being searched. The reason this formulation is incorrect is because it assumes the placement of clutter, non-targets and the real target is re-randomized for each munition. This can never be true for the case of several munitions looking for the same target in the same location. To address the problem correctly requires some consideration of the search path followed by the individual munitions.

The analysis in this section will be limited to the single target/multi-munition scenario. It is based on some unpublished results from Henderson[7] and some previous analysis by this author[8]. For this analysis, we not only need the probability of kill given access by N munitions, but we also need the probability that n munitions will encounter the target, $P_E^{[n]}$. For N munitions searching for the target, we can set up an expression for the probability of killing the target contained within the area A_S,

$$P_{MS}^{[N]}(A_S) = \sum_{n=1}^{N} P_K^{[n]} \cdot P_E^{[n]}(A_S).$$ (13)

While this expression appears simple, the complication arises when we attempt to define $P_E^{[n]}(A_S)$. For this, we will consider the cases of munitions searching over the same path and over opposing paths. For both cases we will limit the analysis to uncorrelated behavior on the part of the munitions. Certainly the assumption of uncorrelated behavior of homogeneous munitions searching over the same path is not valid, but it simplifies the development. For munitions searching over opposing paths there should be much less correlation because the munitions are seeing the targets, and false targets, at very different aspect angles.

4.1.   Multi-Munition, Same Path Formulation

Consider the case of two munitions searching identical paths for a uniformly distributed single target, with a Poisson distribution of false targets in the area A_S. Equation 5 provides the expression for the probability that any given munition will have access to the target, but we need the probability that any combination of the munitions will have access to the target. The probability that one of the two munitions will have access is

$$P_E^{[1]}(A) = 2e^{-\alpha A}(1 - e^{-\alpha A}),$$ (14)

and the probability that both will have access is

$$P_E^{[2]}(A) = e^{-2\alpha A}.$$ (15)

The probability of kill given access by n weapons is $(1 - (1 - P_K P_{TR})^n)$. The two-weapon formula can now be expressed as

$$
\begin{aligned}
P_{MS_S}^{[2]}(A_S) &= P_C \int_0^{A_S} \left(P_K P_{TR} 2e^{-\alpha A}(1 - e^{-\alpha A}) \right. \\
&\quad \left. + (1 - (1 - P_K P_{TR})^2) e^{-2\alpha A} \right) \frac{dA}{A_S} \\
&= P_C P_K P_{TR} \left(\frac{2}{\alpha A_S}(1 - e^{-\alpha A_S}) - \frac{P_K P_{TR}(1 - e^{-2\alpha A_S})}{2\alpha A_S} \right).
\end{aligned}
$$ (16)

The more general expression for N munitions traversing the same search path can be expressed as

$$P_{MS_S}^{[N]}(A_S) = \frac{1}{A_S} \sum_{n=1}^{N} \left[(1 - (1 - P_K P_{TR})^n) \frac{N!}{n!(N-n)!} \right.$$

$$\left. \cdot \int_0^{A_S} (e^{-\alpha A})^n (1 - e^{-\alpha A})^{N-n} dA \right]. \quad (17)$$

4.2. Multi-Munition, Opposing Path Formulation

Now consider the case of two munitions searching opposing paths for a uniformly distributed single target, again with a Poisson distribution of false targets in the area A_S. The probability that the munition conducting the forward search will have access to the target is

$$P_{E_f}(A) = e^{-\alpha A}, \quad (18)$$

and the probability that the munition conducting the reverse search will have access to the target is

$$P_{E_r}(A) = e^{-\alpha(A_s - A)}. \quad (19)$$

The probability that both munitions will have access to the target is

$$P_E^{[2]}(A) = e^{-\alpha A} e^{-\alpha(A_s - A)} = e^{-\alpha A_s}, \quad (20)$$

and we note that it is constant! With these expressions, we can now lay out the expression for two munitions searching opposing paths,

$$P_{MS_O}^{[2]} = P_C \int_0^{A_S} P_K P_{TR} \left(e^{-\alpha A} + e^{-\alpha(A_S - A)} \right.$$

$$\left. - 2e^{-\alpha A_S} + (2 - P_K P_{TR})e^{-\alpha A_S} \right) \frac{dA}{A_S} \quad (21)$$

$$= P_C P_K P_{TR} \left(\frac{2}{\alpha A_S} (1 - e^{-\alpha A_S}) - P_K P_{TR} \, e^{-\alpha A_S} \right).$$

Similar expressions have been derived for four munitions (two searching each direction) and six munitions (three searching each direction) but a general formula for N munitions has yet to be defined.

For the same total number of munitions, the opposing path case will produce the highest mission success value, the same path case will produce the lowest mission success value, and the simple multi-munition roll-up will produce a

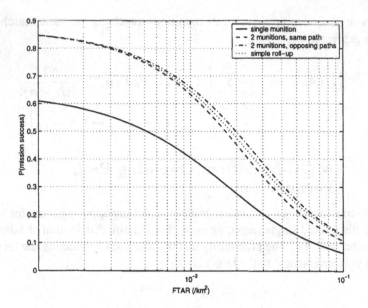

Figure 5.4. Path Considerations for Multi-Munition Case

value in between the other two. The graph shown in Figure 5.4 is for two munitions, but it should be noted that the differences between the curves increases with the number of munitions used in the analysis.

It is worth repeating that the assumption of uncorrelated behavior (at either a real or false target) is not strictly valid, and we should expect a high degree of correlation for the case where the munitions are traversing the same path in the same direction. For scenarios where the potential false targets greatly outnumber the real targets, correlated behavior will degrade the overall mission success rate. For this reason, search patterns should be planned which decrease the degree of correlated behavior at false targets. This can be done through the use of lateral offsets between munitions and/or different approach vectors. While this does not make the assumption of uncorrelated behavior valid, it can reduce the degree of correlation at both targets and false targets. Analytically it becomes intractable to define an expression for arbitrary numbers of munitions executing arbitrarily specified search patterns and degrees of correlation. However, for any realistic effectiveness analysis these are the cases we are most interested in. A numerical simulation with Monte-Carlo runs is the only practical way of performing this more general analysis, and work is currently being done in this area.

5. Implications for Cooperative Behavior

5.1. Cooperative Engagement of Targets

The analysis above indicates that munition success is quite sensitive to FTAR and the target location probability distribution. Cooperative behavior and control has recently become an active area of research, and one of the objectives of the research is to reduce the sensitivity to FTAR and target location error. While there are many aspects of cooperative behavior and control, the two most applicable to the wide area search munition problem are cooperative engagement and cooperative classification. Cooperative engagement is defined as a munition initiating an attack on a target that a second munition has declared. Cooperative classification involves using multiple looks from one or more munitions in order to improve the probability of making a correct target declaration.

Cooperative engagement has potential benefits in several areas. For a target that has been correctly identified, it increases the probability of kill for that target by virtue of multiple warhead events. This increase was described earlier in equation (11). If munitions are chosen for cooperative engagement that are unlikely to find additional targets through continued search, the probability of kill for found targets could be increased without significantly degrading the probability of finding additional targets. Complications arise due to the possibility that declared targets are not real targets (incorrect classification), thus diverting valuable resources for no real gain. Ultimately what is required is a way to compare the probability of success from continued search with the probability of success from attacking a found target. Figure 5.5 depicts all the possible events from searching a Poisson field of targets and non-targets. Starting from the top, the munition can either encounter a target, incorrectly declare (and attack) a false target, or run out of gas prior to the occurrence of either a target encounter or false target attack. For a given target encounter the munition may recognize it as such or incorrectly bypass it. A recognized target will be attacked with an uncertain outcome of the warhead event. A incorrectly bypassed target essentially means that the munition is still in a search mode, but the time remaining for search will be decreased by (on average) the expected time to next target encounter, $E[t_E] = (\eta_T VW)^{-1}$. The basic tree structure repeats at each occurrence of the search state. If we define success as a lethal warhead event on any one of the real targets appearing within the search area, the entire tree can be collapsed to a single level, with the probability of real target attack, false target attack, and running out of gas given as a function of

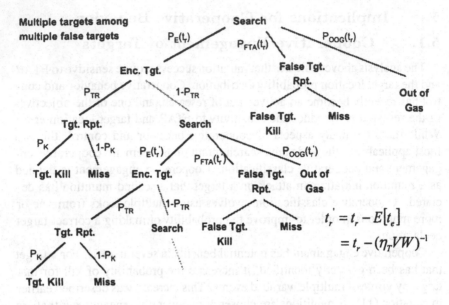

Figure 5.5. Possible Search Outcomes

search time remaining,

$$P_{RT}(t_r) = \frac{P_{TR}\eta_T}{\alpha + P_{TR}\eta_T}\left(1 - e^{-(\alpha + P_{TR}\eta_T)Vt_rW}\right), \qquad (22)$$

$$P_{FTA}(t_r) = \frac{\alpha}{\alpha + P_{TR}\eta_T}\left(1 - e^{-(\alpha + P_{TR}\eta_T)Vt_rW}\right), \qquad (23)$$

$$P_{OOG}(t_r) = e^{-(\alpha + P_{TR}\eta_T)Vt_rW}, \qquad (24)$$

respectively. The term Vt_rW represents the area that can be searched in the time remaining t_r for a given munition velocity V and search width W. The probability for success in search is simply

$$P_{SS}(t_r) = P_K \cdot P_{RT}(t_r). \qquad (25)$$

Figure 5.6 shows a similar tree structure for the event of a declared target to be attacked. Any declared target may or may not be a real target, and any real declared target may or may not be recognized by a second munition being sent to engage it. A correctly found real target again results in a warhead event with uncertain outcome. If the declared target is actually a false target, a second munition may or may not make the same mistake as the first munition making the initial declaration. If it correctly identifies it as a false target it resumes search with the time remaining decreased by the time to arrive at the initially declared target. The analysis in this paper assumes independent events for target/false target declarations, so the probabilities for all individual munitions are the same without regard to order of occurrence. The probability of a real target given that a target declaration has been made can be expressed as

$$P_{RT|TR} = \frac{P_{TR}\eta_T}{P_{TR}\eta_T + P_{FTA|FT}\eta}. \tag{26}$$

With this, we can now define the probability of success from engaging a declared target,

$$
\begin{aligned}
P_{SA} &= P_K P_{TR} P_{RT|TR} + P_{SS}(t_r - t_{ETA}) \cdot (1 - P_{TR}) \cdot P_{RT|TR} \\
&\quad + P_{SS}(t_r - t_{ETA}) \cdot (1 - P_{FTA|FT}) \cdot (1 - P_{RT|TR}). \tag{27}
\end{aligned}
$$

Note that this expression includes success increments from continued search after either missing a real target or bypassing false targets initially declared as real. It should also be noted that once a munition has declared a target, the probability of success in attacking it is

$$P_{SA|TR} = P_K \, P_{RT|TR}. \tag{28}$$

Further note that $P_{SA|TR}$ will always be greater than $P_{SS}(t_r)$, regardless of the amount of search time remaining.

Equations (27) and (28) apply to the case where the target has not yet been attacked by a previous munition. For the case where one or more attacks on the target have previously taken place, the value of attacking the target again should be decreased due to the chance of the target being previously killed. With the assumption of independent warhead events, the probability of a live target $(1 - P_K)^N$ given N previous attacks can be used as a multiplying factor with equation (28) or the first term in (27). Equations (27), (28) with the appropriate multiplying factor for multiple attacks, and equation (25) are quantitative measures of the value associated with either attacking a target or continuing to search, respectively. Although the situation gets quickly complicated when one considers multiple target types of differing priorities, these basic measures can serve as a basis for making decisions on cooperative engagement. Work is progressing to incorporate these quantitative measures into overall schemes for cooperative engagement, and simulation based analysis is being used to evaluate the schemes under more general multi-target scenarios.

5.2. Cooperative Classification of Targets

Cooperative engagement has proven effective in increasing the probability of success for cases where FTAR is low, but it provides no benefit, and is possibly detrimental, for cases where FTAR is higher[2]. The reason for this is that

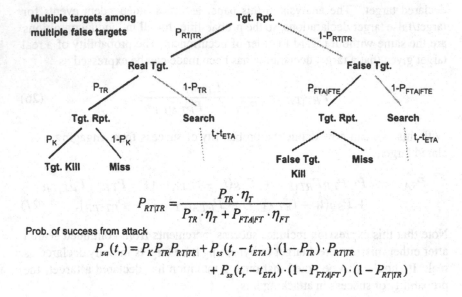

$$P_{RT|TR} = \frac{P_{TR} \cdot \eta_T}{P_{TR} \cdot \eta_T + P_{FTA|FT} \cdot \eta_{FT}}$$

Prob. of success from attack

$$P_{sa}(t_r) = P_K P_{TR} P_{RT|TR} + P_{ss}(t_r - t_{ETA}) \cdot (1 - P_{TR}) \cdot P_{RT|TR}$$
$$+ P_{ss}(t_r - t_{ETA}) \cdot (1 - P_{FTA|FT}) \cdot (1 - P_{RT|TR})$$

Figure 5.6. Possible Engagement Outcomes

cooperative engagement increases the chance that additional munitions will be lost due to false target attacks. Cooperative classification may provide some help for this problem because it can effectively reduce the false target attack rate; however, cooperative classification can potentially increase the chance of missing real targets. For the case of independent classification events, a simple analysis can provide some insight into the potential strengths and weaknesses of cooperative classification. While independent events may not be a realistic assumption, it can be used to provide a first-cut sensitivity analysis.

Figure 5.7 shows a surface plot of P_{MS} vs. FTAR and P_{TR} for two munitions conducting a non-cooperative, opposing path search for a single target in a 100 km^2 area. The extreme sensitivity to FTAR is clearly evident (note the log scale for FTAR), as well as the general insensitivity to P_{TR}. The most basic implementation of cooperative classification would require two subsequent looks with the same classification prior to declaring a real target. For this simple two look scenario the effective probability of false target attack given false target encounter is now the square of the value for non-cooperative classification. This yields an effective FTAR of $\alpha = (P_{FTA|FT})^2 \eta$ which is obviously reduced from the non-cooperative case. The downside of this approach is that the effective probability of correct target report is also the square of the value for the non-cooperative case, thus reducing a value that you would like to keep as close to unity as possible. Figure 5.8 shows a similar surface plot for two munitions conducting a cooperative search along parallel paths. Each muni-

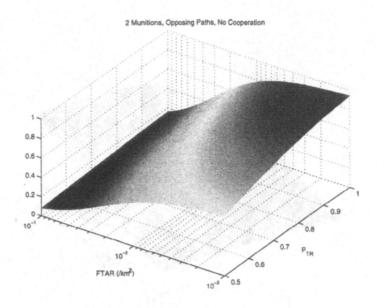

Figure 5.7. Complementary Search w/ 2 Munitions

tion is responsible for searching the total area along the same path, and each can cooperatively classify and attack targets without delay. The sensitivity to FTAR is reduced using cooperative classification, but there is obviously an increase in the sensitivity to P_{TR}. Of great interest is the difference between the two surfaces, thus indicating where cooperative classification improves or degrades the overall probability of success. Figures 5.9 and 5.10 show the difference between the two surface plots, and clearly there are regions where cooperative classification can help ($\Delta P_{MS} > 0$) and hurt ($\Delta P_{MS} < 0$). For low single munition FTAR, the decrease in P_{TR}^2 clearly outweighs any benefit from further decreasing FTAR through cooperative behavior.

A better implementation of cooperative behavior might be to attempt an opposing path formulation. In order to keep the assumption of zero delay for cooperative classification and engagement, we need to assume that each munition is on a parallel track covering half the total area, and at the end of the track they switch lanes and reverse direction. We can add the probability of success for this second pass conditioned on not having engaged the target *or* any false target on the first pass. Assuming independent passes (not completely unreasonable because of the reversed direction), we can assume that the conditional probability of success for the second pass is the same as the unconditional probability of success for the first pass. The probability of not having engaged

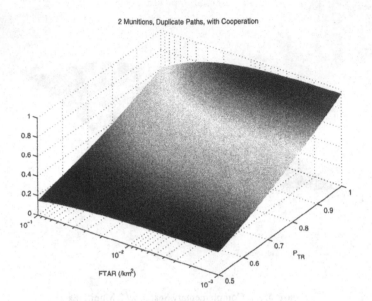

Figure 5.8.   Cooperative Classification/Engagement w/ 2 Munitions

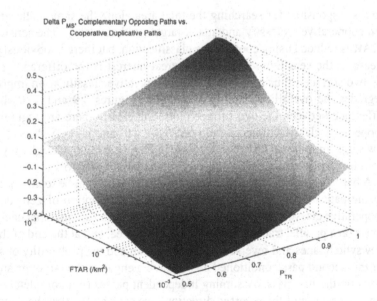

Figure 5.9.   Difference of Cooperative and Complementary Behavior

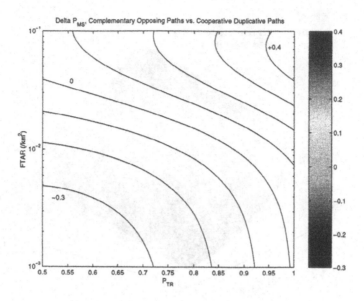

the target or a false target on the first pass is

$$P_{\bar{1}} = e^{-P_{FTA|FT}^2 \eta A_S} \left(1 - P_{TR}^2\right). \tag{29}$$

Figures 5.11 and 5.12 show the ΔP_{MS} for the complementary (non-cooperating) opposing path vs. the cooperative opposing path scenarios. Once again, a positive ΔP_{MS} indicates the cooperative scheme outperforms the non-cooperating scheme. While the situation is improved over the cooperative, same path scheme, there is still a significant operating region for low FTAR's where it is no longer beneficial to employ cooperation. It should be noted that these results are for a very limited single target, two munition scenario, but they highlight some important considerations for the more complex multi-target/multi-munition scenarios we would like to address. Mobile targets will make cooperative classification even more difficult because the target may not be in the same location when the second munition arrives for its confirming look. Ultimately the benefit of analysis such as this may be to provide desired operating regions for the operating characteristic of the ATR algorithm. P_{TR} and $P_{FTA|FT}$ are competing objectives, so system trades will need to be made.

6.   Conclusions

This paper has presented some fundamental analysis of the wide area search munition problem. False target attack rate and the distribution of targets have

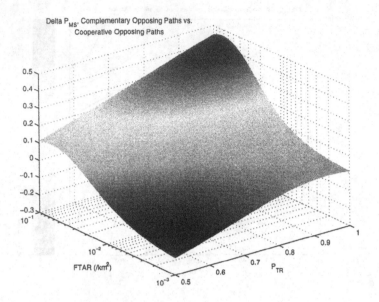

Figure 5.11.   Difference of Cooperative and Complementary Behavior

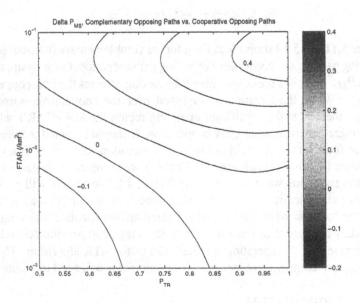

Figure 5.12.   Difference Contour of Cooperative and Complementary Behavior

been identified as critical factors in this problem. Extensions to existing search theory have been presented, specifically in the area of multiple target/false target scenarios. Finally, the implication of this analysis for cooperative behavior has been discussed. Decision factors for cooperative engagement were developed, and the strengths and limitations of cooperative classification were highlighted. Overall, cooperative behavior holds promise for the autonomous wide area search munition problem, but analysis such as has been presented here is required in order to develop behavior algorithms that degrade gracefully in the presence of uncertain target location and/or false targets.

References

[1] P.R. Chandler, M. Pachter, K. Nygard, and Dharba Swaroop, "Cooperative control for target classification", in R. Murphey and P. Pardalos, editors, *Cooperative Control and Optimization*, pages 1-19, Kluwer Academic Publishers, 2002.

[2] D. Gillen and D. R. Jacques, "Cooperative behavior schemes for improving the effectiveness of autonomous wide area search munitions", in R. Murphey and P. Pardalos, editors, *Cooperative Control and Optimization*, pages 95-120, Kluwer Academic Publishers, 2002.

[3] K. Passino, M. Polycarpou, D. Jacques, and M. Pachter, "Distributed cooperation and control for autonomous air vehicles", in R. Murphey and P. Pardalos, editors, *Cooperative Control and Optimization*, pages 233-271, Kluwer Academic Publishers, 2002.

[4] B. O. Koopman, *Search and Screening*, Pergamon Press, New York, 1980.

[5] L. Stone, *Theory of Optimal Search*, Operations Research Society of America (ORSA), Batimore, MD, 1975.

[6] A. Washburn, *Search and Detection, 2nd Ed.*, Institution for Operations Research and the Management Sciences (INFORMS), 1980.

[7] R. Henderson, *Notes on munition search effectiveness*, 1997.

[8] D. R. Jacques and R. Leblanc, "Effectiveness analysis for wide area search munitions", in *Proceedings of the 1998 AIAA Missile Sciences Conference*, Monterey, CA, November 1998.

Chapter 6

PATH PLANNING FOR UNMANNED AERIAL VEHICLES IN UNCERTAIN AND ADVERSARIAL ENVIRONMENTS*

Myungsoo Jun

Sibley School of Mechanical and Aerospace Engineering
Cornell University, Ithaca, NY 14853-7501, USA
mj73@cornell.edu

Raffaello D'Andrea

Sibley School of Mechanical and Aerospace Engineering
Cornell University, Ithaca, NY 14853-7501, USA
rd28@cornell.edu

Abstract   One of the main objectives when planning paths for unmanned aerial vehicles in adversarial environments is to arrive at the given target, while maximizing the safety of the vehicles. If one has perfect information of the threats that will be encountered, a safe path can always be constructed by solving an optimization problem. If there are uncertainties in the information, however, a different approach must be taken. In this paper we propose a path planning algorithm based on a map of the probability of threats, which can be built from *a priori* surveillance data. An extension to this algorithm for multiple vehicles is also described, and simulation results are provided.

Keywords: Unmanned aerial vehicles, path planning, probability map, uncertain adversarial environments, optimization

*Research sponsored by AFOSR Grant F49620-01-1-0361

S. Butenko et al. (eds.), Cooperative Control: Models, Applications and Algorithms, 95-110.
© 2003 Kluwer Academic Publishers.

1.   Introduction

Autonomous robots and vehicles have been used to perform missions in hazardous environments, such as operations in nuclear power plants, exploration of Mars, and surveillance of enemy forces in the battle field. Among these applications is the development of more intelligent unmanned aerial vehicles (UAVs) for future combat in order to reduce human casualties. One of the main challenges for intelligent UAV development is path planning in adversarial environments.

Path planning problems have been actively studied in the robotics community. The problem of planning a path in these applications is to find a collision-free path in an environment with static or dynamic obstacles. Early work focused on holonomic and non-holonomic kinematic motion problems with static obstacles without considering system dynamics. Despite many external differences, most of these methods are based on a few different general approaches: roadmap, cell decomposition, and potential field ([9]). When moving obstacles are involved in planning problems, the time dimension is added to the configuration space ([5]) or state-space of the robot ([6]), and planning is termed motion planning or trajectory planning instead of path planning. Research has recently been performed in motion planning that takes into account dynamic constraints, called *kinodynamic planning* ([10], [8]). All of the aforementioned path or motion planning methods focus on obstacle avoidance issues.

In the UAV path planning problem in adversarial environments, the objective is to complete the given mission — to arrive at the given target within a prespecified time — while maximizing the safety of the UAVs. We can consider adversaries as obstacles and employ similar methods to those used for robot path planning. The main difference between robot path planning and UAV path planning is that a UAV must maintain its velocity above a minimum velocity, which implies that it cannot follow a path with sharp turns or vertices. There has been research on path planning for UAVs in the presence of risk — see [2], [3], [11], [15], and the references therein. The first three approaches are similar. They decompose path planning into two steps: First, a polygonal path is generated from the Voronoi graph by applying Djiktra's algorithm, which is the same as the roadmap and A^* search approaches in robot path planning; the initial polygonal path is then refined to a navigable path by considering the UAV's maneuverability contraints (see [3], for example) or by using the dynamics of a set of virtual masses in a virtual force field emanating from each radar site (see [2], for example), which is similar to the potential field approaches in robotics. By refining this process, one can produce paths without vertices. The main problem with these methods is that the effects of uncertainties in the locations of the radar sites are not considered. Other potential problems are the effects of the decoupling assumption, as stated by the authors; specifically, the value of the cost function after refining may not be minimal, and moving along the Voronoi graph yields suboptimal trajectories. In [15] an optimization problem is solved, where the solution minimizes the integrated risk along the path with a constraint on the total path length. The result is not guaranteed to be optimal if there are uncertainties in the information, such as the locations of the radar sites, etc.. In addition, the computational load grows quickly as the number of radar sites increases.

Uncertainties in the information lead naturally to consider probability models. In [7] a probabilistic map of radar sites are constructed and a safe UAV path is generated by solving a minimization problem. The probability map was constructed using a likelihood function which was defined by considering radar range and other radar characteristics, and by using Bayes' rule. It was assumed, however, that radar sites remain at fixed locations. The algorithm also requires m way-points for trajectory planning, and thus the resulting path is a minimum risk path only among the paths which include the way-points.

In this paper we propose a path planning method for UAVs by using a probability map. The approach in this paper is similar to those in [2] and [3] in that we decompose the problem into two steps — first the generation of a preliminary polygonal path by using a graph, and then a refinement of the path. Our approach differs in that it is based on a map of the probability of threats, and it does not use a Voronoi graph to find a preliminary path. The nodes and links of the graph are based directly on the probability map. We also consider the effects of moving threats, changes in the probability map, and multiple vehicles, and perform an analysis of the effects of refinement on the initial path.

This paper is organized as follows: Section 2 contains some basic definitions in graph theory. Section 3 describes how to build an occupancy probability from measured data and how to calculate a probability of risk. The problem formulation is stated in Section 4. Section 5 provides a method for generating a weighted digraph from the probability map, which enables us to solve the original problem with a shortest path algorithm. The path planning algorithm is described in Section 6, and the extension to multiple vehicles in Section 7. Simulation results are provided in Section 8. Discussion and future problems are found in Section 9.

2. Graph Theory

In this section we will briefly cover some basic material on graph theory. The reader is referred to [1] for more details. We define a *graph*, $G = (\mathcal{N}, \mathcal{A})$, to be a finite nonempty set $\mathcal{N}$ of *nodes* and a collection $\mathcal{A}$ of distinct nodes from $\mathcal{N}$. Each pair of nodes in $\mathcal{A}$ is called a *link* or an *arc*. A *walk* in a graph G is a sequence of nodes $(n_1, n_2, \cdots, n_l)$ such that each of the pairs (n_1, n_2), (n_2, n_3), ..., (n_{l-1}, n_l) are links of G. A walk with no repeated nodes is a *path.* A walk $(n_1, n_2, \cdots, n_l)$ with $n_1 = n_l$, $l > 3$, and no repeated nodes other than $n_1 = n_l$ is called a *cycle*. A graph is *connected* if for each node i there is a path $(i = n_1, n_2, \cdots, n_l = j)$ to each other node j.

A *directed graph* or *digraph* $G = (\mathcal{N}, \mathcal{A})$ is a finite nonempty set $\mathcal{N}$ of nodes and a collection $\mathcal{A}$ of *ordered* pairs of distinct nodes from $\mathcal{N}$; each ordered pair of nodes in $\mathcal{A}$ is called a *directed link*. The definitions of *directed walks*, *directed cycles*, and *connectedness* are analogous to those for graphs.

3.   Probability Map

The basic concept in building the probability map in this paper is similar to the grid-based occupancy maps in map learning methods for autonomous robots in [4], [12], [13] and [14]. Occupancy values for each grid cell are determined based on sensor readings and by applying the conditional probability of occupancy using Bayes' rule. These values are determined by the sensor characteristics, the location of the sensors, the measurement methods, etc..

Assume that the region $\mathcal{R}$ for the mission is given and $\mathcal{R}$ is composed of n cells. Define the following events:

$$\mathcal{D}_i = \text{event that the UAV in cell } i \text{ is detected}$$
$$\text{by the adversary}$$
$$\mathcal{E}_i = \text{event that the adversary is in cell } i$$
$$\mathcal{S}_i = \text{event that the UAV in cell } i \text{ is neutralized}$$
$$\text{by the adversary}$$

Denote the complement of event $\mathcal{A}$ by $\overline{\mathcal{A}}$. If X_i represents the i-th sensor reading, the probability of occupancy of cell i can be expressed as ([13])

$$P(\mathcal{E}_i | X_1, \cdots, X_k) =$$
$$1 - \left(1 + \frac{P(\mathcal{E}_i | X_1)}{1 - P(\mathcal{E}_i | X_1)} \prod_{j=2}^{k} \frac{P(\mathcal{E}_i | X_j)}{1 - P(\mathcal{E}_i | X_j)} \frac{1 - P(\mathcal{E}_i)}{P(\mathcal{E}_i)} \right)^{-1} . \quad (1)$$

The probability that the UAV in cell i is neutralized by the adversary is

$$P(\mathcal{S}_i) = P(\mathcal{S}_i | \mathcal{D}_i) P(\mathcal{D}_i) + P(\mathcal{S}_i | \overline{\mathcal{D}}_i) P(\overline{\mathcal{D}}_i)$$
$$= P(\mathcal{S}_i | \mathcal{D}_i) P(\mathcal{D}_i),$$

where we assume that the probability that the UAV is neutralized when it is not detected by the adversary is 0. Define $p_i \triangleq P(\mathcal{E}_i | X_1, \cdots, X_k)$. Probability

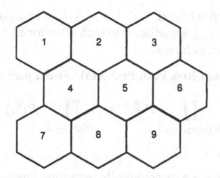

Figure 6.1. The region consists of 9 hexagon cells.

$P(\mathcal{D}_i)$ can be expressed as

$$
\begin{aligned}
P(\mathcal{D}_i) &= P\left(\mathcal{D}_i|(\mathcal{E}_i|X_1,\cdots,X_k)\right) P(\mathcal{E}_i|X_1,\cdots,X_k) + \\
&\quad P\left(\mathcal{D}_i|(\overline{\mathcal{E}}_i|X_1,\cdots,X_k)\right) P(\overline{\mathcal{E}}_i|X_1,\cdots,X_k) \\
&= P\left(\mathcal{D}_i|(\mathcal{E}_i|X_1,\cdots,X_k)\right) p_i + P\left(\mathcal{D}_i|(\overline{\mathcal{E}}_i|X_1,\cdots,X_k)\right)(1-p_i).
\end{aligned}
$$

We thus have

$$
\begin{aligned}
P(\mathcal{S}_i) = P(\mathcal{S}_i|\mathcal{D}_i)\{P\left(\mathcal{D}_i|(\mathcal{E}_i|X_1,\cdots,X_k)\right) p_i + \\
P\left(\mathcal{D}_i|(\overline{\mathcal{E}}_i|X_1,\cdots,X_k)\right)(1-p_i)\}.
\end{aligned} \tag{2}
$$

We assume that we can estimate or calculate the probabilities $P(\mathcal{S}_i|\mathcal{D}_i)$, $P\left(\mathcal{D}_i|(\mathcal{E}_i|X_1,\cdots,X_k)\right)$, $P\left(\mathcal{D}_i|(\overline{\mathcal{E}}_i|X_1,\cdots,X_k)\right)$ and p_i from *a priori* information of the adversary.

The probability $P(\overline{\mathcal{S}})$ that the UAV is NOT neutralized by the adversary when it follows path $(a_1, a_2, \cdots, a_l)$ can be expressed as

$$
P(\overline{\mathcal{S}}) = \prod_{i \in (a_1, a_2, \cdots, a_l)} P(\overline{\mathcal{S}}_i), \tag{3}
$$

where $P(\overline{\mathcal{S}}_i) = 1 - P(\mathcal{S}_i)$. The objective is to find the path $(a_1, a_2, \cdots, a_l)$ that maximizes the probability $P(\overline{\mathcal{S}})$ when cell a_1 is the origin of the mission and cell a_l is the target of the mission.

4. Problem Formulation

Assume that the region of interest consists of n cells, and that the shape of the cells is given. See Figure 6.1, for example. Let $1 \leq O \leq n$ be the cell which contains the origin of the mission, and $1 \leq T \neq O \leq n$ the cell which contains the target. Define a sequence of cells $(a_1, a_2, \cdots, a_l)$, where

$2 \leq l \leq n$ and $a_i \neq a_j$ for all $1 \leq i \neq j \leq l$, to be *a path from cell a_1 to cell a_l* if cell a_i and a_{i+1} are adjacent to each other for all $1 \leq i \leq l - 1$. The problem can be stated as follows:

Problem 1 (Minimum Risk Path Problem). *Find a path from cell O to cell T such that*

$$\prod_{i \in (O, a_2, \cdots, T)} P(\bar{S}_i) \geq \prod_{i \in (O, b_2, \cdots, T)} P(\bar{S}_i) \qquad (4)$$

for all paths $(O, b_2, \cdots, T)$.

Since the logarithm is a monotonically increasing function of its argument, the above expression is equivalent to:

$$\sum_{i \in (O, a_2, \cdots, T)} (-\log(P(\bar{S}_i))) \leq \sum_{i \in (O, b_2, \cdots, T)} (-\log(P(\bar{S}_i))). \qquad (5)$$

5.   Conversion to a Shortest Path Problem

This section describes several methods for generating a digraph from the probability map that was built based on Section 3. After generating the digraph, the Minimum Risk Path Problem is converted to a shortest path problem. The second part of this section describes the Bellman-Ford algorithm which will be used to find the shortest path.

5.1.   Defining Digraph

Define nodes to be cells in the probability map, that is, $\mathcal{N} = \{1, 2, \cdots, n\}$. Define d_{ij}, the weight of link (i, j), as follows:

$$d_{ij} = \begin{cases} -\log(P(\bar{S}_j)) & \text{if cells } i \text{ and } j \text{ are adjacent,} \\ \infty & \text{otherwise.} \end{cases} \qquad (6)$$

For example, $d_{17} = \infty$ for the map in Figure 6.1 since cell 1 and 7 are not adjacent. By defining the weights of the links as in Eq. (6), we can convert the Minimum Risk Path Problem to a shortest path problem.

We may also want to consider path length and include a constant term in the link weight as a penalty:

$$d_{ij} = \begin{cases} -\log(P(\bar{S}_j)) + c & \text{if cells } i \text{ and } j \text{ are adjacent,} \\ \infty & \text{otherwise.} \end{cases} \qquad (7)$$

where c is a constant real number.

With the above definition of nodes and links, the probability map is converted to a digraph. Figure 6.2 is the digraph converted from the probability

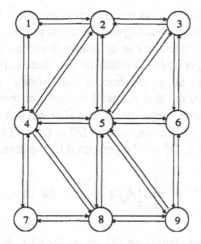

Figure 6.2. The digraph converted from the map of cells in Figure 6.1.

map in Figure 6.1. The Minimum Risk Path Problem has been converted to the problem of finding the shortest path in a network-like routing problem. Our path planning problem is simpler than a network routing problem since we do not have to consider fairness, maximum resource utilization, etc..

We will obtain a preliminary path once the shortest path is found from the converted digraph. For example, if the shortest path from node 9 to node 1 in Figure 2 is $(9, 5, 2, 1)$, the path we should take will pass cells 9, 5, 2, and 1. With this in mind, we can refine the path by using conventional optimization with constraints on the UAV maneuverability.

5.2.   Shortest Path Algorithm

There are many shortest path algorithms. The Djikstra algorithm was used in [3] to find the shortest path in a Voronoi graph. In this paper, we will use the Bellman-Ford algorithm. The Djikstra algorithm is computationally more attractive than the Bellman-Ford algorithm in the worst case: The worst case computation of the Djikstra algorithm is $O(n^2)$ whereas it is $O(n^3)$ for the Bellman-Ford algorithm. There are many problems, however, where the Bellman-Ford algorithm terminates in a very few number of iterations relative to the Djikstra algorithm. Generally, for non-distributed applications, the two algorithms appear to be competitive ([1]).

The main advantage of the Bellman-Ford algorithm is that the iteration process is very flexible with respect to the choice of initial estimates and updates. This allows an asynchronous, real-time distributed implementation of the algorithm, which can tolerate changes in the link lengths — changes in adversary probability in our problem — as the algorithm executes. We can thus update the link lengths without terminating and restarting the algorithm.

The Bellman-Ford algorithm can be briefly described as follows. Suppose that node 1 is the *target* node. We assume that there exists at least one path from every node to the target. A shortest walk from a given node i to node 1, subject to the constraint that the walk contains at most h links and goes through node 1 only once, is referred to as a *shortest*$(\leq h)$ *walk* and its length is denoted by D_i^h. By convention, we take $D_1^h = 0$ for all h. The Bellman-Ford algorithm maintains that D_i^h can be generated by the iteration

$$D_i^{h+1} = \min_j \left[d_{ij} + D_j^h \right], \quad \text{for all } i \neq 1 \tag{8}$$

starting from the initial conditions $D_i^0 = \infty$ for all $i \neq 1$. The assumptions for the distributed asynchronous Bellman-Ford algorithm are i) each cycle has positive length and ii) if (i, j) is a link, then (j, i) is also a link. Our problem satisfies these two assumptions, and thus we can use the distributed asynchronous Bellman-Ford algorithm.

6.   UAV Path Planning

This section describes a path planning algorithm for UAVs by using the shortest path that was given from the digraph and the Bellman-Ford algorithm. The algorithm decomposes the problem into two steps — generation of a polygonal path from the graph, and a refinement of the path. This section also considers the effects of moving adversaries, changes in probabilities, and an analysis of the effects of path refinement.

6.1.   Polygonal Path

After we obtain the shortest path from the digraph described in Section 5, we can build a polygonal path from the origin to the target in the original probability map. Suppose that the shortest path from the digraph is $(a_1, a_2, \cdots, a_l)$. This means that safest path in the probability map should lie within the cells $(a_1, a_2, \cdots, a_l)$, which is called a *channel* in the cell decomposition method. There are an infinite number of paths from the origin to the target which lie within those cells. One might say that it would be best to choose the one with the shortest path length. It is not easy to find the path with shortest path length, however, as this is generally a non-convex optimization problem since the set of cells $(a_1, \cdots, a_l)$ is generally not convex.

We propose the following simple method: Connect the origin in cell a_1 to the center point of cell a_2, connect the center points of cells a_i and a_{i+1}, $2 \leq i \leq l - 1$, connect the center of cell a_{l-1} to the target. See Figure 6.3.

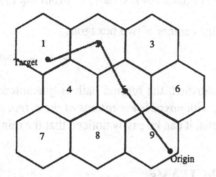

Figure 6.3. A rough path from the origin to the target when the shorted path of the digraph in Figure 2 is $(9, 5, 2, 1)$.

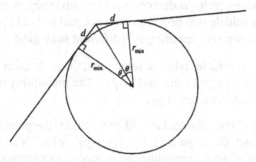

Figure 6.4. Smoothing the vertex of a preliminary path with an arc of a circle of radius r_{min}.

6.2. Path Refinement

In this section we describe how to refine the preliminary path of line segments generated by the Bellman-Ford algorithm. The main idea is similar to what was described in [3]: make the vertices of the line segments *smooth* by tangential arcs of a circle of radius r_{min}, where r_{min} is the minimum radius that a UAV can execute.

The length d between the vertex point and the tangential point (see Figure 6.4) can be expressed as

$$d = r_{min} \cdot \tan \theta. \qquad (9)$$

The valid size of a cell thus depends on r_{min}. If the shape of a cell is hexagon, as is shown in Figure 6.1, the minimum angle between vertices is $2\pi/3$. The minimum value of θ in Figure 6.4 is thus $\pi/6$. From Eq. (9) we have

distance between the centers of two hexagons

$$\geq r_{min} \cdot \tan \frac{\pi}{6} = \frac{r_{min}}{\sqrt{3}}. \quad (10)$$

If this condition is satisfied, the refined path is guaranteed to lie within the predetermined cells, *with possible exceptions at the origin and target cells*. In the case of square cells, it can be easily noticed that the minimum value of θ is $\pi/4$.

7.   Multiple UAVs

When planning for multiple vehicles, it may be desirable for the vehicles to take different paths. To illustrate this point, consider the following simplified scenario: Two paths are available, where the probability that an adversary will be encountered along path i is denoted p_i; if an adversary is encountered, the probability that a vehicle will be neutralized is equal to 1. Let $p_1 \leq p_2$. Let us plan the paths for two vehicles using two different strategies:

Same Path If both vehicles take the same path, the probability that both vehicles reach the target is simply $1 - p_1$. The probability that none of the vehicles reach the target is p_1.

Different Paths If the vehicles take different paths, the probability that both vehicles reach the target is $(1 - p_1)(1 - p_2)$, which is less than the Same Path scenario; we are providing the adversary more opportunities to neutralize at least one of our vehicles. Note, however, that the probability that none of the vehicles reach the target is $p_1 p_2$, which is less than p_1.

The point of this example is that if the success of a mission does not require that all the vehicles reach the target, the probability of success can be enhanced by requiring that the vehicles take different routes to the target. In particular, in an adversarial environment this may decrease the probability that all the vehicles are detected by the adversary at the same time and neutralized.

We propose a simple extension of the single UAV planning problem that results in different vehicle paths. The paths are planned sequentially by adding a penalty weight to the path of the previous vehicles. The cells traversed by one vehicle have thus a high penalty and the other vehicles will avoid these cells. The detailed algorithm is as follows:

  i) Plan a path for vehicle 1 based on the initial probability map;

 ii) Build a map by adding a penalty to the path generated for vehicle 1;

iii) Plan a path for vehicle 2 based on the newly generated map;

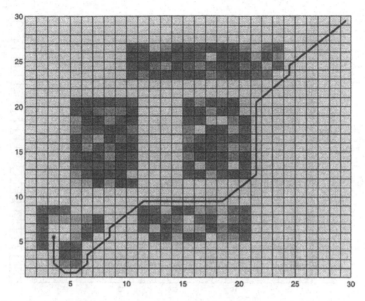

Figure 6.5. Simulation result: the path of the UAV when the path length is not penalized.

iv) Repeat the procedure for the other vehicles.

This method can also be applied to the case of dynamic environments by updating the probability maps and by removing penalties in the links that the vehicles have not yet passed.

8. Simulation

We considered a region $\mathcal{R}$ consisting of 30 × 30 square cells. We allowed oblique moves through the vertex of the square connecting two centers in addition to vertical and horizontal moves.   We generated random numbers with a uniform distribution in certain areas (5 areas in this simulation) and assigned them to $P(S_i)$. In practice, probability $P(S_i)$ should be computed by using Eq. (2) with gathered information, measurement data, etc. The origin is cell $(30, 30)$ and the target is located at cell $(3, 5)$, which is marked by '$*$' in the plots. In the maps, darker regions represents higher value of $P(S_i)$ (more dangerous areas).

The example in Figure 6.5 is the case when weight (6) was used. In this case, the path planning algorithm produced the safest path. The generated path takes a non-direct route for safety reason as can be seen in Figure 6.5. The second example in Figure 6.6 used the same probability map but adopted a different strategy: We used the weight in Eq. (7) with $c = 0.1$. As can be

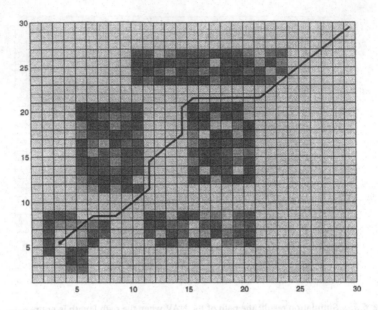

Figure 6.6. Simulation result: the path of the UAV when the path length is penalized.

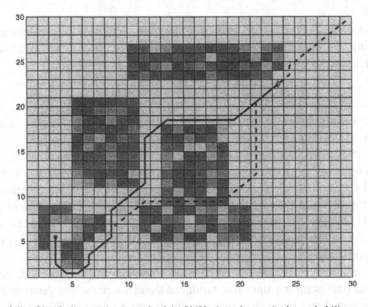

Figure 6.7. Simulation result: the path of the UAV when changes in the probability map were reported during path traversal.

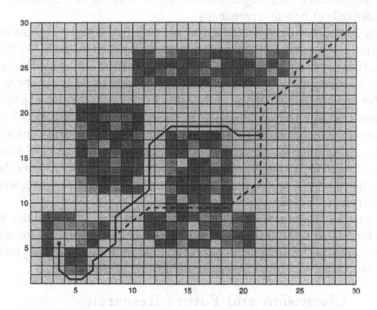

Figure 6.8. Simulation result: the path of the UAV when changes in the probability map were reported during path traversal.

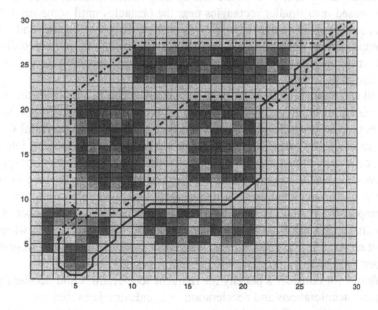

Figure 6.9. Simulation result: paths for multiple UAVs.

seen the generated path length is shorter than that in Figure 6.5, but the path included cells of non-zero probability.

The next scenario considered a dynamic situation. The movements of the adversaries or changes in the probabilities are reported to the UAV. We assumed that it was reported when the UAV was in cell $(23, 22)$, which is marked by '$\Diamond$' in Figure 6.7. The initial probability map is the same as the one in Figure 6.5 and Figure 6.6. We used the same weight as in the first example — without a penalty on path length. The dashed-line path in Figure 6.7 is the initial path, which is same as the one in Figure 6.5. After being reported the changes in the probabilities at the position marked by '$\Diamond$', the UAV took the solid-line path instead of the initial dashed-line path. Figure 6.8 shows the result when the UAV was reported the changes in the probability map when it was in cell $(21, 17)$.

Figure 6.9 shows path planning for multiple vehicles. The solid line is the path for vehicle 1, the dashed line for vehicle 2, and the dotted line for vehicle 3. Referring to the discussion in Section 7, the solid-line path was produced first, the dashed-line path second, and the dotted line path third.

9.   Discussion and Future Research

Our approach is similar to the approximate cell decomposition methods in robot path planning in that it uses discretized cells for path planning. The differences are that i) cells in the cell decomposition method are recursively decomposed into smaller rectangles near the obstacles until some predefined resolution is attained, and thus the sizes of the cells are not same, and ii) the decomposition is performed in the configuration space, not in the two dimensional map, and thus the dimension of the cells is dependent on the degrees of freedom of the robot.

Cell decomposition in robot path planning is accomplished with the initial information on the locations and shapes of the obstacles. If the robot detects an unexpected obstacle inside the channel, a sensory-based potential field is used to guide the robot. In our method, we decompose the region into uniform cells, and change the values of probabilities when we detect unexpected changes during the mission. This can be done in a short time since the map is only two dimensional while the dimension of configuration space in the cell decomposition method is same as the degrees of freedom of the robot. Once the initial graph has been built, we do not have to rebuild the graph when we detect changes in the probabilities; we simply have to modify the weights of the pertinent links.

We did not consider a penalty for frequent accelerations and decelerations. Frequent accelerations and decelerations are undesirable as they require more fuel consumption. The number of accelerations and decelerations can be estimated by the number of heading angle changes. We can thus put a penalty in sharp heading angle changes. In order to capture this penalty, each node should have information of the previously traversed node, and the graph becomes a tree. This tree expands exponentially with the number of cells, which makes it difficult to plan paths in a real time environment. A future research direction is to find a fast method for path planning which considers sharp heading angle changes.

Our proposed algorithm for multiple UAVs is based on a simple heuristic: different paths are desirable, for the reasons outlined in Section 7. The advantage of this approach is that it is very fast, and amenable to distributed computation. A more direct approach would explicitly optimize the probability that a given number of vehicles reach the target; it may be possible to simplify the resulting conditions to yield computational costs comparable to what has been presented in this paper, but with enhanced performance.

References

[1] D. Bertsekas and R. Gallager, *Data Network*, Prentice-Hall, Upper Saddle River, NJ, 1992.

[2] S. A. Bortoff, "Path planning for UAVs", In *Proc. of the American Control Conference*, pages 364–368, Chicago, IL, 2000.

[3] P. R. Chandler, S. Rasmussen and M. Pachter, "UAV cooperative path planning", In *Proc. of AIAA Guidance, Navigation and Control Conference*, Denver, CO, 2000.

[4] A. Elfes, "Sonar-based real-world mapping and navigation", *IEEE Trans. on Robotics and Automation*, RA-3(3):249–265, 1987.

[5] M. Erdmann and T. Lozano-Perez, "On multiple moving objects", *Algorithmica*, 2:477–521, 1987.

[6] T. Fraichard, "Trajectory planning in a dynamic workspace: A 'state-time space' approach", *Advanced Robotics*, 13(1):75–94, 1999.

[7] J. P. Hespanha, H. H. Kizilocak Y. S. and Ateşkan, "Probabilistic map building for aircraft-tracking radars", In *Proc. of the American Control Conference*, pages 4381–4386, Arlington, VA, 2001.

[8] D. Hsu J. C. Kindel, J. C. Latombe and S. Rock, "Randomized kinodynamic motion planning with moving obstacles", In *Proc. Workshop on Algorithmic Foundation of Robotics*, Hanover, NH, 2000.

[9] J. C. Latombe, *Robot Motion Planning*. Kluwer Academic Press, 1990.

[10] S. M. LaValle and J. J. Kuffner, "Randomized kinodynamic planning", In *Proc. IEEE Int. Conf. on Robotics and Automation*, pages 473–479, Detroit, MI, 1999.

[11] T. W. McLain and R. W. Beard, "Trajectory planning for coordinated rendezvous of unmanned air vehicles", In *Proc. of AIAA Guidance, Navigation and Control Conference*, Denver, CO, 2000.

[12] H. P. Moravec, "Sensor fusion in certainty grids for mobile robots", *AI Magazine*, pages 61–74, 1988.

[13] S. Thrun, "Learning metric-topological maps for indoor mobile robot navigation", *Artificial Intelligence*, 99:21–71, 1998.

[14] B. Yamauchi and P. Langley, "Place recognition in dynamic environments", *J. of Robotic Systems*, 14:107–120, 1997.

[15] M. Zabarankin, S. Uryasev and P. Pardalos, "Optimal risk path algorithm", In R. Murphey and P. Pardalos, editors, *Cooperative Control and Optimization*, pages 273-303, Kluwer Academic Publishers, 2002.

Chapter 7

AUTOMATICALLY SYNTHESIZED CONTROLLERS FOR DISTRIBUTED ASSEMBLY

Partial Correctness

Eric Klavins *
Computer Science Department
California Institute of Technology
Pasadena, CA   91125
klavins@caltech.edu

Abstract   We consider the task of assembling a large number of self controlled parts (or robots) into copies of a prescribed assembly (or formation). In particular, we describe a computationally tractable way to synthesize, from a specification of the desired assembly, local controllers to be used by each part, which when taken together, have the global effect of assembling the parts. We then prove that the controlled discrete dynamics of the system are correct with respect to a simplified model of the dynamics— meaning that a maximal number of parts are correctly assembled into copies of the desired assembly.

Keywords: Controller synthesis, distributed control, self-assembly

1.   Introduction

We consider the problem of controlling hundreds or thousands of robots to perform a task in concert. This problem presents many fundamental issues to robotics, control theory and computer science. With a great number of robots, decentralization is critical due to the cost of communication and the need for fault tolerance. In decentralized control, each robot should act based only on information local to it. It then becomes difficult, however, to guarantee or even

*This research is supported in part by the DARPA SEC program under grant number F33615-98-C-3613 and by AFOSR grant number F49620-01-1-0361.

S. Butenko et al. (eds.), Cooperative Control: Models, Applications and Algorithms, 111-127.
© 2003 Kluwer Academic Publishers.

derive the behavior of the entire system given the behaviors of the individual components. In this paper we address this difficulty in a novel way: We begin with a specification of an assembly and develop methods that allow us to automatically *synthesize* individual behaviors so that they are guaranteed to produce the desired global behavior.

Specifically, we consider the task of assembling many disk-shaped parts in the plane into copies of a prescribed assembly (formation), which is specified by a graph with n vertices. We do not allow the parts to collide, making the task more difficult due to the non-trivial topology of the resulting $2n$ dimensional configuration space. As shown in Figure 7.1 we suppose that each part can move itself and can play any role in an assembly, which makes the task particularly rich. We first demonstrate a means of synthesizing from the specified assembly, a set of identical controllers for the parts to run which have the net effect of moving the parts to form copies of the specified assembly without colliding. The idea is that parts should join together into subassemblies which should in turn join together to make larger assemblies and so on. To achieve this, some theory is developed along with algorithms that compile a specified assembly into a list of allowable subassemblies. Next we show how to produce a lookup table from the list which can be used as a discrete event controller (Figure 7.2) that guides parts through a "soup" of other parts and subassemblies. Then we add a continuous motion controller based on the assembly rules represented by the lookup table and a (provably correct) method for deadlock avoidance.

Finally, we show formally that the discrete dynamics given by the lookup table and the deadlock avoidance mechanism (and employed in the control of each part) are correct. The proof assumes a certain logical model of the dynamics which accounts for the discrete interactions between parts (forming neighbor relationships) but neglects the continuous dynamics. A formal analysis of the complete, highly nonlinear hybrid dynamics is not yet forthcoming.

1.1.   Related Research

We are most strongly inspired by the work of Whitesides and his group ([3, 4]) in meso-scale self-assembly. In this work, small, regular plastic tiles with hydrophobic or hydrophylic edges are placed on the surface of some liquid and gently shaken. Tiles with hydrophobic edges are attracted along those edges while hydrophylic edges repel. Striking "crystals" emerge as larger structures self assemble. By using different shapes and edge types, different gross structures can be created. A similar idea is used on a much smaller scale in ([12]) where strands of DNA are attached to tiny gold balls in solution. Complementary strands attract and a gross structure is revealed. By choosing which strands go where, the "programmer" has some control over the resulting emergent structure. At least two next steps are apparent. First, these and similar ([2]) methods generally produce arrays or lattices of parts, meaning that there is no general way to *terminate* a regular pattern at, say, a 5 × 5 array of parts (There has been work on changing the function of parts as they combine ([15]). Second, there is no known formal method of starting with a *specification* of the desired emergent structure and devising the structure of the individual parts. In this paper we address both of these issues by supposing that each part can run a *program* that tells it when to join with another part, and when to repel it, based on some state information. Of course, this is a far way away from the reality of small plastic parts or gold balls, but our ideas could easily be implemented with teams of robots and may even, when developed further, present the chemist with new tools.

The motivation for considering disk shaped parts in the plane and for the potential field construction in Section 4 comes from the work of Koditschek and others ([10, 7]) in assembly. There, a global artificial potential function over the configuration space of n disk shaped parts is used to guide the parts to their assembled state, corresponding to the unique minimum of the potential function. The approach is not distributed, however, because it requires that each part knows the full state of the system to act. Other work has applied similar ideas, in a distributed fashion ([13, 1]), although without a means of synthesizing the desired behavior. Still other approaches to the control of a group of robots ([5]) assume a leader. In contrast, the present paper commits to the synthesis idea and to a strong degree of decentralization, using decentralized potential fields merely as a *primitive* in a more sophisticated hybrid control scheme.

The ideas in this paper also grow from our own work in controller synthesis in manufacturing systems ([9, 8]). Our approach to manufacturing has been to synthesize a decentralized automated factory description from a description of a product. The description includes the layout of the factory and the control programs the robots should run to produce the product. In that sense, the present work is an extension of the idea, although it assumes fewer constraints on the topology of the workspace.

2. The Problem

We consider a simple form of assembly process by assuming that parts are programmable and able to sense the position and state of other nearby parts. We start with m disk-shaped parts (of radius r) confined to move in $\mathbb{R}^2$. Denote the position of part i by the vector x_i. We desire that each part move smoothly, without colliding with other parts, so that all parts eventually take some role in an *assembly* or *formation*. This is shown graphically in Figure 7.1. For simplicity, we assume that the dynamics of each disk are given by $\ddot{x}_i = u_i$. We

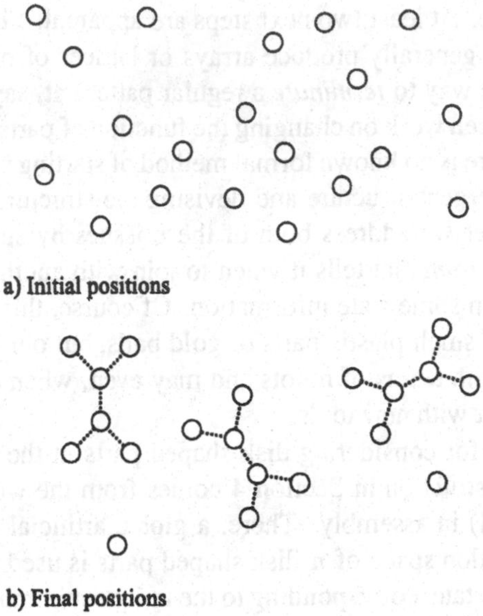

a) Initial positions

b) Final positions

Figure 7.1. The goal of the assembly problem. Each disk shaped part must move from its initial position (a) to a position in a a copy of the specified assembly (b). Dashed lines show the resulting adjacency relationship E. There may be leftover parts.

believe that control of parts with more complicated dynamics can be based on the control algorithms we develop for this simple situation. In this section we describe the goal of assuming a role in a formation formally.

Let $G = (V, E)$ be a finite undirected, acyclic graph. Thus, V is a finite set of nodes (in this paper, $V = \{1, ..., n\}$) and E is a collection of edges of the form $\{a, b\}$ with $a, b \in V$ and $a \neq b$. In this paper, we will call such a graph an *assembly* and only consider the case where G is a tree (i.e., contains no cycles). There are technical details, which are solvable but not addressed in this paper except briefly, that prevent the direct application of the methods in this paper to general graphs.

Given an assembly $G = (V, E)$ with $|V| = n$, consider the case where $m = n$. The problem is to produce a control algorithm to be used by each part that will control the m parts to move, without colliding, from arbitrary initial conditions to positions such that there exists a permutation h of $\{1, ..., m\}$ such that

1 If $\{h(i), h(j)\} \in E$ then $d_{nbr} - \epsilon < ||x_i - x_j|| < d_{nbr} + \epsilon$;

2 If $\{h(i), h(j)\} \notin E$ then $||x_i - x_j|| > d_{nbr}$.

Here $d_{nbr} > 0$ and $\epsilon > 0$ are parameters. The image $h(i)$ of i is called the *role* of i in the assembly. We furthermore require that these assemblies be stable to disturbances in the sense that the set of points $x_1, ..., x_m$ satisfying the above conditions is an attractor of the closed loop dynamics we will construct. If $m = kn$ for some $k \in \mathbf{Z}$ then we still require the above except now with respect to a disjoint union of k copies of G. And of course, if m is not a multiple of n, then we require that as many parts as possible form assemblies in the obvious way.

We note that not all trees can be embedded in the plane in such a way that neighbors are distance d_{nbr} apart and non-neighbors are distance greater than d_{nbr} apart. For simplicity in what follows, we restrict the assemblies we specify to those that can be so embedded.

2.1. Controller Structure

In general we will assume that parts have limited sensing and communication capabilities and we allow them to store a discrete state, s_i, along with their control programs. In particular, we assume that part i can sense its own position and the positions and discrete states of other parts within some range $d_{max} > 0$ of x_i.

The methods we develop below will, given a description of the desired assembly structure, *synthesize* a hybrid controller H_i of the form shown in Figure 7.2. The goal is that when each part runs a copy of H_i (from different initial conditions), the parts will self assemble.

The controller H_i is described by a continuous control law F_i, a predicate $\mathcal{A}$ called the *attraction predicate* and a discrete update rule g. F_i describes the force that the part should apply to itself. $\mathcal{A}(s_i, s_j) \in \{true, false\}$ determines whether parts i and j with states s_i and s_j should try to become neighbors, thereby forming a larger assembly. The update rule $g(s_i, s_j, s_k)$ determines the new discrete state of part k based on the joining of parts i and j. Loosely, the operation of H_i is as follows. Part i starts with some initial position $x_i(0)$, the initial state $s_i(0) = (1,1)$ and no neighbors. It then applies the control force $F_i(\mathbf{x}, \dot{\mathbf{x}}, \mathbf{s})$ to itself until either a new neighbor is detected or it receives a state update from a neighbor. Here $\mathbf{x}$, $\dot{\mathbf{x}}$ and $\mathbf{s}$ are m dimensional vectors describing the complete state of the system. However, F_i may only use the states of the parts within distance d_{max} of part i. The force F_i is computed based on the position, velocity and discrete state of part i and on the discrete states of the sensed parts.

The task of an automatic synthesis procedure, performed by what we are calling a *compiler*, is to take a description of a desired assembly and produce H_i — in this case, F_i, $\mathcal{A}$ and g. The construction of $\mathcal{A}$ and g are described next and the construction of F_i, which requires $\mathcal{A}$, is discussed after that.

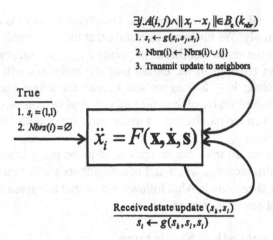

Figure 7.2. The structure of the hybrid controller that is constructed by the compilation scheme in this paper. Arcs denote transitions and are labeled by a predicate/action pair. When an arc's predicate becomes true, the action is taken and control transfers from the source of the arc to the target of the arc.

3. Compilation of Assembly Rules from Specifications

In this paper, an assembly can be specified simply by listing which roles in the assembly are adjacent — that is, by a graph. As mentioned above, we restrict ourselves to the situation where the adjacency graphs are trees, leaving the detail of arbitrary graphs to future work. In any case, we believe that assembling an arbitrary graph will start with the assembly of a spanning tree of that graph.

The goal of this section is to produce the attraction predicate $\mathcal{A}$ and the update rule g from a specified assembly $G_{spec} = (V_{spec}, E_{spec})$, which we assume is a tree. This requires first generating a set of subassemblies of G_{spec} and then compiling $\mathcal{A}$ and g from the set.

3.1. Discrete State of a Part

We intend that the parts control themselves to first form subassemblies of G_{spec}, and from those subassemblies form larger subassemblies and so on until G_{spec} is finally formed. The discrete state of a part must, therefore, include a reference to the subassembly in which it currently plays a role. To this end, we build a list (in Section 3.2) of the particular (connected) subassemblies we will allow: $\mathcal{G} = \{G_1, ..., G_p\}$. We require that each $G_i \in \mathcal{G}$ is of the form (V_i, E_i) where $V_i = \{1, ..., |V_i|\}$ and $E_i \subseteq V_i \times V_i$. Although this representation of subgraphs in $\mathcal{G}$ is arbitrary, because the vertices in V_i could have been named in other ways, some common scheme is required for a graceful definition of the states of parts.

Now, the discrete state of a part consists of a pair $s_i = (j, k) \in \mathbf{Z}^2$ where j is the index of a subassembly in $\mathcal{G}$ and $k \in V_i$ is a *role* in that subassembly.

3.2.   Generating Assembly Sequences

Define an operation on assemblies G_1 and G_2 as follows

Definition 1. *The* **join** *of* G_1 *and* G_2 *via vertices* $u \in V_1$ *and* $v \in V_2$, *denoted* $G_1.u \oplus G_2.v$, *is defined as* $G_1.u \oplus G_2.v = (V, E)$ *where*

$$V = \{1, ..., |V_1| + |V_2|\} \text{ and}$$

$$E = E_1 \cup \{\{a + |V_1|, b + |V_1|\} \mid \{a, b\} \in E_2\} \cup \{u, v + |V_1|\}.$$

For example

$$(\{1, 2\}, \{\{1, 2\}\}).2 \oplus (\{1, 2\}, \{\{1, 2\}\}).1$$

$$= (\{1, 2, 3, 4\}, \{\{1, 2\}, \{2, 3\}, \{3, 4\}\}).$$

We will also use the notations $i.j \oplus k.l$ and $(i, j) \oplus (k, l)$ to mean the join of the assemblies with indices i and k in a given $\mathcal{G}$ via the vertices with indices j and l.

The set of subassemblies $\mathcal{G}$ must have the following property:

Property 1. *For all* $G \in \mathcal{G}$ *there exist* $G_1, G_2 \in \mathcal{G}$, $u \in V_1$ *and* $v \in V_2$ *such that* $G_1.u \oplus G_2.v \simeq G$ *unless* $G = \{\{1\}, \emptyset\}$ *and there does not exist a* $G' \in \mathcal{G} - \{G\}$ *with* $G \simeq G'$.

Here "$\simeq$" means isomorphic in the usual sense: $(V_1, E_1) \simeq (V_2, E_2)$ if there exists a function $h : V_1 \to V_2$ such that $(u, v) \in E_1$ if and only if $(h(u), h(v)) \in E_2$. Such an h is called a *witness* of the isomorphism. Witnesses are used in this paper to "translate" the representation of the join of two graphs to the representation of that graph in $\mathcal{G}$. Property 1 assures that any assembly can be constructed from exactly two other assemblies, so that only pairwise interactions between parts need to be considered by the ultimate controller, and that there is only one representation of each subassembly in the list. (We will also require another property, Property 2 in Section 5, later).

The simplest means of automatically constructing $\mathcal{G}$ from G_{spec} is to simply set $\mathcal{G}$ to be all possible connected subgraphs of G up to isomorphism, producing a set of size $O(2^n)$. This set can be computed using a simple exhaustive search. Since A and g will be obtained from a table constructed from $\mathcal{G}$, this may be an impractically large set for large G_{spec}, although for small assemblies the set of all subassemblies is quite reasonable and produces good controllers. A $\mathcal{G}$ thus constructed trivially satisfies Property 1.

Another means of constructing $\mathcal{G}$ is to build subtrees of G_{spec} one node at a time, starting at some base node and adding nodes to the leaves of subtrees. This algorithm, which we call A_1, requires an assembly G_{spec} and a base node i. It produces a set $\mathcal{G}_{A_1,i}$ of size exactly n, there being one subassembly for each size 1 to n. The set $\mathcal{G}_{A_1,i}$ constructed using A_1 satisfies Property 1 easily since each subassembly (except the singleton assembly) can be obtained by joining the next smallest subassembly with $\{\{1\}, \emptyset\}.1$. Richer subassembly sets can be made by calling A_1 again, starting with a different base node, and combining it with the first set. In this manner a set of size $O(cn)$ can be constructed from a set of c nodes $U \subseteq V_{spec}$. Call this set $\mathcal{G}_{A_1,U}$. It satisfies Property 1 because each of the sets $\mathcal{G}_{A_1,i}$ for $i \in U$ do. The process of combining the sets requires some computation, however, because we must maintain the second part of Property 1. To combine the list $\mathcal{G}_{A_1,i}$ with list $\mathcal{G}_{A_1,j}$ we must compare each element of the first list with each element of the second list to make sure they are not isomorphic. If they are, we keep only one of them for the combined list. Although there is no known polynomial time algorithm for checking the isomorphism of two graphs, checking the isomorphism of two trees of size n takes $O(n^{3.5})$ steps ([14]). Thus, combining two size n lists takes time $O(n^{5.5})$. The reader can check that the combination of sets satisfying Property 1 also satisfies Property 1.

3.3.   Generating Update Rules

From an assembly set $\mathcal{G}$ satisfying Property 1, we can state the definition of $\mathcal{A}$ simply:

Definition 2. *Given $\mathcal{G}$ satisfying Property 1, the* **attraction predicate** *$\mathcal{A}$ is defined as*

$$\mathcal{A}(s_i, s_j) = true \Leftrightarrow \exists\, G \in \mathcal{G} \text{ such that } s_i \oplus s_j \simeq G.$$

We can also define the update rule g.

Definition 3. *Given $\mathcal{G}$ satisfying Property 1 and states s_i and s_j with $\mathcal{A}(s_i, s_j) = true$, the* **update rule** *$g$ is defined as follows. Suppose $G \simeq s_i \oplus s_j$ has index k in $\mathcal{G}$, suppose $h : s_i \oplus s_j \rightarrow G$ witnesses this isomorphism and suppose $s_l = (a, b)$. Then*

$$g(s_i, s_j, s_l) \doteq (k, h(b')),$$

where $b' \in V(s_i \oplus s_j)$ is the name of vertex b after taking disjoint unions in Definition 1 of the join operation. If $\mathcal{A}(s_i, s_j) = false$ then the update rule is not defined: $g(s_i, s_j, s_l) \doteq \perp$.

Algorithm A_2:

Input: $\mathcal{G}$, a list of subgraphs with
          Property 1
Output: T, a tabular representation
          of $\mathcal{A}$ and g

For $i = 1$ to $|\mathcal{G}| - 1$
  For $k = i$ to $|\mathcal{G}| - 1$
    For $j = 1$ to $|V_i|$
      For $l = 1$ to $|V_k|$
        If $\exists\, G \in \mathcal{G}$ with $i.j \oplus k.l \simeq G$
        Let h be the witness
        $T_{i,j,k,l} = (\; index(G'),\; \langle h \rangle\,)$
        Else $T_{i,j,k,l} = \bot$

Figure 7.3. The procedure for constructing a table of size $O(|\mathcal{G}|^2 n^3)$ from a list of sub-assemblies $\mathcal{G}$ of a specified tree G_{spec}. The predicate $\mathcal{A}$ and the update rule g can be read off the resulting table in constant time.

The procedure for determining the values of $\mathcal{A}$ and g require determining tree isomorphisms — which is likely too time consuming to be done online. We can, however, perform all the necessary computations offline by compiling $\mathcal{G}$ into a table. The result is that H_i can make all discrete transitions essentially instantaneously because all that is required is a table look-up. Furthermore, the size of the table is $O(|\mathcal{G}|^2 n^3)$. As was shown, $|\mathcal{G}|$ can be taken to be cn, so that even complicated assemblies require only $O(n^5)$ storage.

The construction proceeds in two steps. First, we determine a representation of the update function g resulting from a join of $G_i.j$ with $G_k.l$. Second we build a table of all possible joins between all possible pairs of distinct graphs taken from $\mathcal{G} - G_{spec}$. The result is a four dimensional table T where each entry $T_{i,j,k,l}$ is the representation of $G_i.j \oplus G_k.l$.

Given $G_i.j$ and $G_k.l$, let $G = (V, E) = G_i.j \oplus G_k.l$. We must first determine whether there exists a $G' \in \mathcal{G}$ such that $G \simeq G'$ then, we require a witness h of this isomorphism because we must have a means of translating the new roles of each part in the new assembly into their representations in $\mathcal{G}$. Suppose such an h exists. Then we represent the table entry $T_{i,j,k,l}$ as a pair

$$(\; index(G'),\; \langle h(1), ..., h(|V_i| + |V_j|)\rangle\;).$$

Otherwise, set $T_{i,j,k,l} = \bot$. The procedure for constructing T is shown in Figure 7.3, it takes time $O(|\mathcal{G}|^3 n^{6.5})$ because of the added complexity of finding a witness for each join.

To summarize, given G_{spec}, constructing A and g, the discrete part of the controller H_i, proceeds in two steps. First, a list of subassemblies $\mathcal{G}$ is build from G_{spec} using one of the methods discussed in Section 3.2. Second, using algorithm A_2, a table T is built from the $\mathcal{G}$. $A(s_i, s_j)$ can be computed simply by checking whether $T_{s_i,s_j} \neq \bot$ and $g(s_i, s_j, (a, b))$ can be determined by looking up T_{s_i,s_j} and reading off $h(b)$.

4.   Implementation of Assembly Rules

Completing the controller H_i shown in Figure 7.2 requires a definition of F_i as well as some method by which parts can communicate. In this section we define an F_i and assume a simple communications scheme that works in simulation and about which we have a preliminary analytical understanding.

We suppose that parts can only communicate with their neighbors. The difficulty is then that two parts playing roles in the same subassembly might try to update the state of that subassembly simultaneously. Thus, such an update requires a means of obtaining consensus among all parts in the subassembly. Consensus can be difficult or even impossible if the processing is asynchronous and there are process failures or link failures ([11]), although approximate algorithms exist for these situations ([6]). In what follows, we assume a good consensus algorithm with no process or communication failures. Consideration of the many complications we may add, although important, would take us too far afield of the present topic and are somewhat independent of methods we have so far described.

4.1.   An Example Implementation

For each part i, we can decide, using A, whether part i should move toward j or not. To this end define

$$S(i) = \{j \mid \|x_i - x_j\| < d_{max}\}$$
$$Attract(i) = (\{j \mid A(s_i, s_j)\} \cup Nbrs(i)) \cap S(i)$$
$$Repel(i) = (\{j \mid \neg A(s_i, s_j)\} - Nbrs(i)) \cap S(i).$$

$S(i)$ is the set of parts that i can sense. Note that these sets are easily computed from a table compiled from a given G_{spec}. One way of forming the control law F_i is to sum, for each $j \in Attract(i)$ a vector field F_{att} which has an equilibrium set at distance d_{nbr} from x_j and for each $j \in Repel(i)$ a vector field F_{rep} which has x_j as a repellor. We can construct these fields from the potential functions defined by

$$V_{att}(x_i, x_j) = \left(\frac{\|x_i - x_j\| - d_{nbr}}{\|x_i - x_j\| - r} \right)^2$$

$$V_{rep}(x_i, x_j) = \left(\frac{1}{\|x_i - x_j\| - r} \right)^2.$$

Recall that r is the radius of the (disk shaped) parts. Then we set

$$F_{att}(x_i, x_j) = -\frac{1}{\|x_i - x_j\|} \frac{\partial V_{att}}{\partial x_i}(x_i, x_j)$$

$$F_{rep_1}(x_i, x_j) = -\frac{1}{\|x_i - x_j\|} \frac{\partial V_{rep}}{\partial x_i}(x_i, x_j)$$

$$F_{rep_2}(x_i, x_j) = \begin{cases} 0 \text{ if } \|x_i - x_j\| > d_{nbr} + \delta \\ F_{rep_1}(x_i, x_j) \text{ } otherwise, \end{cases}$$

where $\delta > 0$ is some small constant. We have scaled the gradients of the potential functions by $\|x_i - x_j\|^{-1}$ so that the "influences" of parts nearest i are felt most strongly. We have also defined two versions of the repelling field. We use F_{rep_2} because it is only active when parts violate condition (2) from Section 2. We will see the reason for this shortly.

For the complete control law we use

$$F_i(\mathbf{x}, \dot{\mathbf{x}}, \mathbf{s}) = \sum_{j \in Attract(i)} F_{att}(x_i, x_j)$$

$$+ \sum_{j \in Repel(i)} F_{rep_2}(x_i, x_j) - b\dot{x}_i,$$

where $b > 0$ is a damping parameter. In practice we assume a maximum actuator force, setting $u_i = max\{u_{max}, F_i(\mathbf{x}, \dot{\mathbf{x}}, \mathbf{s})\}$.

We have built a simulation environment for the above system. We have investigated a variety of initial conditions, with varying numbers of agents (from tens to hundreds), and various specifications of the desired assembly G_{spec}. Some simulations can be viewed at http://www.cs.caltech.edu/~klavins/rda/.

Two deadlock situations arose in our initial simulations. First, F may have spurious stable equilibriums which prevent attracting pairs from moving toward each other. Second, it is possible that the set of currently formed subassemblies admit no joins in $\mathcal{G}$. That is, it may be that at some time there do not exist parts i and j such that $\mathcal{A}(s_i, s_j)$ is true.

To avoid these situations, we employ a simple deadlock avoidance method. For each subassembly $G_k \in \mathcal{G}$ we define a *stale time* $stale(k) \in \mathbb{R}$. Any subassembly that has not changed state within $stale(i)$ seconds of its formation time should (1) break apart, setting the state of each part in it to $(1,1)$ and (2) have each part "ignore" other parts from that same assembly for $stale(k)$ seconds. If k_{spec} is the index of G_{spec} in $\mathcal{G}$, we set $stale(k_{spec}) = \infty$. The result is a new controller $H_{d,i}$ that checks for staleness and implements (1) and (2) above, but is otherwise similar to H_i in Figure 7.2. We also change the definitions of $Attract(i)$ and $Repel(i)$. Suppose that $Ignore(i)$ is the set of all part indices that part i is presently ignoring due to a staleness break-up. Then

$$Attract_d(i) = Attract(i) - Ignore(i)$$
$$Repel_d(i) = Repel(i) - Ignore(i).$$

F_i is then changed accordingly. Using this deadlock avoidance measure, we have not yet seen a set of initial conditions for any G_{spec} we tried for which our simulation did not converge upon the maximum possible number of parts playing roles in a final assembly, except when partial assemblies were out of sensor range of parts needed to complete them.

5.   Partial Correctness of the Assembly Process

In this section, we describe a *discrete* model of the assembly process that allows to prove properties of the rules generating by the compiler, independent of the physical setting in which the assembly process is implemented. This is only half the story, of course. A complete proof of correctness of the hybrid assembly system is not yet available.

For each part $i \in \{1, ..., m\}$ we define its state at step k by a triple

$$q_i(k) = (s_i(k), N_i(k), \vartheta_i(k)).$$

Here $s_i(k)$ is the assembly index, role pair as in the description of the compiler, $N_i(k) \in \{1, ..., m\}$ is the set of neighbors of i at step k and $\vartheta_i(k)$ is the "ignore"set of part i at step k. Initially,

$$q_i(0) = (s_i(0), N_i(0), \vartheta_i(0)) = ((1,1), \emptyset, \emptyset).$$

Let $q(k) = \{q_1(k), ..., q_m(k)\}$.

Define $G(k) = (\{1, ..., m\}, E(k))$ to be the graph induced by the neighbor sets: $\{a, b\} \in E(k)$ if and only if $a \in N_b(k)$. Define $S_i(k)$ to be the connected component of G_k containing i. Thus, $S_i(k)$ is the subassembly of agent i at step k. Define $R(k)$ to be the binary relation on $\{1, ..., n\}$ defined by

$$\begin{aligned}
i \, R(k) \, j \quad &\Leftrightarrow \quad \mathcal{A}(s_i(k), s_j(k)) \\
&\wedge \quad S_i(k) \neq S_j(k) \\
&\wedge \quad (s_i(k) = s_j(k) = (1,1) \Rightarrow i \notin \vartheta_j(k)).
\end{aligned}$$

To advance the system from state $q(k)$ proceed as follows. There are three cases.

1 There exist i and j such that $i \, R(k) \, j$. In this case, set

$$s_l(k+1) = \begin{cases} g(s_i, s_j, s_l) \text{ if } l \in S_i(k) \cup S_j(k) \\ s_l(k) \text{ otherwise} \end{cases}$$

and set

$$N_l(k+1) = \begin{cases} N_l(k) \cup \{j\} \text{ if } l = i \\ N_l(k) \cup \{i\} \text{ if } l = j \\ N_l(k) \text{ otherwise} \end{cases}$$

for all $l \in \{1, ..., n\}$. The ignore sets do not change.

2 $R(k) = \emptyset$ and there are at least two components of $G(k)$ that are not isomorphic to a final subassembly. In this case, we advance the system by breaking up a subassembly. First, choose $X(k)$ to be a minimally sized assembly of $G(k)$. And set

$$s_l(k+1) = \begin{cases} (1,1) \text{ if } l \in X(k) \\ s_l(k) \text{ otherwise} \end{cases}$$

and put

$$N_l(k+1) = \begin{cases} \emptyset \text{ if } l \in X(k) \\ N_l(k) \text{ otherwise} \end{cases}$$

and finally

$$\vartheta_l(k+1) = \begin{cases} X(k) \text{ if } l \in X(k) \\ \emptyset \text{ otherwise} \end{cases}$$

for all $l \in \{1, ..., n\}$.

3 $R(k) = \emptyset$ and there are zero or one components that are not isomorphic to a final assembly.

The system is assembled. Set $s_l(j) = s_l(k)$ and $N_l(j) = N_l(k)$ for all $j > k$.

We prove two properties about the system defined above. The first is a *safety* property, asserting that only subassemblies in $\mathcal{G}$ form during executions of the system. The second is a *progress* property, asserting essentially that the number of components of $G(k)$ decreases as k increases. From this property we can conclude that every run of the system ends with a maximum number of final subassemblies being formed.

Theorem 5.1. *For all $k \in \mathbb{N}$, every component of $G(k)$ is isomorphic to some graph $G' \in \mathcal{G}$.*

Proof: This is true of $G(0)$ since all components are singletons. Suppose it is true of $G(k)$. Either rule 1 or rule 2 above is used. In the first case, suppose that i and j are chosen so that $iR(k)j$. Then $G(k+1)$ is the same as $G(k)$ except that $G(k+1)$ contains the additional edge (i,j) joining $S_i(k)$ and $S_j(k)$ together. Since $iR(k)j \Rightarrow \mathcal{A}(s_i(k), s_j(k)) \Rightarrow S_i(k).i \oplus S_j(k).j \in \mathcal{G}$, the new component is isomorphic to a graph in $\mathcal{G}$. In the second case, suppose that X is chosen as a minimal component of $G(k)$. Then $G(k+1)$ has the same components as $G(k)$ except that it does not contain X and it does contain $|X|$ singletons. $\square$

In the next theorem we suppose that $G(k_1)$ contains no copies of the final assembly. The does not reduce the generality of our arguments because final assemblies do not play any part in the execution of a system. Thus, once a final assembly is built, we can remove the nodes in it from consideration.

Before we state the theorem, we have a definition. Let the predicate $P(k)$ be defined by

$$P(k) \iff R(k) = \emptyset$$
$$\wedge \quad G(k) \text{ contains at least two nonfinal assemblies.}$$

Thus, $P(k)$ is equivalent to the condition for rule two in the definition of execution. We also define a new property on $\mathcal{G}$ that is that we require of assembly sequences in addition to Property 1:

Property 2. $\{\{1\}, \emptyset\} \in \mathcal{G}$ and for all $G \in \mathcal{G}$ there is a $u \in V(G)$ such that $G.u \oplus \{\{1\}, \emptyset\}.1$ is isomorphic to some graph in $\mathcal{G}$, unless G is the final assembly.

Lemma 5.2. *Suppose Property 2 holds for $\mathcal{G}$. Let $k_1 < k_2$ be such that*

1 $G(k_1)$ contains no final assemblies;

2 $P(k_1)$ is true;

3 $P(k_2)$ is true;

4 There is no j such that $k_1 < j < k_2$ and $P(j)$ is true.

Then the number of components of $G(k_2)$ is less than then number of components of $G(k_1)$.

Proof: Suppose that $G(k_1)$ has x components. To obtain the next state, rule 2 is chosen, by condition 2 in the assumptions of the theorem. Suppose that a minimal assembly of size y is chosen to be broken up. Then $G(k+1)$ has $x - 1 + y$ components.

In steps $k_1 + 1, ..., k_2 - 1$, rule 1 is used exclusively by assumption 4 in the theorem. Also, since the singletons added at step k cannot join with each other, by definition of their ignore sets in rule 2, they must each combine with assemblies descendant from one of the $x-1$ non singletons in $G(k_1+1)$. There is always one available since all non-final assemblies can join with singletons. Thus, at step k_2 there are $x - 1$ or fewer components in $G(k_2)$. $\square$

Corollary 1. *Every sequence of states ends with a maximum number of final assembles being formed.*

Proof: Let $c(k)$ be the number of components in $G(k)$. By Lemma 5.2 and the fact that rule 1 always decreases the number of components in the neighbor graph, we can find a sequence of states $k_1, ..., k_r$ such that $c(k_1), ..., c(k_r)$ is a decreasing sequence. At some point, $c(k_i) \leq m/n$ and thus some component of $G(k_i)$ has n elements. Since the only assembly with n elements that is allowed is the final assembly, $G(k)$ has a final assembly in it. We can remove it and start the process again with $m - n$ parts and no non-final assemblies. $\square$.

6. Conclusion

The ideas in this paper represent only the first steps toward understanding and realizing specifiable, programmable self assembly. Many relatively unexplored and apparently fruitful issues remain. First, although simulations and the proof in Section 5 suggest that the implementation (particular choice of F_i) combined with the deadlock avoidance procedure produces controllers that assemble a maximum number of parts safely (without collisions), this must be verified analytically using the tools in Section 5 and tools from non-linear dynamical systems.

Arbitrary graphs (as opposed to trees) require certain embeddings of their subassemblies in order to assemble themselves. For example, suppose we assemble a graph by first assembling a spanning tree of the graph and then "closing" it. If we require the closing procedure to respect the d_{nbr} distance requirements we have used, then the tree can not *cross over* itself while closing. This means the tree must assemble to an appropriate embedding class — a constraint we do not yet deal with, but plan to address soon.

Many variations on the theme presented here should also be explored: hierarchical assembly with intermediate goal assemblies, three dimensional assembly (which has fewer "closing" problems than in two dimensions), assembly of non-homogeneous parts, assembly of parts with complex dynamics (e.g. non-holonomic), and so on. Finally, we are exploring hardware implementations of these algorithms so that the issues of asynchronous processing, inaccurate sensors and faulty communications may be realistically addressed.

Acknowledgments

I thank Dan Koditschek with whom I have discussed many of the ideas. The research is supported in part by DARPA grant numbers JCD.61404-1-AFOSR.614040 and RMM.COOP-1-UCLA.AFOSRMURI.

References

[1] T. Balch and M. Hybinette, "Social potentials for scalable multirobot formations", In *IEEE International Conference on Robotics and Automation*, San Francisco, 2000.

[2] E. Bonabeau, S. Guerin, D. Snyers, P. Kuntz, and G. Theraulaz, "Three-dimensional architectures grown by simple 'stigmergic' agents", *BioSystems*, 56:13–32, 2000.

[3] N. Bowden, L. S. Choi, B. A. Grzybowski, and G. M. Whitesides, "Mesoscale self-assembly of hexagonal plates using lateral capillary forces: Synthesis using the "capilary" bond", *Journal of the American Chemical Society*, 121:5373–5391, 1999.

[4] T. L. Breen, J. Tien, S. R. J. Oliver, T. Hadzic and G. M. Whitesides, "Design and self-assembly of open, regular, 3D mesostructures", *Science*, 284:948–951, 1999.

[5] J. P. Desai, V. Kumar and J. P. Ostrowski, "Control of changes in formation for a team of mobile robots", In *IEEE International Conference on Robotics and Automation*, Detroit, 1999.

[6] M. Franceschetti and J. Bruck, "A group membership algorithm with a practical specification", To appear in *IEEE Transactions on Parallel and Distributed Systems*, 2001.

[7] S. Karagoz, H. I. Bozma and D. E. Koditschek, "Event driven parts moving in 2d endogenous environments", In *Proceedings of the IEEE Conference on Robotics and Automation*, pages 1076–1081, San Francisco, CA, 2000.

[8] E. Klavins, "Automatic compilation of concurrent hybrid factories from product assembly specifications", In *Hybrid Systems: Computation and Control Workshop, Third International Workshop*, Pittsburgh, PA, 2000.

[9] E. Klavins and D. Koditschek, "A formalism for the composition of concurrent robot behaviors", In *Proceedings of the IEEE Conference on Robotics and Automation*, 2000.

[10] D. E. Koditschek, and H. I. Bozma, "Robot assembly as a noncooperative game of its pieces", *Robotica*. to appear, 2000.

[11] N. Lynch, *Distributed Algorithms*, Morgan Kaufmann, 1996.

[12] C. A. Mirkin, "Programming the assembly of two- and three-dimensional architectures with DNA and nanoscale inorganic building blocks" *Inorganic Chemistry*, 39(11):2258–2272, 2000.

[13] H. Reif and H. Wang, "Social potential fields: A distributed behavioral control for autonomous robots", In *Proceedings of the 1994 Workshop on the Algorithmic Foundations of Robotics*. A.K.Peters, Boston, MA, 1995.

[14] S. W. Reyner, "An analysis of a good algorithm for the subtree problem", *SIAM Journal on Computing*, 6:730–732, 1977.

[15] K. Saitou, "Conformational switching in self-assembling mechanical systems", *IEEE Transactions on Robotics and Automation*, 15(3):510–520, 1999.

Chapter 8

THE SEARCH FOR THE UNIVERSAL CONCEPT OF COMPLEXITY AND A GENERAL OPTIMALITY CONDITION OF COOPERATIVE AGENTS

Victor Korotkich

Faculty of Informatics and Communication

Central Queensland University

Mackay, Queensland 4740, Australia

v.korotkich@cqu.edu.au

Abstract   There are many different notions of complexity. However, complexity does not have a generally accepted universal concept. It is becoming more clear that the search for the universal concept must be done within a final theory. In this chapter a concept of structural complexity for the first time suggests the real opportunity to search the universal concept of complexity within a final theory. Experimental facts given in this chapter allow to suggest a general optimality condition of cooperative agents in terms of structural complexity. The optimality condition says that cooperative agents show their best performance for a particular problem when their structural complexity equals the structural complexity of the problem. According to the optimality condition to control a complex system efficiently means to equate its structural complexity with the structural complexity of the problem.

Keywords: Complexity, cooperative agents, optimality condition, travelling salesman problem

1.   Introduction

The concept of complexity is fundamental in our understanding of reality. There exist many different notions of complexity, which are useful in numerous applications [39, 25]. Many new notions proposed recently indicate that complexity is seen as the key to a variety of problems [15, 9]. However, complexity does not have a generally accepted universal concept and it is not agreed what complexity precisely means.

S. Butenko et al. (eds.), Cooperative Control: Models, Applications and Algorithms, 129-163.

This situation, of course, is not satisfactory and yet is rooted in the main unresolved problems of complexity theory. It aims to define complexity of all physical systems within a single framework [13] and thus actually to become a theory of everything [14]. But there is still no theory that gives us such knowledge about reality.

Therefore, the search for the universal concept of complexity has a strong requirement. It must be done within a final theory [42]. In this case the theory does not allow a deeper explanatory base as it is already built on irreducible notions. Otherwise, a concept of complexity can not escape a reductionist chain of explanations. This casts doubts on the potential universality of the concept since it is possible to seek an explanation of the theory in terms of a more profound theory and so on.

The main problem of finding a final theory is to identify irreducible notions to be based on. There are not many options available for the task. Even notions originated in spacetime are not adequate as it is an approximation to reality [44, 45]. Description of physical systems in terms of a final theory is given in [19]. This theory captures a new type of processes. They are hierarchical formations of integer relations. The theory is based on integers only, which act as the ultimate building blocks of the hierarchical formations. It is final because integers are irreducible to anything more basic.

The concepts of integer and integer relation are an integral part of our mental equipment. They are a set of abstract symbols subject to the arithmetic laws. Usually, integer relations appear as solutions to Diophantine equations. However, integer relations do not seem to us rooted in firm reality. They do not have the power to evoke in our minds images of objects that form composite objects much as atoms form molecules. The notion of integer relation is too abstract to give us insight what we think the formation is all about. It is difficult to imagine how integer relations can be really formed from other integer relations, because it is not possible to visualize how this may actually happen.

In the final theory a new perception of integer relations as geometrical objects that can form into each other is presented [19]. This gives a geometrical meaning to the integer relations and visualizes their hierarchical formations. The visualization reveals self-similarity, local and non-local symmetries in the hierarchical formations. A hierarchical formation of the integer relations resembles formations of physical systems we describe in terms of complexity. This is in favor to interpret the hierarchical formation as a building-up process that combines elements of one level to form more complex elements of the next level. This interpretation helps us to see the hierarchical formations in terms of complexity and results in a quantitative concept of structural complexity [19]. This concept of complexity suggests for the first time the real opportunity to search the universal concept of complexity within a final theory.

With this concept at hand we begin to wonder whether it might have implications for complex systems. An important step in this search would be to show that the structural complexity determines the functioning of a complex system. Most complex systems can be modeled by using interacting or cooperative agents. The agents, which can range from billiard balls in a random interaction to organisms that adapt and learn, offer the quickest route to building models of complex systems. The following question about the functioning of cooperative agents is considered.

- *Is it possible to have a general optimality condition specifying when cooperative agents show their best performance for a particular problem?*

The idea is to demonstrate that for complex systems such an optimality condition could be formulated terms of structural complexity. The huge difficulty of the problem gives us no other options but to start with extensive computational experiments and see whether they lead us to a consistent picture that we can use to answer the question. Such computational experiments made during last years [20, 22] are given in the chapter. In the experiments the structural complexity of cooperative agents is increased in order to see how their performance changes in solving a problem of a class. A remarkable result always appears for each problem tested. Namely, the performance of the agents unimodally peaks at some point in an interval as their structural complexity increases. Moreover, it turns out that the structural complexity at which the performance peaks is the characteristic of the problem itself.

In general, results of the experiments give facts to believe that indeed such an optimality condition can be suggested in terms of structural complexity. The optimality condition is formulated in the following universal form:

- *cooperative agents show their best performance for a particular problem when their structural complexity equals the structural complexity of the problem.*

Importantly, the optimality condition is about how to improve performance of a complex system. In particular, it says that if the structural complexity of a complex system is less than the structural complexity of a problem then the first one must be increased until it reaches the structural complexity of the problem. Only then the system shows the best performance for the problem. Moreover, if the structural complexity of a complex system is greater than the structural complexity of a problem then for the same purpose the first one must be decreased till it becomes equal to the second one.

Therefore, the optimality condition says clearly how to harness effects of complex systems. According to the optimality condition to make good use of a complex system means to equate its structural complexity with the structural complexity of the problem.

The sections of the chapter specify the content and contain details that are supposed to be enough for the reader to be convinced in the course of the chapter. More detail about a particular result may be found in references they point to.

In Section 2 an integer code series (ICS) [20] is given. It is an instrument used to make a connection between sequences in spacetime and the hierarchical formations of integer relations. The integer code series expresses integrals of piecewise constant functions in a form that contains the sequence and powers of integers. ICS allows us to extract powers of integers in integer relations and view the integer relations as geometrical objects that like physical ones can form into each other.

In Section 3 a global description of sequences in terms of structural numbers is given. Structural numbers are simply definite integrals but expressed in terms of the integer code series and thus have the geometrical interpretation of the integral. Structural numbers combine integers and geometry. From one side ICS gives a definite integral in terms of powers of integers and can pack them in integer relations. From another side the geometrical interpretation of the definite integral links the integer relations with a geometrical pattern whose area is measured by the integral.

In Section 4 sequences are described in terms of structural numbers. This description shows a remarkable system of linear equations that in turn reveals specific integer relations. This type of integer relations is in the focus of our consideration. An important aim of this section is to see these integer relations as objects that like physical ones can form into each other.

In Section 5 a notion of integer pattern is given. The notion is used to develop a new perception of the integer relations. Integer patterns provide a geometrical meaning of the integer relations and visualize their hierarchical formations.

In Section 6 an isomorphism between the hierarchical formations of integer relations and hierarchical formations of integer patterns is shown. By the isomorphism the integer relations appear as two-dimensional geometrical objects that can form into each other. In particular, an integer relation acquires a shape of a corresponding integer pattern and can be measured by its area. A hierarchical formation of integer patterns visually reveals self-similarity, local and non-local symmetries. A connection between sequences in spacetime and the hierarchical formations of integer relations or equivalently integer patterns is presented. This suggests that the laws of a physical system in spacetime are coded in the hierarchical formations.

In Section 7 due to the isomorphism the hierarchical formations of both integer relations and integer patterns are presented as one reality, called a web of relations. The web of relations provides rich properties in the description of physical systems. Firstly, the web of relations is a final theory as it is based on integers only and does not use any notions of spacetime. Secondly, the hierarchical formations can be interpreted as a building-up process that combines elements of one level to form more complex elements of the next level. This pictures the hierarchical formations in terms of complexity. Thirdly, the geometrical meaning of the integer relations reveals self-similarity, local and non-local symmetries in the hierarchical formations.

In Section 8 an analogy between the hierarchical formation of integer relations and formations of physical systems is discussed. This analogy is in favor to interpret the hierarchical formation of integer relations as a building-up process that combines elements of one level to form more complex elements of the next level. The interpretation helps us to see the hierarchical formations in terms of complexity and results in a quantitative concept of structural complexity. The concept of complexity suggests for the first time a real opportunity to search the universal concept of complexity within a final theory.

In Section 9 optimality condition is discussed as a common property of fitness landscapes. There are no indications that fitness landscapes may have such a common property. Many important classes of problems are NP-hard and their fitness landscapes are rugged. This gives serious doubts on the possibility to have optimality condition within a general setting.

It is worth mentioning that fitness landscapes appear rugged in the representation based on Cartesian order. In this order the nearest proximity of an element consists of elements that are different from the element by a minimal value in one coordinate. In Section 10 a different representation of fitness landscapes is considered. In this representation the nearest proximity of an element consists of elements that are the closest to the element in terms of structural complexity.

In Section 11 an agent-based method to computationally investigate how performance of cooperative agents changes as their structural complexity increases is presented. A class of benchmark TSP problems [36] is used in computational experiments for this investigation. In the experiments the structural complexity of cooperative agents is increased by a control parameter. This allows us to see how the agents' performance changes as they solve the same problem for different values of the parameter. A remarkable result always appears for all the problems tested. Namely, the performance of the agents unimodally peaks for some value of the parameter as their structural complexity increases. This means that fitness landscapes of the problems appear in the unimodal form when represented in terms of structural complexity.

Moreover, the experiments give evidence to suggest that for each problem a corresponding value of structural complexity, i.e., where performance peaks, is a characteristic of the problem itself. The characteristic clearly stands to be defined as the structural complexity of the problem. Finally, this allows to formulate an optimality condition with a general setting: the cooperative agents show their best performance for a particular problem when their structural complexity equals the structural complexity of the problem.

Main results of the chapter are summarized in conclusions.

2.   Sequences and Integer Code Series

When we consider physical systems in spacetime sequences appear naturally. They realize in a discreet form our understanding that physical systems can interact and move in space as time goes in the ordered fashion. The sequence gives approximation to the trajectory of a physical system by recording the space position in successive time instants. Sequences are used also as the main carriers of information and objects of computation.

Then it is not surprising that in order to measure complexity of physical systems many notions of complexity in the first place are defined to measure complexity of sequences. A concept of structural complexity [19] presented in this chapter applies to sequences as well. The following facts are responsible for its introduction.

There is a connection between sequences as a description of physical systems in spacetime and hierarchical formations of integer relations as a description of physical systems in a new reality. This suggests that the laws of a physical system in spacetime are coded in the hierarchical formations. The most important fact about our description is that it is given in terms of a final theory. This theory is final because it is based on integers only and they are irreducible to anything more basic.

The hierarchical formation is a process that resembles formations of physical systems characterized by concept of complexity. This is in favor to interpret the hierarchical formation of integer relations as a building-up process that combines elements of one level to form more complex elements of the next level. This interpretation helps us to see the hierarchical formations in terms of complexity and results in a quantitative concept of structural complexity.

Let I be an integer alphabet and

$$I_n = \{s = s_1...s_n, \ s_i \in I, i = 1, ..., n\}$$

be the set of all sequences of length $n \geq 2$ with symbols in I. If $I = \{-1, +1\}$ then I_n is the set of all binary sequences of length n. Let $\delta > 0$ and $\varepsilon > 0$ be respective spacings of a spacetime lattice (δ, ε) in $1 + 1$ dimensions. A sequence $s = s_1...s_n \in I_n$ encodes on the lattice (δ, ε) the dynamics of a physical system that has a space position $s_i \delta$ at a time instant $i\varepsilon, i = 1, ..., n$.

An integer code series (ICS) [19] plays the key role in the connection between the descriptions of physical systems. It expresses integrals of piecewise constant functions in a form that contains the sequence and powers of integers. ICS allows us to extract powers of integers in integer relations and view them as geometrical objects that like physical ones can form into each other.

Let $W_{\delta\varepsilon}([t_m, t_{m+n}])$ be a set of piecewise constant functions $f : [t_m, t_{m+n}] \to \Re^1$ such that each function f belonging to the set is constant on

$$(t_{i-1}, t_i], \quad i = m+1, ..., m+n$$

and equals

$$f(t_m) = s_1\delta, \quad f(t) = s_i\delta, \; t \in (t_{i-1}, t_i], \; i = m+1, ..., m+n,$$

$$t_i = i\varepsilon, \; i = m, ..., m+n,$$

where m is an integer and s_i, $i = 1, ..., n$ are real numbers.

A piecewise constant function of the set $W_{\delta\varepsilon}([t_m, t_{m+n}])$ can be completely characterized by a sequence of real numbers. Namely, let a function $f \in W_{\delta\varepsilon}([t_m, t_{m+n}])$ and $s_i\delta$ be its value on an interval $(t_{i-1}, t_i]$, $i = m+1, ..., m+n$ then a sequence $s = s_1...s_n$ is called a code of the function, denoted $c(f)$, as it completely determines the function in the set. By this way sequences have a representation in terms of piecewise constant functions (see Figure 8.1).

Let $f^{[k]}$ denote the kth, $k = 1, 2, ...,$ integral of a function $f \in W_{\delta\varepsilon}([t_m, t_{m+n}])$, i.e., a family of functions whose k^{th} derivative equals f, and $f^{[0]} = f$. The notation $f^{[k]}$ in what follows stands for a function of the family implying that values $f^{[i]}(t_m)$, $i = 1, ..., k$ specify the function. The integer code series gives the kth $k = 1, 2, ...,$ integral of a function $f \in W_{\delta\varepsilon}([t_m, t_{m+n}])$ in terms of its code $c(f)$ and importantly in *powers of integers*.

Integer Code Series. *Let a function* $f \in W_{\delta\varepsilon}([t_m, t_{m+n}])$, $c(f) = s_1...s_n$ *and values* $f^{[i]}(t_m)$, $i = 1, ..., k$, *be known, where* $k = 1, 2, ...$ *Then the value of the kth integral* $f^{[k]}$ *of the function f at a point* t_{m+l+1}, $l = 0, ..., n-1$ *is completely specified by the code* $c(f) = s_1...s_n$ *of the function and is given by* [19]

$$f^{[k]}(t_{m+l+1}) = \sum_{i=0}^{k-1} \alpha_{kmi}((m+l+1)^i s_1 + ... + (m+1)^i s_{l+1})\delta\varepsilon^k$$

$$+ \sum_{i=1}^{k} \beta_{k,l+1,i} f^{[i]}(t_m)\varepsilon^{k-i}, \tag{1}$$

where coefficients

$$\alpha_{kmi} = \binom{k}{i}((-1)^{k-i-1}(m+1)^{k-i} + (-1)^{k-i}m^{k-i})/k!,$$

$i = 0, ..., k-1$ *are independent of the code* $c(f)$ *and the point* t_{m+l+1}, $\binom{k}{i}$, $i = 0, ..., k-1$ *is the binomial coefficient and coefficients*

$$\beta_{k,l+1,i} = \frac{(l+1)^{k-i}}{(k-i)!}, \quad i = 1, ..., k$$

are independent of the code $c(f)$ *and integer m.*

3. Structural Numbers of Sequences and their Geometrical Interpretation

The integer code series is applied to make the connection between spacetime and the hierarchical formations. For this purpose it is useful to consider a sequence $s = s_1...s_n \in I_n$ on a lattice (δ, ε) in terms of its **Structural Numbers** $\vartheta_1(s), ..., \vartheta_k(s), ...$

$$s = s_1...s_n \implies \vartheta_1(s), ..., \vartheta_k(s), ...$$

defined for $k = 1, 2, ...$ by the formula [19]

$$\vartheta_k(s) = \sum_{i=0}^{k-1} \alpha_{kmi}((m+n)^i s_1 + (m+n-1)^i s_2 + ... + (m+1)^i s_n)\delta\varepsilon^k, \quad (2)$$

which has a recognizable ICS character. This character follows by comparing (1) and (2). Let a mapping $\rho_{m\delta\varepsilon}$ take a sequence $s = s_1...s_n, s_i \in \Re^1, i = 1, ..., n$ to a function $f \in W_{\delta\varepsilon}[t_m, t_{m+n}]$, denoted $f = \rho_{m\delta\varepsilon}(s)$, such that $c(f) = s$ (see Figure 8.1) and whose k^{th} integral satisfies $f^{[k]}(t_m) = 0, k = 1, 2, ...$. It follows from (2) that the kth structural number $\vartheta_k(s), k = 1, 2, ...$ of a sequence $s \in I_n$ on a lattice (δ, ε) is the value of the kth integral of a function $f = \rho_{m\delta\varepsilon}(s) \in W_{\delta\varepsilon}[t_m, t_{m+n}]$ at the point t_{m+n}

$$\vartheta_k(s) = f^{[k]}(t_{m+n}), \quad k = 1, 2, ... \quad (3)$$

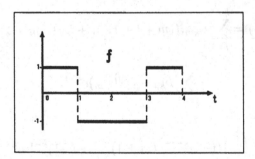

Figure 8.1. Graph of a function $f = \rho_{011}(s)$, $s = c(f) = s_1...s_4 = +1 - 1 - 1 + 1$, $\delta = 1$, $\varepsilon = 1$, $m = 0$, $n = 4$, $t_{m+n} = 4$.

To illustrate let us consider a function f such that (see Figure 8.1)

$$f = \rho_{011}(s), \quad s = c(f) = s_1...s_4 = +1 - 1 - 1 + 1.$$

Symmetry of the function can be easily seen and we can sketch, thinking in geometrical images, the graph of its first $f^{[1]}$ and second $f^{[2]}$ integrals as shown in Figure 8.2. From the figure we conclude that

$$f^{[1]}(t_4) = 0, \quad f^{[2]}(t_4) = 0$$

and, by using (3),

$$\vartheta_1(s) = 0, \quad \vartheta_2(s) = 0.$$

Importantly, by definition structural numbers are simply definite integrals but expressed in terms of the integer code series. Indeed, it is clear that the kth, $k = 1, 2, ...$, structural number of a sequence s on a lattice (δ, ε) is the definite integral of a function $f^{[k-1]}$ on the interval $[t_m, t_{m+n}]$, where $f = \rho_{m\delta\varepsilon}(s)$. From this perspective structural numbers of a sequence carry information about the sequence as a whole and can be seen as its global properties.

Therefore, structural numbers of sequences have the geometrical interpretation of the definite integral. For example, the third structural number of the sequence $s = +1 - 1 - 1 + 1$ (see Figure 8.1) is given by

$$\vartheta_3(s) = \int_{t_0}^{t_4} f^{[2]} dt$$

and is equal to the area of the geometrical pattern located under the graph of the function $f^{[2]}$ and above the t-axis (see Figure 8.2).

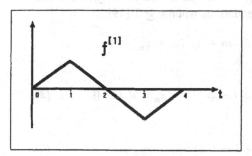
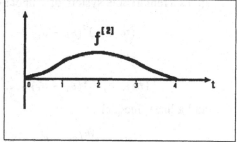

Figure 8.2. Graph of the first $f^{[1]}$ and second $f^{[2]}$ integrals (sketched) of the function f in Figure 1.

In this context the integer code series reveals a new role of the definite integral. It turns out that the definite integral equates the integer relations and geometrical patterns as well as their hierarchical formations. Namely, from one side ICS can express a definite integral in terms of powers of integers and pack them in integer relations. From another side the geometrical interpretation of the definite integral links the integer relations with a geometrical pattern whose area is measured by the integral. This fact allows us to show that the integer relations have a geometrical meaning as spatial objects and can form into each other.

4.   Structural Numbers and Integer Relations

Structural numbers can describe a sequence uniquely. This description shows a remarkable system of linear equations that in turn reveals specific integer relations. This type of integer relations is in the focus of our consideration. An important aim of this section is to see these integer relations as objects that like physical ones can form into each other. These formations are organized hierarchically with integers as the ultimate building blocks.

The following notation plays an essential role in the further presentation of the concept of structural complexity. For a pair of different sequences $s, s' \in I_n$ let $C(s, s')$ denote an integer $k \geq 1$ such that the first k structural numbers of the sequences are equal

$$\vartheta_1(s) = \vartheta_1(s'), ..., \vartheta_k(s) = \vartheta_k(s'),$$

whereas the $(k+1)^{\text{st}}$ structural numbers are not $\vartheta_{k+1}(s) \neq \vartheta_{k+1}(s')$. If $\vartheta_1(s) \neq \vartheta_1(s')$ then $C(s, s') = 0$. Clearly, a sequence $s \in I_n$ can be specified uniquely with respect to another sequence $s' \in I_n$ by using the first $C(s, s')$ of their structural numbers.

It is proved for sequences $s, s' \in I_n$, $s \neq s'$ that $C(s, s') = k \leq n$ and this gives a remarkable system of k linear equations when $k \geq 1$ [19]

$$(m + n)^0(s_1 - s'_1) + ... + (m + 1)^0(s_n - s'_n) = 0$$

$$\cdot \qquad \quad \cdot \qquad \quad \cdot \qquad \quad \cdot$$

$$(m + n)^{k-1}(s_1 - s'_1) + ... + (m + 1)^{k-1}(s_n - s'_n) = 0 \qquad (4)$$

and a linear inequality

$$(m + n)^k(s_1 - s'_1) + ... + (m + 1)^k(s_n - s'_n) \neq 0. \qquad (5)$$

in integers $(s_i - s'_i)$, $i = 1, ..., n$. System (4) has the Vandermonde matrix when $k = n$.

The system (4) shows up in many different contexts and under various guises (for example, [18, 43]). It sets links between structural numbers and many areas of mathematics and physics [19]. However, we consider the system (4) in a new way by interpreting it as a record of a *process*. This process is a hierarchical formation of integer relations.

Notice that system (4) can be viewed as a system of integer relations, since the equations are in essence specific relations between powers i, $i = 0, ..., k-1$

of integers $m + n, ..., m + 1$. This is especially clear in concrete cases. For example, consider sequences

$$s = +1 - 1 - 1 + 1 - 1 + 1 + 1 - 1 - 1 + 1 + 1 - 1 + 1 - 1 - 1 + 1,$$

$$s' = -1 + 1 + 1 - 1 + 1 - 1 - 1 + 1 + 1 - 1 - 1 + 1 - 1 + 1 + 1 - 1,$$

which are the initial segment of length 16 PTM sequences [34, 29] starting with $+1$ and -1 respectively and $n = 16, m = 0$. Then it can be shown that $C(s, s') = 4$ and (4) becomes a system of specific integer relations

$$+16^0 - 15^0 - 14^0 + 13^0 - 12^0 + 11^0 + 10^0 - 9^0 - 8^0 + 7^0 + 6^0 - 5^0 + 4^0 - 3^0 - 2^0 + 1^0 = 0$$

$$+16^1 - 15^1 - 14^1 + 13^1 - 12^1 + 11^1 + 10^1 - 9^1 - 8^1 + 7^1 + 6^1 - 5^1 + 4^1 - 3^1 - 2^1 + 1^1 = 0$$

$$+16^2 - 15^2 - 14^2 + 13^2 - 12^2 + 11^2 + 10^2 - 9^2 - 8^2 + 7^2 + 6^2 - 5^2 + 4^2 - 3^2 - 2^2 + 1^2 = 0$$

$$+16^3 - 15^3 - 14^3 + 13^3 - 12^3 + 11^3 + 10^3 - 9^3 - 8^3 + 7^3 + 6^3 - 5^3 + 4^3 - 3^3 - 2^3 + 1^3 = 0, \quad (6)$$

whereas (5) does not follow the character in (6),

$$+16^4 - 15^4 - 14^4 + 13^4 - 12^4 + 11^4 + 10^4 - 9^4 - 8^4 + 7^4 + 6^4 - 5^4 + 4^4 - 3^4 - 2^4 + 1^4 \neq 0. \quad (7)$$

For clarity common factor 2 is cancelled out in (6) and (7).

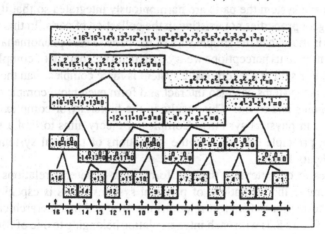

Figure 8.3. Represented in this form system (6) and inequality (7) appear as a hierarchical formation of integer relations with integers 16, ..., 1 as the ultimate building blocks. In the formation all integer relations share the same organizing principle.

System (4) opens a way to the reality of hierarchical formations. Integer relations behave there in an interesting manner that we could rather expect from physical systems. We show and explain a picture of this reality by using (6) and (7). There are integer relations of the similar character that are not immediately evident from system (6) and can be identified by a more careful analysis. These integer relations demonstrate connections, indicated by edges (see Figure 8.3). Connections between an integer relation and a number of integer relations are in place if the integer relation can be produced from the integer relations by a mathematical operation.

The operation takes a number of integer relations with the same power and produces from them an integer relation in the following way. Firstly, it modifies the left part of the integer relations by increasing the power of the integers by 1. Secondly, the operation combines the modified parts together. The fact they result in zero indicates that the parts are harmoniously integrated in a new whole(see Figure 8.3).

An interesting picture of formations appears when the integer relations are interpreted as objects that like physical ones combine and form a composite object according to the operation. In this case the operation can be seen as the organizing principle of these formations as it is shared by all the integer relations.

We are not used to see that an integer relation can be formed from another integer relations. This process is unusual but it is quite familiar in the context of physical systems. Physical systems can interact. When certain conditions are in place they combine as parts and form a composite system as a new whole. In the composite system the parts are harmoniously integrated so that it exhibits new emergent properties not existing in the collection of parts. In this situation the whole is more than the simple sum of the parts. These phenomena of physical systems give us perceptions we try to explain by concept of complexity. In particular, it is said that a composite system is more complex than the parts.

Composite systems can also interact and form even more complex systems as new wholes and so on. This results in the hierarchy of complexity levels we observe in physical systems. Complexity theory aims to find a universal organizing principle that governs the formations of physical systems across the complexity levels.

It comes to our attention that the formations of integer relations in some sense resemble the formations of physical systems. This is especially clear when the formations of integer relations are arranged in a hierarchical formation (as in Figure 8.3) with each integer relation belonging to one of four levels. We can observe that following the organizing principle elements of each level, except the top one, combine and form elements of the next level.

The two complementary levels added below complete the picture and make the hierarchical formation consistent (see Figure 8.3). It is worth to note that integers 16, ..., 1 act as the ultimate building blocks of these formations. The first complementary level, called the ground level, may be thought as a reservoir that generates integers in two states, i.e., the negative and the positive, for the second complementary level, called the ground level. For example, integer 3 is generated in the negative state while integer 15 in generated in the positive state on the zero level. The elements and their states generated on the zero level for this hierarchical formation are shown in Figure 8.3. The elements of the zero level combine and form elements of the first level, which are the integer relations with power 0. In these formations the state of an element of the zero level is translated into the arithmetic sign in the integer relation.

The hierarchical formation with all these levels in place can be represented by a diagram as in Figure 8.3. The diagram reads that integers in certain states are generated on the zero level, which in turn form integer relations of the first level, which in turn form integer relations of the second level and so on till the fourth level. There is only one element at the top of the diagram, which can not form an element of the next level by itself because of (7). The diagram illustrates how level by level each integer relation is formed.

Therefore, by considering the sequences s, s' in terms of structural numbers and the analysis of system (6) and inequality (7) we reveal a process. This process is a hierarchical formation of integer relations. This sets up a connection between the sequences and the hierarchical formation, which can be used to characterize the sequences s, s' themselves. For example, we may say that the complexity of sequence s with respect to sequence s' is quite enough to generate (6), which is intuitively perceived as a complex phenomenon. At the same time it may be said that this complexity is not enough, as we observe in (7), to make this phenomenon by one unit more complex.

5. Integer Patterns: Geometrical Meaning of the Integer Relations and their Formations

In the previous section system (6) and inequality (7) are interpreted as a record of a hierarchical formation of integer relations. The concepts of integer and integer relations are an integral part of our mental equipment. They are a set of abstract symbols subject to the laws of arithmetic. Usually, integer relations appear as solutions to Diophantine equations.

However, integer relations do not appear to us rooted in firm reality. They do not have the power to evoke in our minds images of objects that like physical ones can form into each other. The concept of integer relation is too abstract to give us insight what we think the formation is all about. It is difficult to imagine how integer relations can be really formed from other integer relations, because it is not possible to visualize how this may actually happen.

To extend the scope of integer relations and develop a new perception of them as geometrical objects that can form into each other a notion of integer pattern is introduced [19]. Integer patterns provide a geometrical meaning and visualize the integer relations and their hierarchical formations. We begin to present integer patterns by a notion of

Pattern of a Function. *Let* $f : [a, b] \to \Re^1$ *be a piecewise constant or continuous function. Let*

$$P(f, t) = \{(t, x) \in \Re^2, \min\{0, f(t)\} \leq x \leq \max\{0, f(t)\}\},$$

if function f is continuous at a point $t \in [a, b] \subset \Re^1$, and $P(f, t) =$

$$\{(t, x) \in \Re^2, \min\{0, f(t-0), f(t+0)\} \leq x \leq \max\{0, f(t-0), f(t+0)\}\},$$

if function f is discontinuous at the point $t \in (a, b)$. Moreover, if function f is discontinuous at the point $t = a$ then

$$P(f, t) = \{(t, x) \in \Re^2, \min\{0, f(a+0)\} \leq x \leq \max\{0, f(a+0)\}\},$$

and if function f is discontinuous at the point $t = b$ then

$$P(f, t) = \{(t, x) \in \Re^2, \min\{0, f(b-0)\} \leq x \leq \max\{0, f(b-0)\}\}.$$

The pattern of a function f is defined by

$$P(f, [a, b]) = \{P(f, t), t \in [a, b]\}.$$

To illustrate the definition the pattern of a function f is brought out by shading in Figure 8.4.

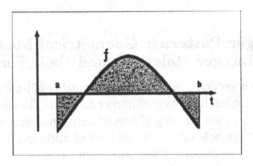

Figure 8.4. The pattern $P(f, [a, b])$ (shaded) of a function f. The shaded area allows us to see the pattern as a spatial object.

Furthermore, we describe integer patterns as patterns with a special character. For a function

$$f \in W_{\delta\varepsilon}([t_m, t_{m+n}]), \quad f = \rho_{m\delta\varepsilon}(s), \quad s = s_1 \dots s_n \in I_n \tag{8}$$

let $T(f, [t_m, t_{m+n}])$ be a set of intervals $[t_{m+i}, t_{m+i+1}] \subset [t_m, t_{m+n}]$ such that

$$s_{i+1} \neq 0, \quad i = 0, \dots, n-1.$$

Let $T(f^{[k]}, [t_m, t_{m+n}])$, $k = 1, 2, \ldots$ be a set of intervals

$$[t_i, t_j] \subseteq [t_m, t_{m+n}], \quad m \le i, \quad i+1 < j \le m+n$$

such that

$$f^{[r]}(t_i) = 0, \quad f^{[r]}(t_j) = 0, \quad r = 1, \ldots, k,$$

and there exists no t_l, $l = i+1, \ldots, j-1$ such that $f^{[r]}(t_l) = 0$ for all $r = 1, \ldots, k$ at one time.

Integer Pattern. *The pattern* $P(f^{[k]}, [t_i, t_j])$, $k = 0, 1, \ldots$ *of the kth integral of a function f given by* (8) *is called an integer pattern if*

$$[t_i, t_j] \in T(f^{[k]}, [t_m, t_{m+n}]) \ne \emptyset.$$

We illustrate the definition of integer pattern by examples.

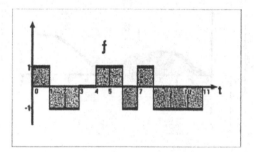 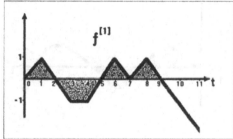

Figure 8.5. Integer patterns (shaded) of a function f and its first integral $f^{[1]}$, where $f = \rho_{011}(s)$, $s = +1 - 1 - 1\,0 + 1 + 1 - 1 + 1 - 1 - 1 - 1 \in I_{11}$, $I = \{-1, 0, +1\}$.

For instance, Figure 8.5 shows integer patterns

$$P(f, [t_0, t_1]), \quad P(f, [t_1, t_2]), \quad P(f, [t_2, t_3]), \quad P(f, [t_4, t_5]), \quad P(f, [t_5, t_6]),$$

$$P(f, [t_6, t_7]), \quad P(f, [t_7, t_8]), \quad P(f, [t_8, t_9]) \quad P(f, [t_9, t_{10}]), \quad P(f, [t_{10}, t_{11}])$$

of a function $f = \rho_{011}(s)$, where

$$s = +1 - 1 - 1\,0 + 1 + 1 - 1 + 1 - 1 - 1 - 1 \in I_{11}, \quad I = \{-1, 0, +1\},$$

and integer patterns

$$P(f^{[1]}, [t_0, t_2]), \quad P(f^{[1]}, [t_2, t_5]), \quad P(f^{[1]}, [t_5, t_7]), \quad P(f^{[1]}, [t_7, t_9])$$

of its first integral $f^{[1]}$.

The integer pattern $P(f^{[2]}, [t_0, t_7])$ of the second integral $f^{[2]}$ and the integer pattern $P(f^{[3]}, [t_0, t_7])$ of the third integral $f^{[3]}$ are brought out by shading in Figure 8.6. Integer pattern $P(f^{[2]}, [t_0, t_7])$ consists of two patterns

$$P(f^{[2]}, [0, 3.5]), \quad P(f^{[2]}, [3.5, 7]), \tag{9}$$

which are not integer patterns because two conditions are not satisfied,

$$f^{[1]}(3.5) \neq 0$$

and 3.5 is not an integer. These two patterns are identical in shape and like opposite to each other. They may be viewed coupled to give the integer pattern $P(f^{[2]}, [t_0, t_7])$. Integer pattern $P(f^{[2]}, [t_0, t_7])$ forms into integer pattern $P(f^{[3]}, [t_0, t_7])$ under the integration of $f^{[2]}$. This may be pictured as if patterns (9) merged together in the formation of the integer pattern $P(f^{[3]}, [t_0, t_7])$.

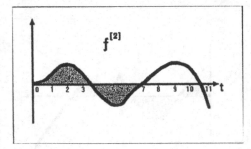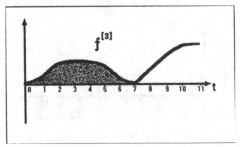

Figure 8.6. The integer pattern $P(f^{[2]}, [t_0, t_7])$ of the second integral $f^{[2]}$ and the integer pattern $P(f^{[3]}, [t_0, t_7])$ of the third integral $f^{[3]}$ of the function f depicted in Figure 5. The integer patterns are brought out by shading. Integer pattern $P(f^{[2]}, [t_0, t_7])$ forms into integer pattern $P(f^{[3]}, [t_0, t_7])$ under the integration of $f^{[2]}$.

From general results presented in the next section it follows that integer pattern $P(f^{[2]}, [t_0, t_7])$ corresponds to an integer relation

$$+7^1 - 6^1 - 5^1 + 3^1 + 2^1 - 1^1 = 0 \tag{10}$$

and integer pattern $P(f^{[3]}, [t_0, t_7])$ corresponds to another integer relation

$$+7^2 - 6^2 - 5^2 + 3^2 + 2^2 - 1^2 = 0 \tag{11}$$

Clearly, the organizing principle forms integer relation (11) from integer relation (10). The correspondence between the integer patterns and the integer relations illustrates the geometrical interpretation of the organizing principle. Namely, the organizing principle forms (11) from (10) as the integration of $f^{[2]}$ forms $P(f^{[3]}, [t_0, t_7])$ from $P(f^{[2]}, [t_0, t_7])$.

6. Isomorphism between Hierarchical Formations of Integer Relations and Integer Patterns

The key idea about system (4) and inequality (5) is to interpret them as a record of a hierarchical formation of integer relations [19].

Firstly, it is shown that system (4)

$$(s_1 - s_1')(m+n)^0 + ... + (s_n - s_n')(m+1)^0 = 0$$

$$\vdots \qquad \vdots \qquad \vdots \qquad \vdots \qquad \vdots$$

$$(s_1 - s_1')(m+n)^{C(s,s')-1} + ... + (s_n - s_n')(m+1)^{C(s,s')-1} = 0 \quad (12)$$

can be associated with a hierarchical set $WR(s, s', n, m, I_n)$ of $C(s, s') \geq 1$ levels whose elements of k level $k = 1, ..., C(s, s')$ are integer relations of the form

$$\boxed{\lambda_1 d_1^{k-1} + ... + \lambda_l d_l^{k-1} = 0} \quad (13)$$

where $\lambda_i, d_i, i = 1, ..., l$ are integers such that $d_i > d_{i+1}, i = 1, ..., l-1$ and k is the power of integers $d_i, i = 1, ..., l$.

Secondly, it is interpreted that elements of k level $k = 2, ..., C(s, s')$ of the set $WR(s, s', n, m, I_n)$ are formed from elements of $(k-1)$ level according to the organizing principle. The principle can be described mathematically as follows. If $r \geq 1$ integer relations

$$\boxed{\lambda_{i1} d_{i1}^{k-1} + ... + \lambda_{il(i)} d_{il(i)}^{k-1} = 0} \quad (14)$$

of k level $k = 1, 2..., C(s, s') - 1$ with the ith, $i = 1, ..., r$, relation containing $l(i)$ terms satisfy

$$\boxed{\sum_{i=1}^{r} \lambda_{i1} d_{i1}^k + ... + \lambda_{il(i)} d_{il(i)}^k = 0} \quad (15)$$

and the inclusion of each of the relations (14) is important for (15), then it is said that integer relation (15) is formed from integer relations (14). In general, integer relations of type (14) do not form an integer relation, because integer relation (15) is more than their simple sum

$$\sum_{i=1}^{r} \lambda_{1i} d_{i1}^{k-1} + ... + \lambda_{il(i)} d_{il(i)}^{k-1} = 0. \quad (16)$$

The power of integers $d_{ij}, i = 1, ..., r, j = 1, ..., l(i)$ in (15) is increased by 1 compared to (16). This means that integer relations (14) must have a special property to form integer relation (15).

The interpretation of the set $WR(s, s', n, m, I_n)$ as a hierarchical formation becomes complete when two complementary levels are added below the levels of the set. The functioning of these two levels can be made consistent with the integer relations and their formations (see Figure 8.3).

In particular, the first complementary level, called the ground level, consists of integers. It may be thought as a reservoir that can generate integers in any amount at the second complementary level, called the zero level, in two states, i.e., the negative and the positive. For the set $WR(s, s', n, m, I_n)$ integer $(m + n - i), i = 0, ..., n - 1$ as an element is generated in an amount of $(s_i - s'_i)$ in a state specified by the sign of $(s_i - s'_i)$. The elements of the zero level combine and form elements of the first level of $WR(s, s', n, m, I_n)$. In these formations the state of an element of the zero level is translated into the arithmetic sign in the integer relation. Then the elements of the first level form the elements of the second level and so on till level $C(s, s')$. In such a manner all the elements on all the levels, starting with the integers at the ground one, become unified into one hierarchical formation $WR(s, s', n, m, I_n)$. The remarkable property is that the integers act as the ultimate building blocks of this hierarchical formation.

System (12) stands as a record of the hierarchical formation. Inequality (5) in this case becomes

$$(s_1 - s'_1)(m + n)^{C(s,s')} + ... + (s_n - s'_n)(m + 1)^{C(s,s')} \neq 0 \qquad (17)$$

tells us that the hierarchical formation can not propagate to the level higher than level $C(s, s')$. The elements of this level can not all form elements on the next level because of (17).

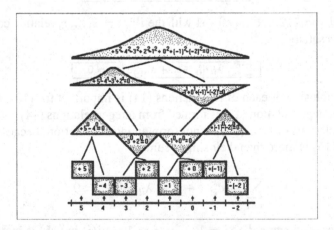

Figure 8.7. A hierarchical formation of integer relations and a corresponding hierarchical formation of integer patterns are displayed together to underline their unified character. The figure allows us to see how the integer relations form into each other. Note that an integer relation can be measured by the area of a corresponding integer pattern.

Usually, the left side of integer relation

$$\lambda_1 d_1^{k-1} + ... + \lambda_l d_l^{k-1} = 0 \tag{18}$$

is the recipe for the manipulations of the symbols based on the laws of arithmetic. The integer relation (18) reads that the quantity on the left side, i.e., the result of the manipulations, is equal to the quantity on the right, i.e., zero.

A new meaning of (18) appears in the context of the hierarchical formations. In our case the integer relation

$$\boxed{\lambda_1 d_1^{k-1} + ... + \lambda_l d_l^{k-1} = 0} \tag{19}$$

is an element of k level $k = 1, ..., C(s, s')$ of a hierarchical formation. It is formed from elements of the previous level and is a building block for elements of the next level. The left side of (19), called the code of the integer relation, contains information about how the element is formed. The fact that the code of an integer relation equals zero is the criterion of its existence. To show the new meaning of the integer relations as elements of the hierarchical formations they are placed in boxes like in (19).

We may ask the question: is there something "physical" happening when the formation of a new integer relation takes place? Integer relations do not appear to us rooted in firm reality and have no power to evoke in our minds images of objects like physical ones. As a result the notion of formation of integer relations is too abstract to give us insight what we think the formation is all about and tells nothing about its nature. It is difficult to imagine how integer relations can be really formed from other integer relations, because it is not possible to visualize how this may actually happen. For example, formations of elements in Figure 8.3 appear as a mystery because the laws involved are not clear.

Remarkably, by using the integer code series it is proved that there exists an isomorphism $\psi_{\delta\varepsilon}$ between elements of a set $WR(s, s', n, m, I_n)$, i.e., integers and integer relations, and integer patterns of integrals $f^{[k]}$, $k = 0, ..., C(s, s')$ of a function

$$f = \rho_{m\delta\varepsilon}(s - s') \in W_{\delta\varepsilon}([t_m, t_{m+n}]) \tag{20}$$

as well as between their hierarchical formations [19]. As a result a hierarchical formation of integer patterns $WP_{\delta\varepsilon}(s, s', n, m, I_n)$

$$\psi_{\delta\varepsilon} : WR(s, s', n, m, I_n) \Longleftrightarrow WP_{\delta\varepsilon}(s, s', n, m, I_n).$$

is defined. The isomorphism $\psi_{\delta\varepsilon}$ gives a geometrical meaning to the integer relations and their formations in terms of integer patterns and their formations. In particular, it interprets the organizing principle as the integration of the function.

The isomorphism visualizes a hierarchical formation of integer relations

$$WR(s, s', n, m, I_n)$$

by a corresponding hierarchical formation of integer patterns $WP_{\delta\epsilon}(s, s', n, m, I_n)$ when it is represented as follows. The integer patterns of the kth, $k = 0, ..., C(s, s')$, integral $f^{[k]}$ of a function f given by (20) are viewed as elements of k level of the hierarchical formation

$$WP_{\delta\epsilon}(s, s', n, m, I_n).$$

The $(k + 1)^{\text{st}}$ level of this hierarchical formation has a coordinate system with the graph of the $(k + 1)^{\text{st}}$ integral $f^{[k+1]}$, $k = 0, ..., C(s, s') - 1$. This coordinate system is placed at some distance above the kth level which has a coordinate system with the graph of the kth integral $f^{[k]}$. There are edges between integer patterns on k level and integer patterns on $(k + 1)$ level. These edges show how integer patterns of k level form integer patterns of $(k + 1)$ level under the integration of the kth integral $f^{[k]}$ (see Figure 8.7).

The isomorphism is illustrated in Figure 8.7, where a hierarchical formation of integer relations and a corresponding hierarchical formation of integer patterns are shown together. We can see that the integer relations appear as two-dimensional geometrical objects that like physical ones form into each other. In particular, an integer relation acquires a shape of a corresponding integer pattern and can be measured by its area. The hierarchical formation of integer patterns reveals self-similarity, local and non-local symmetries. Figure 8.7 gives a static picture of the hierarchical formations and can be thought as a sequence of snapshots, one at each level of the formation.

Results of the previous sections can be summarized by a diagram that shows a connection between sequences and hierarchical formations of the integer relations and integer patterns:

$$\{ \text{ Sequences in Spacetime } \}$$

$$\Downarrow$$

$$\{ \text{ Structural Numbers of Sequences } (2) \}$$

$$\Downarrow$$

$$\left\{ \begin{array}{c} \text{System of Linear Equations } (12) \\ \text{and Inequality } (17) \end{array} \right\}$$

$$\Downarrow$$

$$\left\{ \begin{array}{c} \text{Hierarchical Formation} \\ \text{of Integer Relations} \end{array} \right\} \Longleftrightarrow \left\{ \begin{array}{c} \text{Hierarchical Formation} \\ \text{of Integer Patterns} \end{array} \right\}$$
$$\text{isomorphism}$$

7.   The Hierarchical Formations as a Final Theory

In the last section we presented a connection between sequences as description of physical systems in spacetime and the hierarchical formations of integer relations and integer patterns. This connection opens a way to describe physical systems and suggests to consider the hierarchical formations as a new reality.

This reality, called a web of relations, consists of the hierarchical formations of integer relations where all pairs of different sequences $s, s' \in I_n$ are considered, $n \to \infty$ and integer alphabet I is the set of all integers $\mathbf{Z}$ [19]

$$WR(I_n) = \bigcup_{m \in \mathbf{Z}} \bigcup_{s \in I_n} \bigcup_{s' \in I_n} WR(s, s', m, I_n),$$

$$WR(I) = \lim_{n \to \infty} WR(I_n), \quad WR = WR(\mathbf{Z}).$$

The isomorphism $\psi_{\delta\varepsilon}$ defines a web of integer patterns that for a given δ and ε incorporates the hierarchical formations of integer patterns corresponding to WR into one whole

$$\psi_{\delta\varepsilon} : WR \to WP_{\delta\varepsilon}.$$

Due to the isomorphism WR and $WP_{\delta\varepsilon}$ can be viewed as one structure, i.e., called the web of relations, whose elements have the dual character. They are the integer relations as well as integer patterns specified by δ and ε. The web of relations provides rich properties in the description of physical systems.

Firstly, the web of relations is a final theory because it is based on integers only. This theory is about the integer relations and their formations, whose existence is completely specified by the laws of arithmetic.

For example, consider two statements about integers written as integer relations

$$+5^1 - 4^1 - 3^1 + 2^1 = 0 \quad -1^1 + 0^1 + (-1)^1 - (-2)^1 = 0. \tag{21}$$

These statements are true since in each of them the left side indeed equals the right side. These integer relations do exist as they express properties of the integers which follow from the laws of arithmetic. Moreover, integer relations (21) are elements

$$\boxed{+5^1 - 4^1 - 3^1 + 2^1 = 0}$$

$$\boxed{-1^1 + 0^1 + (-1)^1 - (-2)^1 = 0} \tag{22}$$

of the second level of the web of relations because they are formed from elements of the first level (see Figure 8.7), i.e., the first one from elements

$$\boxed{+5^0 - 4^0 = 0}$$

$$\boxed{-3^0 + 2^0 = 0}$$

the second one from elements

$$\boxed{-1^0 + 0^0 = 0}$$

$$\boxed{+(-1)^0 - (-2)^0 = 0}$$

The existence of integer relations (21) gives the solid ground for the existence of the elements (22).

Can elements (22) combine and form an element of the third level? The answer to this question is "yes" and we have (see Figure 8.7)

$$\boxed{+5^2 - 4^2 - 3^2 + 2^2 - 1^2 + 0^2 + (-1)^2 - (-2)^2 = 0} \qquad (23)$$

This formation is completely specified by the laws of arithmetic. However, in its turn two elements

$$\boxed{+5^1 - 4^1 - 3^1 + 2^1 = 0}$$

$$\boxed{-1^1 + (-1)^1 + (-2)^1 - (-4)^1 = 0} \qquad (24)$$

of the second level of the web of relations can not form an element of the third level because according to the same laws

$$+5^2 - 4^2 - 3^2 + 2^2 - 1^2 + (-1)^2 + (-2)^2 - (-4)^2 \neq 0.$$

Thus, the laws of arithmetic forbid the formation of an element from the elements (24) and we can state that this element does not exist.

This is how formations happen in the web of relations. They are based on the simple organizing principle and are controlled completely by the laws of arithmetic. The organizing principle has a universal character as it applies equally to all elements and all levels.

Secondly, the hierarchical formation resembles formations of physical systems. Many physical phenomena can be interpreted as hierarchical formations of more elementary phenomena that occur at different scale levels. These phenomena are characterized in terms of complexity. The hierarchical formation has the building-block character of the elements on each of its levels. This is in favor to interpret the hierarchical formation as a process that combines elements of one level to form more complex elements of the next higher level. This interpretation helps us to see the hierarchical formations in terms of complexity and suggests a quantitative concept of structural complexity presented in the next section.

Thirdly, the geometrical meaning of the integer relations brings "physics" into the description. Measurable quantities of the integer patterns give us a way to characterize the hierarchical formations quantitatively. For example, the area of the integer pattern can be considered as a natural quantity of the element. The geometrical meaning of the hierarchical formation reveals self-similarity, local and non-local symmetries (as in Figure 8.7). These notions are among the key concepts of physical systems.

8. Structural Complexity

The most important fact about the hierarchical formations is that they describe physical systems in terms of a final theory. To understand the meaning of the hierarchical formations in the previous sections we discussed an analogy between them and formations of physical systems. This suggests a cyclic connection between physical systems and the hierarchical formations. Physical systems can be described by the hierarchical formations, which in their turn have the analogy with the formations of physical systems.

The analogy is in favor to interpret the hierarchical formation of integer relations as a building-up process that combines elements of one level to form more complex elements of the next higher level. This helps us to see the hierarchical formations in terms of complexity. In this section we continue the discussion, which results in a concept of structural complexity [19].

Physical systems can interact and when certain conditions are in place form a composite system as a new whole. In the composite system the parts are harmoniously integrated. Hence, the composite system exhibits new emergent properties that do not exist in the collection of parts. In this situation the whole is more than the simple sum of the parts. These phenomena of physical systems give us perceptions we try to explain in terms of complexity. In particular, it is said that a composite system is more complex than the parts. Composite systems also can interact and form even more complex systems as new wholes and so on. This results in the hierarchy of complexity levels we observe in physical systems.

From this perspective, superstring theory proposes strings as the ultimate building blocks. The familiar elementary particles can be thought as vibrations of strings. The elementary particles interact with each other to form the entire range of atoms on the next level. There are about one hundred naturally occurring atoms with emergent properties that do not exist in the particles themselves. The parts represented by the atoms form molecules on the next level. The molecules as new wholes display emergent properties not present in the atoms themselves. For example, the properties in salt, or sodium chloride, are emergent and do not exist in sodium or chloride per se [30]. Starting from the infinitesimal level this process can be continued to the cosmological one.

Complexity theory aims to find a universal organizing principle that governs the formations of physical systems across the complexity levels [13]. In this capacity the complexity theory becomes a theory of everything and thus must be sought as a final theory. It also requires two important ingredients: the elements and a universal organizing principle telling us how elements of one level form composite elements of the next level. It would be a real wonder to know the identity of these fundamental elements as they are supposed to model the formations of all physical systems. However, the question whether such a complexity theory exists has certain arguments of its own.

The web of relations meets the above requirements and thus could be useful in a quest of the complexity theory. In particular, it is a final theory. The web of relations has the integer relations as its elements and the organizing principle specifying how integer relations of one level compose integer relations of the next level. Moreover, we see that the hierarchical formations of integer relations resemble the formations of physical systems and in some sense mirror their complexity levels.

Indeed, there are physical systems of different complexity levels on one side and the integer relations of different levels on the other. In addition, from one side physical systems, by virtue of an interaction, can form more complex systems, which are *qualitatively* more than the simple sum of the parts. From the other side integer relations, owing to the organizing principle, can form new integer relations, which are *quantitatively* more than their simple sum.

There is more information about some aspects of the web of relations than about corresponding parts in physical systems. Firstly, the ultimate building blocks of the hierarchical formations are known. They are integers. At the same time in physical reality the fundamental question about the ultimate building blocks is still debated. Strings are current candidates for this role. Secondly, the integer relations follow the same organizing principle, which applies to all levels. This principle is well-defined and has a clear geometrical interpretation. In physical systems the character of interaction can be different from one level to another and unknown. This means that the web of relations has a potential in understanding these aspects of physical systems.

The interpretation of the hierarchical formations in terms of complexity can be naturally expressed by a quantitative concept of complexity. Namely, in a hierarchical formation elements of level $k \geq 1$ can be viewed more complex by 1 than elements of level $k - 1$ from which they are formed. This defines that the complexity of an element on k level equals k assuming that integers at the ground level have zero complexity.

In its turn the complexity of a hierarchical formation can be defined by the maximum complexity of its elements. A hierarchical formation $WR(s, s', n, m, I_n)$ starts at the ground level and successively forms all its elements of k level from all its elements of $(k - 1)$ level until it reaches the highest level $C(s, s')$. The elements of level $C(s, s')$ can be viewed as the result of the hierarchical formation and have the maximum complexity.

The complexity of a hierarchical formation $WR(s, s', n, m, I_n)$ is a characteristic of a corresponding pair of sequences s, s'. We can simply say that the pair of sequences s, s' generates the hierarchical formation $WR(s, s', n, m, I_n)$. This allows us to introduce for sequences a new concept of complexity called

Structural Complexity. *The structural complexity of a sequence $s \in I_n$ with respect to a sequence $s' \in I_n$ is the complexity of the hierarchical formation*

$$WR(s, s', n, m, I_n)$$

the sequences generate and equals $C(s, s')$. The structural complexity of a sequence $s \in I_n$ is the maximum complexity of the hierarchical formations generated by the sequence together with sequences from I_n and is defined by

$$C(s) = \max_{s' \in I_n} C(s, s').$$

It is conjectured that for a binary sequence s of length n [19]

$$C(s) \le C(\eta(n)) = \lfloor log_2 n \rfloor, \tag{25}$$

where $\eta(n)$ is the initial segment of length n of the PTM sequence.

The concept of structural complexity suggests for the first time a real opportunity to search the universal concept within a final theory.

9. On General Optimality Condition of Agents Best Performance

We consider a group of $N \ge 2$ cooperative agents that is involved in a process consisting of $n \ge 2$ steps. At each step an agent can choose among two different strategies, labeled by $+1$ and -1.

Let **S** be the set of all $N \times n$ binary matrices, called the strategy space, with elements in $\{-1, +1\}$. The agents can be described by a binary $N \times n$ matrix, called a strategy matrix,

$$S = \{s_{ij}\} \in \mathbf{S}, \quad i = 1, ..., N, j = 1, ..., n,$$

where a binary sequence $s_i = s_{i1}...s_{in} \in B_n$ specifies the ith, $i = 1, ..., N$, agent and s_{ij} is a strategy used by the agent at the jth step of the process.

Let P be a class of problems the agents consider. Let $F_f(S)$ be the value of a functional F_f that describes the performance of the agents when they deal with a problem $f \in P$ and come up with a strategy matrix $S \in \mathbf{S}$ to minimize the functional. The group of agents demonstrates its best performance for a problem $f \in P$ by producing strategies that are the global minimum

$$G(F_f) = \{S : S \in \mathbf{S}, F_f(S) = \min_{S' \in \mathbf{S}} F_f(S')\}$$

of the functional F_f. It is assumed that the agents are given in terms of rules specifying how they choose strategies, cooperate and exchange information.

We consider the following question. Is there a general optimality condition specifying how properties of the cooperative agents must be connected with properties of a particular problem f so that the agents as a group show their best performance for the problem? The key point of the question is what concepts could be involved in the optimality condition to describe these properties of the agents and the problem.

It is well-known that many important classes of practical problems are NP-hard. A corresponding functional F_f of a problem f from such a class, depending on the context called a fitness landscape [33], potential energy surface [7] or cost function [27], has a very rugged and complicated nature. The optimality condition for a class of problems P, if existed, must be a constructive representation of a common property of the class of functionals $\{F_f, f \in P\}$ with respect to the global minimum. However, there are no indications that fitness landscapes have such a common property. This gives serious doubts on the possibility of the optimality condition within such a general setting.

Surprisingly then, that for a benchmark class of traveling salesman problems (TSP) [36] we have found such a common property. In particular, the consideration of the strategy space in terms of structural complexity reveals fitness landscapes in a regular way and shows their common property. This common property is that each fitness landscape takes a unimodal form when the strategy space is viewed in terms of structural complexity.

10.   Representation of Fitness Landscapes in terms of Structural Complexity

We associate a strategy matrix $S \in \mathbf{S}$ with a corresponding complexity matrix S_C defined by

$$S_C = \{C(s_i, s_j)\}, \ i, j = 1, ..., N,$$

where $C(s_i, s_j)$ is the structural complexity of a sequence $s_i \in B_n$ with respect to a sequence $s_j \in B_n$. The structural complexity of a strategy matrix $S \in \mathbf{S}$ is defined by

$$C(S) = \sum_{i=1}^{N} \sum_{j=1}^{N} C(s_i, s_j).$$

The structural complexity of agents resulting in a strategy matrix S is defined by $C(S)$.

Let $S(lower), S(upper) \in \mathbf{S}$ be strategy matrices such that the first one is composed of only $+1$'s

$$S(lower) = \begin{pmatrix} +1 & +1 & ... & +1 & +1 \\ +1 & +1 & ... & +1 & +1 \\ . & . & ... & . & . \\ +1 & +1 & ... & +1 & +1 \\ +1 & +1 & ... & +1 & +1 \end{pmatrix}$$

and the second one is composed of the initial segment of length n of the PTM sequence starting with $+1$

$$S(upper) = \begin{pmatrix} +1 & -1 & ... & -1 & +1 \\ +1 & -1 & ... & -1 & +1 \\ . & . & ... & . & . \\ +1 & -1 & ... & -1 & +1 \\ +1 & -1 & ... & -1 & +1 \end{pmatrix}.$$

These two matrices play in an important role because they give the lower and upper bound for the structural complexities of the strategy matrices. In particular, $N \times N$ complexity matrices $S_C(lower), S_C(upper)$ of $N \times n$ matrices $S(lower), S(upper)$ are

$$S_C(lower) = \begin{pmatrix} 0 & 0 & ... & 0 \\ 0 & 0 & ... & 0 \\ . & . & ... & . \\ 0 & 0 & ... & 0 \end{pmatrix}$$

because $C(s,s) = 0$, when $s = +1 + 1... + 1$ and

$$S_C(upper) = \begin{pmatrix} \lfloor log_2 n \rfloor & \lfloor log_2 n \rfloor & ... & \lfloor log_2 n \rfloor \\ \lfloor log_2 n \rfloor & \lfloor log_2 n \rfloor & ... & \lfloor log_2 n \rfloor \\ . & . & ... & . \\ \lfloor log_2 n \rfloor & \lfloor log_2 n \rfloor & ... & \lfloor log_2 n \rfloor \end{pmatrix}$$

because $C(s,s) = \lfloor log_2 n \rfloor$, when s is the initial segment of length n of the PTM sequence. Due to the extreme property of the PTM sequences the structural complexity of a strategy matrix $S \in \mathbf{S}$ varies within the bounds given by the matrices $S(lower)$ and $S(upper)$

$$0 = C(S(lower)) \leq C(S) \leq C(S(upper)) = N^2 \lfloor log_2 n \rfloor.$$

We define a partial order in the strategy space $\mathbf{S}$ that compares matrices in terms of structural complexity. Namely, if $S, S' \in \mathbf{S}$ are matrices

$$S_C = \{C_{ij}\}, \quad S'_C = \{C'_{ij}\}, \quad i,j = 1,...,N$$

such that $C_{ij} \leq C'_{ij}$ for all $i,j = 1,...,N$ and there exists at least one pair of their elements such that $C_{ij} < C'_{ij}$, then it is said that matrix S' is more complex than matrix S, denoted $S \prec S'$. The definition of the partial order is consistent with the fact that if $S \prec S'$ then the structural complexity of S' is greater then the structural complexity of S, i.e.,

$$C(S) < C(S').$$

A set of matrices $S_0, S_1, ..., S_k \in \mathbf{S}$, $k \geq 1$ is called a complexity trajectory in the strategy space $\mathbf{S}$ if

$$S_0 \prec S_1 \prec ... \prec S_{k-1} \prec S_k.$$

The complexity trajectories are distinguished in $\mathbf{S}$ because each next element $S_i, i = 1, ..., k$ in a complexity trajectory $S_0, S_1, ..., S_k \in \mathbf{S}$ is more complex than the previous one S_{i-1}. The structural complexity of the agents increases if they move along such a complexity trajectory. We are interested to know how the performance of the agents to solve a particular problem changes as their structural complexity increases. In other words, the idea is to consider corresponding fitness landscapes with the strategy space represented in terms of structural complexity.

It is worth mentioning that fitness landscapes appear rugged in the representation of the strategy space based on Cartesian order. In this representation the closest proximity of a strategy matrix consists of strategy matrices that are different from the strategy matrix by a minimum value in one coordinate. The partial order presented above gives a different representation of the strategy space. In this representation the closest proximity of a strategy matrix consists of strategy matrices that are different from the strategy matrix by minimum values in terms of structural complexity. Remarkably, computational experiments show that in this representation fitness landscapes appear regular as they take the unimodal form.

11.   A General Optimality Condition of Cooperative Agents

In this section we present an agent-based method to computationally investigate how performance of the cooperative agents changes as their structural complexity increases. In the investigation a class of benchmark TSP problems [36] is used.

The computational experiments for each problem of the class show a remarkable result. The structural complexity of the agents solving a TSP problem increases, i.e., they move along a complexity trajectory in the strategy space, there is a value of structural complexity such that the performance of the agents increases till this value and then decreases after it. This means that fitness landscapes of the problems appear in the unimodal form when the strategy space is represented in terms of structural complexity.

Moreover, the experiments give evidence to suggest that for each problem a corresponding value of structural complexity, i.e., where the performance of the method peaks, is a characteristic of the problem itself. The characteristic can be defined as the structural complexity of the problem. Finally, this allows us to formulate an optimality condition: the cooperative agents show their best performance for a problem when their structural complexity equals the structural complexity of the problem.

The method admits the following description. In a group of N cooperative agents each agent solves the TSP problem, i.e., to find the shortest closed path between $n + 1$ cities going through each city exactly once. The agents start in the same city and each of them first chooses the next city at random. Then at each city an agent chooses between two strategies, i.e., to pick up a next city purely at random or travel to the next city which is currently closest to the agent. In other words, at each city an agent uses a random strategy, labeled $+1$ or the greedy strategy, labeled -1 in order to get to the next city. Thus, an agent can be described by a binary sequence of length n as the agent step by step chooses between these two strategies to visit the cities.

The method is designed to have a control parameter that can increase the structural complexity of cooperative agents. This is realized by an algorithm that each agent uses to choose a next strategy. The algorithm, called the PTM algorithm, is about the next strategy: "win - stay, lose - consult PTM generator".

The PTM algorithm allows to organize cooperation between the agents and control them to increase their structural complexity. This is done in the following way. After each step distances traveled by the agents so far become known to everyone. Each agent takes this information into account to evaluate whether he/she wins or loses at the step, i.e., the last strategy is successful or unsuccessful. If according to the information an agent finds out that the distance he traveled so far deviates from the shortest one to date by not more than a value of a parameter v then his last strategy is successful and unsuccessful otherwise.

The parameter v is defined as follows. Let D_{ij} be the distance traveled by the ith, $i = 1, ..., N$, agent after j steps $j = 1, ..., n$ and

$$D_j^- = \min_{i=1,...,N} D_{ij}, \quad D_j^+ = \max_{i=1,...,N} D_{ij}.$$

All distances D_{ij} traveled by the agents after j steps belong to the interval $[D_j^-, D_j^+]$. The parameter v specifies a point

$$D_j(v) = D_j^+ - v(D_j^+ - D_j^-), \quad 0 \le v \le 1 + \varepsilon, \quad \varepsilon > 0$$

dividing the interval into two parts and determines what the agents view as a successful strategy. If a distance traveled by an agent after j steps $j = 1, ..., n$ belongs to the interval $[D_j^-, D_j(v)]$, called successful, then the agent's last strategy is successful, i.e., the agent wins. If the distance traveled by an agent belongs to the interval $(D_j(v), D_j^+]$, called unsuccessful, then the agent's last strategy is unsuccessful, i.e., the agent loses.

The method for the two limiting values of the parameter v comes up with the following strategy matrices.

1. When $v = 0$, then $D_j(v) = D_j^+$. This means that the successful interval $[D_j^-, D_j(v)]$ coincides with the whole interval $[D_j^-, D_j^+]$ and an agent always wins at each step irrespective of distances traveled the agents. Therefore, an agent starting with a random strategy and following the PTM algorithm to choose strategies never changes the random strategy all the way. A corresponding binary sequence of each agent consists completely of $+1$'s and the agents for this value of the parameter v come up with the strategy matrix

$$S(lower) = \begin{pmatrix} +1 & +1 & \dots & +1 & +1 \\ +1 & +1 & \dots & +1 & +1 \\ \cdot & \cdot & \dots & \cdot & \cdot \\ +1 & +1 & \dots & +1 & +1 \\ +1 & +1 & \dots & +1 & +1 \end{pmatrix}.$$

2. In the opposite limit, when $v = 1 + \varepsilon$, then $D_j(v) < D_j^-$. This means that the unsuccessful interval $(D_j(v), D_j^+]$ covers the whole interval $[D_j^-, D_j^+]$ and an agent at each step always loses irrespective of distances traveled by the agents. Therefore, each agent following the PTM algorithm at each step always asks PTM generator which strategy to choose next. A corresponding binary sequence of each agent is the initial segment of length n of the PTM sequence starting with $+1$ and the agents for this value of the parameter v come up with the strategy matrix

$$S(upper) = \begin{pmatrix} +1 & -1 & \dots & -1 & +1 \\ +1 & -1 & \dots & -1 & +1 \\ \cdot & \cdot & \dots & \cdot & \cdot \\ +1 & -1 & \dots & -1 & +1 \\ +1 & -1 & \dots & -1 & +1 \end{pmatrix}.$$

The PTM algorithm provides a very important property of the method. The parameter v increases from 0 to $1 + \varepsilon$, then the successful interval gets smaller. Thus, an agent is urged more to ask PTM generator which strategy to choose next. This brings corresponding binary sequences of the agents in terms of structural complexity more closer to the PTM sequence and consequently increases their structural complexity (for more details see [19]).

Therefore, the method for any TSP problem, as the parameter v is varied from 0 to $1 + \varepsilon$, produces a complexity trajectory in the strategy space $\mathbf{S}$

$$S(lower) \prec S_1 \prec \dots \prec S_k \prec S(upper)$$

and increases the structural complexity of the agents from 0 to $N^2 \lfloor log_2 n \rfloor$

$$0 = C(S(lower)) < C(S_1) < \dots < C(S_k) < C(S(upper)) = N^2 \lfloor log_2 n \rfloor.$$

Let $F_f(v, i)$ be a distance traveled by the ith agent, $i = 1, ..., N$, when the method is used with the value v of the parameter for a problem f. The performance of the agents as a group, when the method is used with the value v of the parameter and results in a strategy $S(v) \in \mathbf{S}$, is characterized by the average distance traveled by the agents

$$F_f(S(v)) = \frac{\sum_{i=1}^{N} F_f(v, i)}{N}.$$

This average distance is sought to be minimized.

Extensive computational experiments by using the benchmark class of TSP problems have been done to investigate how the performance of the agents changes as their structural complexity increases. In the experiments the interval $[0, 1 + \varepsilon]$ of the parameter v is subdivided by points

$$v_i = i\Delta v, \quad i = 0, 1, ..., l(\Delta v),$$

with resolution Δv, where $l(\Delta v) = \lfloor \frac{1+\varepsilon}{\Delta v} \rfloor + 1$. For all problems of the class we used the resolution

$$\Delta v = 0.1, \ 0.05, \ 0.01, \ 0.005, \ 0.001$$

and the number of agents

$$N = 100, 200, ..., 10000.$$

The method is used for each TSP problem f of the class successively for values v_i of the parameter and its performance $F_f(S(v_i)), i = 1, ..., l(\Delta v)$ is recorded for these values. As a result a discrete function

$$F_f(S(v_i)) = F_f'(v_i), \quad i = 1, ..., l(\Delta v)$$

of the parameter v is obtained.

The computational experiments for all the problems have demonstrated clearly that as the number N of the agents increases and the resolution Δv decreases the function

$$F_f'(v_i), \quad i = 1, ..., l(\Delta v)$$

converges to a smooth unimodal function F_f^* of the parameter v. Figure 8.8 gives an example of this result for a problem bier127 when $N = 10000$.

This means that for each TSP problem there is a value v^* of the parameter such that the performance of the agents increases till this point and then decreases after it. The method for a problem f, as the parameter v is varied from 0 to $1 + \varepsilon$, produces a complexity trajectory

$$S(lower) \prec ... \prec S_f(v^*) \prec ... \prec S(upper), \tag{26}$$

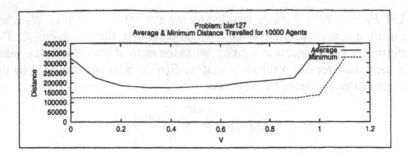

Figure 8.8. The figure shows the main result of the experiments for a problem bier127 when N = 10000. The performance of the agents as a function of the parameter v converges to a smooth unimodal function with one global minimum.

depending on the problem f. It is assumed that a limiting procedure is in place in (26) to eliminate dependence of the trajectory on the number of the agents.

Moreover, the experiments give strong facts to suggest that for each problem f tested a corresponding value of structural complexity $C(S_f(v^*))$, i.e., where the performance of the method peaks, is a characteristic of the problem itself. The performance of the agents for the problem increases as long as their structural complexity is less then $C(S_f(v^*))$ and decreases as soon as it becomes greater. This characteristic can be defined as the structural complexity of the problem.

Finally, this allows us to formulate a general optimality condition: the agents show their best performance for a TSP problem when their structural complexity equals the structural complexity of the problem. Importantly, the optimality condition is a guide how to control the structural complexity of the agents in improving their performance. Namely, if their structural complexity is less than the structural complexity of a problem then it must be increased till this value. If their structural complexity is greater than the structural complexity of the problem then it must be decreased till this value for the agents to show their best.

12.   Conclusions

Complex systems can not be understood without clear explanations what complexity precisely means. There are many different notions of complexity. However, complexity does not have a generally accepted universal concept. It is becoming more and more clear that the search for the universal concept of complexity has a strong requirement. It must be done within a final theory. In this chapter a concept of structural complexity for the first time suggests the real opportunity to search the universal concept of complexity within a final theory.

The concept of structural complexity is put forward as a key in understanding and control of complex systems. Experimental facts given in the chapter allow to suggest a general optimality condition of cooperative agents in terms of structural complexity. The optimality condition says that cooperative agents show their best performance for a particular problem when their structural complexity equals the structural complexity of the problem. The optimality condition specifies clearly how to harness effects of a complex system. According to the optimality condition to control a complex system efficiently means to equate its structural complexity with the structural complexity of the problem.

References

[1] C. Bennett, "On the nature and origin of complexity in discrete, homogeneous, locally-interacting systems", *Foundations of Physics*, 16: 585-592, 1986.

[2] L. Brillouin, *Science and Information Theory*, Academic, London, 1962.

[3] G. J. Chaitin, "On the length of programs for computing binary sequences", *Journal of ACM*, 13: 547-569, 1966.

[4] G. J. Chaitin, *Exploring Randomness*, Springer-Verlag, 2001.

[5] J. P. Crutchfield, "The calculi of emergence", *Physica D*, 75: 11-54, 1994.

[6] H. C. Fogedby, *J. Stat. Phys.*, 69: 411, 1992.

[7] M. Garey and D. Johnson, *Computers and Intractability: A Guide to the Theory of NP Completeness*, Freeman, San Franisco, 1979.

[8] M. Gell-Mann, *The Quark and the Jaguar: Adventures in the Simple and Complex*, W. H. Freeman and Company, New York, 1994.

[9] M. Gell-Mann and S. Lloyd, *Complexity*, 2: 44, 1996.

[10] P. Grassberger, *Int. J. Theor. Phys.*, 25: 907, 1986.

[11] J. P. Crutchfield and D. P. Feldman, *Phys. Rev. E*, 55: 1239, 1997.

[12] J. H. Holland, *Hidden Order: How Adaptation Builds Complexity*, Addison-Wesley, Reading, MA, 1995.

[13] J. H. Holland, *Emergence: From Chaos to Order*, Perseus Books, Reading, Massachusetts, 1998.

[14] J. Horgan, *The End of Science*, Broadway Books, New York, 1996.

[15] B. A. Huberman and T. Hogg, *Physica D*, 22: 376, 1986.

[16] S. Kauffman, *At Home in the Universe: the Search for Laws of Self-organization and Complexity*, Oxford University Press, New York, 1995.

[17] A. Kolmogorov, "Three approaches to the definition of the concept 'Quantity of Information'", *Problems of Information Transmission*, 1: 1-7, 1965.

[18] N. M. Korobov, *Trigonometric Sums and their Applications*, Nauka, Moscow, 1989.

[19] V. Korotkich, *A Mathematical Structure for Emergent Computation*, Kluwer Academic Publishers, 1999.

[20] V. Korotkich, "On complexity and optimization in emergent computation", In P. Pardalos, editor, *Approximation and Complexity in Numerical Optimization*, pp. 347-363, Kluwer Academic Publishers, 2000.

[21] V. Korotkich, "On optimal algorithms in emergent computation", In A. Rubinov and B. Glover, editors, *Optimization and Related Topics*, pp. 83-102, Kluwer Academic Publishers, 2001.

[22] V. Korotkich, "On self-organization of cooperative systems and fuzzy logic", In V. Dimitrov and V. Korotkich, editors, *Fuzzy Logic: A Framework for the New Millennium*, pp. 147-167, Springer-Verlag, 2002.

[23] R. Landauer, *IBM J. Res. Dev.* 3: 183, 1961.

[24] C. Langton, *Physica D*, 42: 12, 1990.

[25] M. Li and P. Vitanyi, *An Introduction to Kolmogorov Complexity and its Applications*, Springer-Verlag, 1997.

[26] S. Lloyd and H. Pagels, "Complexity as thermodynamic depth", *Annals of Physics*, 188: 186-213, 1988.

[27] P. G. Mezey, *Potential Energy Hypersurfaces*, Elsevier, Amsterdam, 1987.

[28] L. Mordell, "On a sum analogous to a Gauss' sum", *Quart. J. Math.*, 3: 161-167, 1932.

[29] M. Morse, "Recurrent geodesics on a surface of negative curvature", *Trans. Amer. Math. Soc.*, 22: 84, 1921.

[30] R. Nadeau and M. Kafatos. *The Non-local Universe*, Oxford University Press, New York, 2001.

[31] G. Nicolis and I. Prigogine, *Exploring Complexity*, New York, Freeman, 1989.

[32] C. Papadimitriou, *Computational Complexity*, Addison-Wesley, Reading, MA, 1994.

[33] A.S. Perelson and S.A. Kauffman, editors, *Molecular Evolution on Rugged Landscapes: Proteins, RNA, and the Immune System*, vol. 9 of Santa Fe Studies, Reading, MA, Addison-Wesley, 1991.

[34] E. Prouhet, "Memoire sur quelques relations entre les puissances des nombres", *C.R. Acad. Sci.*, Paris, 33: 225, 1851.

[35] S. Ramanujan, "Note on a set of simultaneous equations", *J. Indian Math. Soc.*, 4: 94-96, 1912.

[36] G. Reinelt, *TSPLIB 1.2* [Online]
Available from: URL ftp://ftp.wiwi.uni-frankfurt.de/pub/TSPLIB 1.2, 2000.

[37] C. E. Shannon, "A mathematical theory of communication", *Bell System Technical Journal*, July and October, 1948, reprinted in C. E. Shannon and W. Weaver, editors, *A Mathematical Theory of Communication*, University of Illinois Press, Urbana, 1949.

[38] R. Solomonoff, "A formal theory of inductive inference", *Information and Control*, 7:1-22, 1964.

[39] L. Szilard, *Z Phys.*, 53: 840, 1929.

[40] A. Thue, "Uber unendliche zeichenreihen", *Norske vid. Selsk. Skr. I. Mat. Nat. Kl. Christiana*, 7: 1, 1906.

[41] J. Traub, G. Wasilkowski and H. Wozniakowski, *Information-based Complexity*, Academic Press, 1988.

[42] S. Weinberg, *Dreams of a Final Theory*, Pantheon, New York, 1992.

[43] H. Weyl, "Uber die gleichverteilung von zahlen mod", *Eins. Math. Ann.*, 77: 313-352, 1915/16.

[44] J. A. Wheeler, *A Journey into Gravity and Spacetime*, Scientific American Library, New York, 1990.

[45] E. Witten, "Reflections on the fate of spacetime", *Physics Today*, 49: 24-30, 1996.

[46] S. Wolfram, *Cellular automata and complexity*, Addison-Wesley, 1994.

[47] W. H. Zurek, "Algorithmic randomness and physical entropy", *Physical Review A*, 40: 4731-4751, 1989.

Chapter 9

ROBUST DECISION MAKING: ADDRESSING UNCERTAINTIES IN DISTRIBUTIONS*

Pavlo Krokhmal
Risk Management and Financial Engineering Lab,
ISE Dept., University of Florida, USA
krokhmal@ufl.edu

Robert Murphey
Air Force Research Laboratory, Eglin AFB, Florida, USA
murphey@eglin.af.mil

Panos Pardalos
Center for Applied Optimization, ISE Dept., University of Florida, USA
pardalos@ufl.edu

Stanislav Uryasev
Risk Management and Financial Engineering Lab,
ISE Dept., University of Florida, USA
uryasev@ufl.edu

Grigory Zrazhevski
Risk Management and Financial Engineering Lab,
ISE Dept., University of Florida, USA,
and Taras Shevchenko State University, Kiev, Ukraine
zgrig@mechmat.univ.kiev.ua

*This work was partially supported by the Air Force grant F49620-01-1-0338.

S. Butenko et al. (eds.), Cooperative Control: Models, Applications and Algorithms, 165-185.
© 2003 *Kluwer Academic Publishers.*

Abstract   This paper develops a general approach to risk management in military applications involving uncertainties in information and distributions. The risk of loss, damage, or failure is measured by the Conditional Value-at-Risk (CVaR) measure. Loosely speaking, CVaR with the confidence level α estimates the risk of loss by averaging the possible losses over the $(1 - \alpha) \cdot 100\%$ worst cases (e.g., 10%). As a function of decision variables, CVaR is convex and therefore can be efficiently controlled/optimized using convex or (under quite general assumptions) linear programming. The general methodology was tested on two Weapon-Target Assignment (WTA) problems. It is assumed that the distributions of random variables in the WTA formulations are not known with certainty. The total cost of a mission (including weapon attrition) was minimized, while satisfying operational constraints and ensuring destruction of all targets with high probabilities. The risk of failure of the mission (e.g., targets are not destroyed) is controlled by CVaR constraints. The case studies conducted show that there are significant qualitative and quantitative differences in solutions of deterministic WTA and stochastic WTA problems.

Keywords: Risk management, Conditional Value-at-Risk, uncertainty, military applications, stochastic programming.

1.   Introduction

This paper develops a general approach to managing risk in military applications involving stochasticity and uncertainties in distributions. Various military applications such as intelligence, surveillance, planning, scheduling etc., involve decision making in dynamic, distributed, and uncertain environments. In a large system, multiple sensors may provide incomplete, conflicting, or overlapping data. Moreover, some components or sensors may degrade or become completely unavailable (failures, weather conditions, battle damage). Uncertainties in combat environment induce different kinds of risks that components, sensors or armed units are exposed to, such as the risk to be damaged or destroyed, risk of mission incompleteness (e.g., missing a target) or failure, risk of false target attack etc. Therefore, planning and operating in stochastic and uncertain conditions of a modern combat require robust decision-making procedures. Such procedures must take into account the stochastic nature of risk-inducing factors, and generate decisions that are not only effective on average (in other words, have good "expected" performance), but also safe enough under a wide range of possible scenarios. In this regard, risk management in military applications is similar to practices in other fields such as finance, nuclear safety, etc., where decisions targeted only at achieving the maximal expected performance may lead to an excessive risk exposure. However, in contrast to other applications, distributions of the stochastic risk-inducing factors are often unknown or uncertain in military problems. Uncertainty in distributions of risk parameters may be caused by a lack of data, unreliability of data, or the specific nature of a risk factor (e.g., in different circumstances a risk factor may exhibit different stochastic behavior). Therefore, decision making in military applications must account for uncertainties in distributions of stochastic parameters and be robust with respect to these uncertainties.

In this paper, we propose a general methodology for managing risk in military applications involving various risk factors as well as uncertainties in distributions. The approach is tested with several stochastic versions of the Weapon-Target Assignment problem.

The paper is organized as follows. Section 2 presents key theoretical results on risk management using Conditional Value-at-Risk (CVaR) risk measure, and describes the general approach to controlling risk when distributions of risk factors are uncertain. Section 3 develops various formulations of the stochastic Weapon-Target Assignment (WTA) problem with CVaR constraints. Results of numerical experiments for one-stage and two-stage stochastic WTA problems are presented in Section 4. The Conclusions section summarizes the obtained results and outlines the directions of future research. Finally, the Appendix presents formal definitions and results concerning the risk management using the CVaR risk measure.

2.   The General Approach

Presence of uncertainty in a decision-making model leads to the problem of estimation and managing/controlling of risk associated with the stochastic parameters in the model. Over the recent years, risk management has evolved into a sophisticated discipline combining both rigorous and elegant theoretical results and practical effectiveness (this especially applies to the risk management in finance industry). Generally speaking, risk management is a set of activities aimed at reducing or preventing *high losses* incurred from an incorrect decision. The losses (e.g., damages, failures) in a system are quantified by a *loss function* $L(x, \xi)$ that depends upon decision vector x and a stochastic vector ξ standing for uncertainties in the model. Assuming for now that a distribution of the parameter ξ is known, it is possible to determine the distribution of the loss function $L(x, \xi)$ (see Figure 9.1). Then, the problem of preventing high losses is a problem of controlling and shaping the loss distribution and, more specifically, its right tail, where the high losses reside. To estimate and quantify the losses in the tail of the loss distribution, a *risk measure* has to be specified. In particular, a risk measure introduces the ordering relationships for risks, so that one is able to discriminate "less risky" decisions from the "more risky" ones[1]. The appropriate choice of a risk measure is, in most cases,

---

[1] Artzner et al., 1999, have introduced a concept of "ideal", or *coherent*, risk measure. A *coherent* risk measure, which satisfies to a set of axioms developed in this paper, is expected to produce "proper" and "consistent" estimates of risk.

dictated by the nature of uncertainties and risks in the problem at hand. In military applications, for example, one usually deals with the probabilities of events, such as the probability to hit a target, the probability to detect the enemy's aircraft, and so on. Therefore *percentile* risk measures that represent the risk in terms of percentiles of the loss distribution are particularly suitable for the risk management in military applications. Popular percentile risk measures include Value-at-Risk (VaR), Conditional Value-at-Risk, Maximum Loss, and Expected Shortfall. Figure 9.1 displays some of these measures; Value-at-Risk with confidence level α (α-VaR), which is the α-percentile of loss distribution, Maximum Loss ("1.0-percentile" of loss distribution), and α-CVaR, which approximately equals to the expectation of losses exceeding α-VaR.

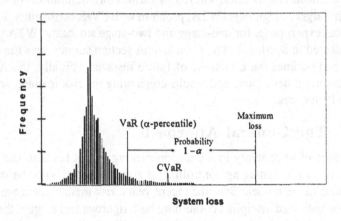

Figure 9.1. Loss function distribution and different risk measures.

We build our approach for risk management in military applications on the CVaR methodology, which is a relatively new development ([7], [8]). This section presents the general framework of risk management using Conditional Value-at-Risk, and extends it to the case when the distributions of stochastic parameters are not certain.

2.1.　Risk Management Using Conditional Value-at-Risk

Suppose that the uncertain future is represented by a finite number of future outcomes (scenarios). Then, approximately, Value-at-Risk with confidence level α (α-VaR) is defined as the loss that can be exceeded only in $(1 - \alpha) \cdot 100\%$ of worst scenarios. Similarly, one may think of α-CVaR (i.e., Conditional Value-at-Risk with confidence level α) as of the average loss over $(1 - \alpha) \cdot 100\%$ of worst cases (see Figure 9.2). We say "approximately" because $(1 - \alpha) \cdot 100\%$ may not be an integer number. Exact definition of α-VaR and α-CVaR is given in the Appendix.

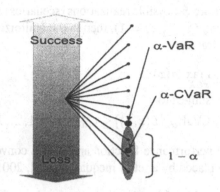

Figure 9.2.   A visualization of VaR and CVaR concepts.

These intuitive definitions are correct if, for example, all scenarios are equally probable, and $(1 - \alpha) \cdot 100$ is an integer number. The formal definitions of α-VaR and α-CVaR that apply to any loss distribution and value of confidence level α are more complex, and the reader can find them in the Appendix. The Appendix also contains the key formal results concerning calculation and controlling of the Conditional Value-at-Risk for general loss distributions. Here we mention only the most important properties of CVaR and their practical implications.

The Conditional Value-at-Risk function $\text{CVaR}_\alpha[L(x, \xi)]$ has the following properties ([7], [8], [1]):

- CVaR is continuous with respect to confidence level α (other percentile risk measures like VaR, Expected Shortfall, etc., may be discontinuous in α);

- CVaR is convex in α and x, provided that the loss function $L(x, \xi)$ is convex in x (VaR, Expected Shortfall are generally non-convex in x);

- CVaR is *coherent* in the sense of Artzner et al., 1999;

- in case of a continuous loss distribution CVaR equals the conditional expectation of losses exceeding the VaR level.

From the viewpoint of managing and controlling of risk, the most important property of CVaR, which distinguishes it from all other percentile risk measures, is the convexity with respect to decision variables, which permits the use of convex programming for minimizing CVaR. If the loss function $L(x, \xi)$

can be approximated by a piecewise linear function, the procedure of controlling or optimization of CVaR is reduced to solving a Linear Programming (LP) problem.

Assume that there are S possible realizations (scenarios) $\xi_1, ..., \xi_S$ of vector ξ with probabilities π_s ($\sum_{s=1}^{S} \pi_s = 1$), then in the optimization problem with multiple CVaR constraints

$$\max_{x \in X} \ g(x)$$
subject to
$$\text{CVaR}_{\alpha_n}[L(x, \xi)] \leq C_n, \quad n = 1, ..., N,$$

where $g(x)$ is some performance function and X is a convex set, each CVaR constraint may be replaced by a set of inequalities ([7], 2001)

$$
\begin{aligned}
L(x, \xi_s) - \zeta_n &\leq w_{ns}, \quad s = 1, ..., S, \\
\zeta_n + (1 - \alpha_n)^{-1} \sum_{s=1}^{S} \pi_s w_{ns} &\leq C_n, \\
\zeta_n \in \mathcal{R}, \quad w_{ns} \in \mathcal{R}^+, \quad s &= 1, ..., S,
\end{aligned}
\tag{1}
$$

where $\mathcal{R}$ and $\mathcal{R}^+$ are the sets of real and non-negative real numbers correspondingly, and w_{ns} are auxiliary variables. If in the optimal solution the n-th CVaR constraint is active, then the corresponding variable ζ_n is equal to α_n-VaR (i.e., α_n-th percentile of the loss distribution).

In the risk management methodology discussed above the distribution of stochastic parameter ξ is considered to be known. The next subsection extends the presented approach to the case, when the distribution of stochastic parameters in the model is not certain.

2.2. Risk Management Using CVaR in the Presence of Uncertainties in Distributions

The general approach to managing risks in an uncertain environment, where the distributions of stochastic parameters are not known for sure, can be described as follows. Suppose that we have some performance function $F(x, \xi)$, dependent on the decision vector $x \in X$ and some random vector $\xi \in \Xi$, whose distribution is not known for certain. We assume that the actual realization of vector ξ may come from different distributions $\Theta_1, ..., \Theta_N$. The vector ξ stands for the uncertainties in data that make it impossible to evaluate the efficiency $F(x, \xi)$ of the decision for sure. Thus, there always exists a possibility of making an incorrect decision, and, consequently, suffering loss, damage, or failing the mission. If the loss in the system is evaluated by function $L(x, \xi)$, then risk of high losses can be controlled using CVaR constrains. Let formulate the problem of maximizing the expected performance function

$F(x,\xi)$ subject to some operational constraints $Ax \leq b$ and CVaR risk constraints. Due to the unknown distribution of vector ξ, we are unable to find the expectation $\mathsf{E}_{\Theta}[F(x,\xi)]$. Therefore, being on the conservative side, we want the decision x to be optimal with respect to each measure Θ_n, and this leads to the following *max-min* problem:

$$\max_{x \in X} \quad \min_{\Theta_n, \, n=1,...,N} \quad \mathsf{E}_{\Theta_n}[F(x,\xi)] \qquad (2)$$

subject to

$$Ax \leq b,$$

$$\text{CVaR}\alpha[L(x,\xi) \,|\, \Theta_n] \leq C, \quad n = 1, ..., N,$$

where multiple CVaR constraints with respect to different measures Θ_n control the risk for high losses $L(x,\xi)$ to exceed some threshold C. In formulation (2) we assume that the performance function F is concave in x, and the loss function L is convex in x. These assumptions are not restrictive; on the contrary, they indicate that given more than one decision with equal performance one favors safer decisions over the riskier ones.

Model (2) explains how to handle the risk of generating an incorrect decision in an uncertain environment. In military applications, different types of risks and losses may be explicitly involved, for example, along with loss function $L(x,\xi)$ one may consider a loss function $R(x,\xi)$ for the risk of false target attack. Control for this type of risk can also be included in the model by a similar set of CVaR constraints:

$$\max_{x \in X} \quad \min_{\Theta_n, \, n=1,...,N} \quad \mathsf{E}_{\Theta_N}[F(x,\xi)]$$

subject to

$$Ax \leq b,$$

$$\text{CVaR}_{\alpha_1}[L(x,\xi) \,|\, \Theta_n] \leq C_1, \quad n = 1, ..., N,$$

$$\text{CVaR}_{\alpha_2}[R(x,\xi) \,|\, \Theta_n] \leq C_2, \quad n = 1, ..., N.$$

In the next sections we test the presented approach to risk management in military applications with the Weapon-Target Assignment problem.

3.   Example: Stochastic Weapon-Target Assignment Problem

The Weapon-Target Assignment (WTA) problem considers the optimal assignment of weapons to targets so as to minimize the surviving value of targets. The WTA problem is used in planning environment that features a whole spectrum of uncertainties, such as the number and types of targets in the battle space, their positions, and the probability of a weapon to destroy a target (e.g., probability of kill). To generate robust decisions, one must account for these uncertainties and the corresponding risks. In this section we present two formulations of the stochastic Weapon-Target Assignment problem that address the uncertainties in a weapon's probability of kill and in the number of targets.

3.1.   Deterministic WTA Problem

The generic formulation of the Weapon Target Assignment problem is as follows. Given the set of targets and set of available weapons, one must find the optimal assignment of weapons to targets, such that, for example, the damage to the targets is maximized, or the cost of the operation is minimized. The WTA formulation that maximizes the damage to the targets (see, for example, [5], [3], [6]) leads to a non-linear programming problem (NLP), with linear constraints and is the subject of a future paper. In this paper we adopt another setup, where the total cost of the mission (including battle damage or loss) is minimized, while satisfying constraints on mission accomplishment (i.e., destruction of all targets with some prescribed probabilities). We assume that different weapons have different costs and efficiencies, and, in general, each may have a "multishot" capacity so that it may attack more than one target. In the deterministic setup of the problem we include also the constraint that prescribes how many targets a single weapon can attack.

The deterministic WTA problem is

$$\min_{x} \quad \sum_{k=1}^{K}\sum_{i=1}^{I} c_{ik}\, x_{ik} \tag{3a}$$

subject to

$$\sum_{k=1}^{K} x_{ik} \leq m_i, \quad i = 1, ..., I, \tag{3b}$$

$$x_{ik} \leq m_i\, v_{ik}, \quad i = 1, ..., I, \quad k = 1, ..., K, \tag{3c}$$

$$\sum_{k=1}^{K} v_{ik} \leq t_i, \quad i = 1, ..., I, \tag{3d}$$

$$1 - \prod_{i=1}^{I}(1 - p_{ik})^{x_{ik}} \geq d_k, \quad k = 1, ..., K, \tag{3e}$$

$$x_{ik} \in \mathbb{Z}^{+}, \quad v_{ik} \in \{0, 1\},$$

where

x_{ik} is the number of shots to be fired by weapon i at target k;

$v_{ik} = 1$, if weapon i fires at target k, and $v_{ik} = 0$ otherwise;

c_{ik} is the cost (including the battle loss or damage) of firing one shot from weapon i at target k; c_k includes the relative value of target k with respect to all other targets;

m_i is the shots capacity for weapon i;

t_i is the maximal number of targets which can be attacked by weapon i;

p_{ik} is the probability of destroying target k by firing one shot from weapon i;

d_k is the minimal required probability for destroying target k;

$\mathcal{Z}$ is the set of integer numbers, and $\mathcal{Z}^+$ is the set of non-negative integers.

The objective function in this problem equals to the total cost of the mission. The first constraint, (3b), states that the munitions capacity of weapon i cannot be exceeded. The second and the third constraints (3c) and (3d) are responsible for not allowing weapon i to attack more than t_i targets, where $t_i \leq K$. The last constraint (3e) ensures that after all weapons are assigned, target k is destroyed with probability not less than d_k. Note that this non-linear constraint can be linearized:

$$\sum_{i=1}^{I} \ln(1 - p_{ik})\, x_{ik} - \ln(1 - d_k) \leq 0. \qquad (4)$$

In this way the deterministic WTA problem (3a) can be formulated as a linear integer programming (IP) problem.

3.2. One-Stage Stochastic WTA Problem with CVaR Constraints

In real-life situations many of the parameters in model (3a)–(3e) are not deterministic, but stochastic values. For example, the probabilities p_{ik} of destroying target k may depend upon battle situation, weather conditions, and so on, and consequently, may be treated as being uncertain. Similarly, the cost of firing c_{ik}, which includes battle loss/damage, may also be a stochastic parameter. The number of targets K may be uncertain as well.

First, we consider a one-stage Stochastic Weapon-Target Assignment (SWTA) problem, where the uncertainty is introduced into the model by assuming that probabilities p_{ik} are stochastic and dependent on some random parameter ξ:

$$p_{ik} = p_{ik}(\xi).$$

In accordance to the described methodology of managing uncertainties and risks, we model the stochastic behavior of probabilities p_{ik} using scenarios. Namely, probabilities $p_{ik}(\xi)$ take different values $p_{ik}(\xi_s) = p_{iks}$, $s = 1, ..., S$

under S different scenarios. Such a scenario set may be constructed, for example, by utilizing the historical observations of weapons' efficiency in different environments, or by using simulated data, experts' opinions etc.

We now replace the last constraint in (3a) by a CVaR constraint, where the loss function takes a positive value if the probability of destroying target k is less than d_k:

$$L_k(x, \xi) = \sum_{i=1}^{I} \ln\left(1 - p_{ik}(\xi)\right) x_{ik} - \ln(1 - d_k), \qquad (5)$$

and takes a negative value otherwise. Recall that a CVaR constraint with confidence level α bounds the (weighted) average of $(1 - \alpha) \cdot 100\%$ highest losses. In our case, allowing small positive values of loss function (5) for some scenarios implies that for these scenarios target k is destroyed with probability slightly less than d_k, which may still be acceptable from a practical point of view.

Except for the constraint on the target destruction probability, the one-stage Stochastic WTA problem is identical to its deterministic predecessor:

$$\min_x \sum_{k=1}^{K} \sum_{i=1}^{I} c_{ik} x_{ik} \qquad (6)$$

subject to

$$\sum_{k=1}^{K} x_{ik} \leq m_i, \quad i = 1, ..., I,$$

$$x_{ik} \leq m_i v_{ik}, \quad i = 1, ..., I, \quad k = 1, ..., K,$$

$$\sum_{k=1}^{K} v_{ik} \leq t_i, \quad i = 1, ..., I,$$

$$\text{CVaR}\alpha\left[L_k(x, \xi)\right] \leq C_k, \quad k = 1, ..., K.$$

Here α is the confidence level, C_k are some (small) constants, and all other variables and parameters are defined as before. As demonstrated in (1), for the adopted scenario model with probabilities p_{ik}, the CVaR constraint for the k-th target

$$\text{CVaR}\alpha\left[L_k(x, \xi)\right] \leq C_k$$

is represented by a set of linear inequalities:

$$\sum_{i=1}^{I} \ln(1 - p_{iks}) x_{ik} - \ln(1 - d_k) - \zeta_k \leq w_{sk}, \quad s = 1, ..., S,$$

$$\zeta_k + (1 - \alpha_k)^{-1} S^{-1} \sum_{s=1}^{S} w_{sk} \leq C_k, \tag{7}$$

$$\zeta_k \in \mathcal{R}, \quad w_{sk} \geq 0, \quad s = 1, ..., S, \quad k = 1, ..., K.$$

Thus, the one-stage Stochastic WTA problem can be formulated as a mixed-integer programming (MIP) problem:

$$\min_x \sum_{k=1}^{K} \sum_{i=1}^{I} c_{ik} x_{ik} \tag{8}$$

subject to

$$\sum_{k=1}^{K} x_{ik} \leq m_i, \quad i = 1, ..., I,$$

$$x_{ik} \leq m_i v_{ik}, \quad i = 1, ..., I, \quad k = 1, ..., K,$$

$$\sum_{k=1}^{K} v_{ik} \leq t_i, \quad i = 1, ..., I,$$

$$\sum_{i=1}^{I} \ln(1 - p_{iks}) x_{ik} - \ln(1 - d_k) - \zeta_k \leq w_{sk},$$

$$s = 1, ..., S, \quad k = 1, ..., K,$$

$$\zeta_k + (1 - \alpha_k)^{-1} S^{-1} \sum_{s=1}^{S} w_{sk} \leq C_k, \quad k = 1, ..., K,$$

$$x_{ik} \in \mathcal{Z}^+, \quad v_{ik} \in \{0, 1\}, \quad \zeta_k \in \mathcal{R}, \quad w_{sk} \geq 0,$$

$$s = 1, ..., S, \quad i = 1, ..., I, \quad k = 1, ..., K.$$

Note that different values of probability p_{ik} represent the uncertainty in the distributions of stochastic parameters discussed in the previous section. Indeed, different values of probability p_{ik} imply different probability measures for the random variable associated with the event of destroying target k by firing one unit of munitions by weapon i. In effect, CVaR constraint (6) is a risk constraint that incorporates multiple probability measures.

3.3.   Two-Stage Stochastic WTA Problem with CVaR constraints

In this section we consider a more complex, but also more realistic two-stage Stochastic WTA problem, where the uncertain parameter is the number of targets to be destroyed.

This problem is more realistic since it models the effect of target discovery as being dynamic; that is, not all targets are known at any single instance of time. To address this type of uncertainty, we need to modify notations.

Consider I weapons are deployed in some bounded region of interest in interval of time T with the goal of finding targets and then, once found, attacking those targets. If we delay all assignments of weapon shots to targets until the final time T, then we have a deterministic, "static" WTA problem as in (3a)–(3e). If, on the other hand, we assume that weapons have at least 2 opportunities to shoot during the interval T, then the WTA problem is dynamic. In the later case we have the opportunity to avoid expending all our shots at targets discovered early in T by explicitly modeling the number of undiscovered targets in the objective function.

Assume that K now represents the number of *categories* of targets (the targets may be categorized, for example, by their importance, vulnerability, etc).

We will assume the problem has 2 stages. That is, at any given time, we may always partition all targets into those thus far determined and those that we conjecture to exist but have not yet found. Our conjecture may be based on evidence obtained by prior reconnaissance of the region of interest. At some arbitrary time $0 < \tau < T$ assume that there are n_k detected targets and η_k undetected targets in each category $k = 1, ..., K$. Thus we have two clearly identified stages in our problem: in the first stage one has to destroy the targets known at time τ, in the second stage one must destroy the targets that we conjecture will be found by time T. In other words, one needs to make an assignment of weapons that will allow for the destruction of the targets known at time τ while reserving enough munition capacity for destroying the targets we expect to find in $\tau < t < T$.

Setup of the two-stage stochastic WTA problem can be considered as a part of a moving horizon or quasi-multistage stochastic WTA algorithm, where the WTA problem with many time periods is solved by recursive application of a two-stage algorithm ([6]).

To simplify the problem setup, we remove the constraint on the number of targets a single weapon can attack (the second and third constraints in problems (3a)), since this constraint makes the problem combinatorial. Also, we assume that the probabilities p_{ik} are known (*not* random), so that the only stochastic parameters in the two-stage SWTA problem are the numbers of undetected (second-stage) targets $\eta_k, k = 1, ..., K$.

We model the uncertainty in the number of targets at the second stage, by we introducing a scenario model, where under scenario $s \in \{1, ..., S\}$ there are $\eta_k(s) = \eta_{ks}$ undetected targets in category k.

The first- and second-stage decision variables are defined as follows:

x_{ik} is the number of munitions to be fired by weapon i at a single target in category k during the first stage;

$y_{ik}(s)$ is the number of munitions to be fired by weapon i at a single target in category k during the second stage scenario s.

Note that the same decision is made for all targets within a category, i.e., once weapon i fires, say, 2 missiles at a specific target in category k, it must fire 2 missiles at every other target in this category.

The recursive formulation of the two-stage stochastic WTA problem is

$$\min \left\{ \sum_{k=1}^{K} \sum_{i=1}^{I} n_k c_{ik} x_{ik} + E_\eta[Q(x,\eta)] \right\} \tag{9a}$$

subject to

$$\sum_{k=1}^{K} n_k x_{ik} \leq m_i, \quad i = 1, ..., I, \tag{9b}$$

$$\sum_{i=1}^{I} \ln(1 - p_{ik}) x_{ik} - \ln(1 - d_k) \leq \varepsilon_{1k}, \quad k = 1, ..., K, \tag{9c}$$

$$\sum_{k=1}^{K} \varepsilon_{1k} \leq C, \tag{9d}$$

$$x_{ik}, \varepsilon_{1k} \geq 0, \quad i = 1, ..., I, \quad k = 1, ..., K,$$

where the recourse function $Q(x, \eta)$ is the solution of the problem

$$Q(x, \eta) = \min_y \left\{ \sum_{k=1}^{K} \sum_{i=1}^{I} \eta_k(s) c_{ik} y_{ik}(s) + M \sum_{i=1}^{I} \delta_i \right\} \tag{10a}$$

subject to

$$\sum_{k=1}^{K} (n_k x_{ik} + \eta_k(s) y_{ik}(s)) \leq m_i + \delta_i, \quad \forall i, \tag{10b}$$

$$\sum_{i=1}^{I} \ln(1 - p_{ik}) y_{ik}(s) - \ln(1 - d_k) - \zeta_k \leq w_k(s), \quad \forall k, s, \tag{10c}$$

$$\zeta_k + (1 - \alpha_k)^{-1} S^{-1} \sum_{s=1}^{S} w_k(s) \leq \varepsilon_{2k}, \quad \forall k, \tag{10d}$$

$$\sum_{k=1}^{K} (\varepsilon_{1k} + \varepsilon_{2k}) \leq C, \tag{10e}$$

$$y_{ik}(s), \delta_i \in Z^+, \quad w_k(s), \varepsilon_{2k} \geq 0, \quad \zeta_k \in \mathcal{R}, \quad M \gg 1.$$

Let us discuss the recourse problem (9a)–(10e). As before, we minimize the total cost of the mission. The first constraint (9b) is the munitions capacity constraint. The second constraint, (9c), allows a first-stage target in category k to survive with (small) error ε_{1k}, and the third constraint (9d) bounds the sum of errors ε_{1k} by some (small) constant C.

In the recourse function (10a) the first constraint (10b) requires the weapon i to not exceed its munitions capacity while destroying the first- and second-stage targets. The possible infeasibility of the munitions capacity constraint can be relaxed using auxiliary variables δ_i that enter the objective function with cost coefficient $M \gg 1$. The second and third constraints (10c)-(10d) form a CVaR constraint that controls the failure of destroying second-stage targets with the prescribed probabilities d_k. Similarly to the deterministic constraint in (9a), CVaR of failure to destroy a second-stage target in category k is bounded by (small) error variable ε_{2k}. The total sum of errors ε_{1k} and ε_{2k} at both stages is bounded by small constant C, which makes possible a tradeoff between the degree of mission accomplishment at the first and second stages.

The extensive form of the two-stage SWTA problem (9a)–(10a) is

$$\min \left\{ \sum_{k=1}^{K} \sum_{i=1}^{I} n_k\, c_{ik}\, x_{ik} + \frac{1}{S} \sum_{s=1}^{S} \sum_{k=1}^{K} \sum_{i=1}^{I} \eta_{ks}\, c_{ik}\, y_{ik}(s) + M \sum_{i=1}^{I} \delta_i \right\} \quad (11)$$

subject to

$$\sum_{k=1}^{K} \left(n_k\, x_{ik} + \eta_{ks}\, y_{ik}(s) \right) \le m_i + \delta_i, \quad \forall\, i,\, s,$$

$$\sum_{i=1}^{I} \ln(1 - p_{ik})\, x_{ik} - \ln(1 - d_k) \le \varepsilon_{1k}, \quad \forall\, k,$$

$$\sum_{i=1}^{I} \ln(1 - p_{ik})\, y_{ik}(s) - \ln(1 - d_k) - \zeta_k \le w_{ks}, \quad \forall\, k,\, s,$$

$$\zeta_k + (1 - \alpha_k)^{-1} S^{-1} \sum_{s=1}^{S} w_{ks} \le \varepsilon_{2k}, \quad \forall\, k,$$

$$\sum_{k=1}^{K} (\varepsilon_{1k} + \varepsilon_{2k}) \le C,$$

$$x_{ik},\, y_{ik}(s),\, \delta_i \in \mathcal{Z}^+, \quad w_{ks},\, \varepsilon_{1k},\, \varepsilon_{2k} \in \mathcal{R}^+ \quad \zeta_k \in \mathcal{R}, \quad M \gg 1.$$

The two-stage stochastic WTA problem is also a MIP problem.

4.   Numerical Results

In this section we present and discuss numerical results obtained for both one-stage and two-stage stochastic WTA problems. The algorithms for solving deterministic, one- and two-stage stochastic WTA problems were implemented in C++, and we used CPLEX 7.0 Callable Library to solve the corresponding IP and MIP problems. We used simulated data (sets of weapons and targets, the corresponding costs and probabilities etc.) for testing the implemented algorithms.

4.1.   Single-Stage Deterministic and Stochastic WTA Problems

For the deterministic and one-stage stochastic WTA problems we used the following data:

- 5 targets ($K = 5$)

- 5 weapons, each with 4 shots ($I = 5, \ m_i = 4$)

- any weapon can attack any target ($t_i = 5$),

- probabilities p_{ik} and costs c_{ik} depend only on the weapon index i: $p_{ik} = p_i, c_{ik} = c_i$

- all targets have to be destroyed with at least probability 95% ($d_k = 0.95$)

- the confidence levels α_k in CVaR constraint are 0.90

- there are 20 scenarios ($S = 20$) for probabilities $p_{ik}(s)$ in the one-stage SWTA problem; all scenarios are equally probable.

According to the aforementioned, we used simulated data for probabilities p_{iks} and costs c_{ik}. It was assumed that probabilities $p_{iks} = p_{is}$ are uniformly distributed random variables, and the Figure 4.1 displays the relation between the cost of missile of weapon i and its efficiency (i.e., probability to destroy a target):

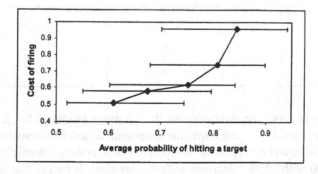

Figure 9.3. Dependence between the cost and efficiency for different types of weapons in one-stage SWTA problem (8) deterministic WTA problem (3a).

On this graph, diamonds represent the average probability of destroying a target by firing one shot from weapon i, and the horizontal segments represent the support for random variable $p_{ik}(\xi) = p_i(\xi)$. The average probabilities

$$\bar{p}_{ik} = \frac{1}{S} \sum_{s=1}^{S} p_{iks}$$

were used for p_{ik} in the deterministic problem (3a).

The efficiency and cost of weapons 1 to 5 increase with the index of weapon, i.e., Weapon 1 is the least efficient and cheapest, whereas Weapon 5 is the most precise, but also most expensive one.

Tables 9.1 and 9.2 represent the optimal solutions (variables x_{ik}) of the deterministic and one-stage stochastic WTA problems.

Table 9.1. Optimal solution of the deterministic WTA problem (3a)

Target	T1	T2	T3	T4	T5	Total shots
Weapon 1	0	2	1	0	1	4
Weapon 2	0	1	2	0	0	3
Weapon 3	1	0	0	1	1	3
Weapon 4	1	0	0	1	1	3
Weapon 5	0	0	0	0	0	0

Table 9.2. Optimal solution of the one-stage stochastic WTA problem (6), (8)

Target	T1	T2	T3	T4	T5	Total shots
Weapon 1	0	1	1	0	1	3
Weapon 2	0	0	1	1	1	3
Weapon 3	2	0	0	1	0	3
Weapon 4	0	1	1	0	1	3
Weapon 5	1	1	0	1	0	3

One can observe the difference in the solutions produced by deterministic and stochastic WTA problems: the deterministic solution does *not* use the most expensive and most precise Weapon 5, whereas the stochastic solution of problem (8) with CVaR constraint uses this weapon. It means that the CVaR-constrained solution of problem (8) represents a more expensive but safer decision.

On a different dataset, we obtained a similar result: the optimal solution of the stochastic problem with CVaR constraints did not use the cheapest and the most unreliable weapon, whereas the deterministic solution used it.

We have also performed testing of the deterministic solution under different scenarios. The deterministic solution failed to destroy more than one target under 13 of 20 scenarios.

This example highlights the importance of using risk management procedures in military decision-making applications involving uncertainties.

4.2. Two-Stage Stochastic WTA Problem

For the two-stage stochastic WTA problems we used the following data:

- 3 categories of targets ($K = 3$)

- 4 weapons, each with 15 shots ($I = 4,\ m_i = 15$)

- probabilities p_{ik} and costs c_{ik} depend only on the weapon index i: $p_{ik} = p_i,\ c_{ik} = c_i$

- all targets have to be destroyed with probability 95% ($d_k = 0.95$)

- the confidence levels α_k in CVaR constraint are equal 0.90

- there are 15 scenarios ($S = 15$) for the number of undetected targets η_{ks} (for each k, the number of undetected targets η_{ks} is a random integer between 0 and 5); all scenarios are equally probable.

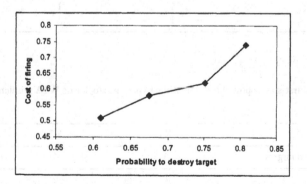

Figure 9.4. Dependence between the cost and efficiency for different types of weapons in two-stage SWTA problem (11).

For the probabilities p_{ik} in the two-stage problem, we used the first four average probabilities from the deterministic WTA problem, and the efficiency-cost dependence is shown in Figure 4.2.

Tables 9.3 to 9.5 illustrate the optimal solution of the problem (11). Table 9.3 contains the first-stage decision variables x_{ik}, and Tables 9.4 and 9.5 display the second-stage variables $y_{ik}(s)$ for scenarios $s = 1$ and $s = 2$, just for illustrative purposes.

Similarly to the analysis of the one-stage stochastic WTA problem, we compared the scenario-based solution of problem (11) with the solution of the "deterministic two-stage" problem, where the number of second-stage targets in each category is taken as the average over 15 scenarios. The comparison shows that the solution based on the expected information leads to significant munitions shortages in 5 of 15 (i.e., 33%) scenarios, and consequently to failing the mission at the second stage. Recall from the analysis of the one-stage SWTA problem that the solution based on the expected information also exhibited poor robustness with respect to different scenarios. Indeed, solutions that use only the *expected* information, are supposed to perform well *on average*, or *in the long run*. However, in military applications there is *no long run*, and therefore such solutions may not be robust with respect to *many possible scenarios*.

Table 9.3.   First-stage optimal solution of the two-stage stochastic WTA problem

Category	K1	K2	K3
# of detected targets	3	5	2
Weapon 1	0	0	0
Weapon 2	0	0	0
Weapon 3	1	1	1
Weapon 4	1	1	1

Table 9.4.   First-stage optimal solution of the two-stage stochastic WTA problem (11) for the first scenario

Category	K1	K2	K3
# of undetected targets	1	4	2
Weapon 1	0	0	2
Weapon 2	0	0	1
Weapon 3	1	1	0
Weapon 4	1	1	0

Thus, solutions of both one-stage and two-stage SWTA problems confirm the general conjecture on the potential importance of exploiting stochastic models and risk management in military applications.

Table 9.5.   Second-stage optimal solution of the two-stage stochastic WTA problem (11) for the second scenario

Category	K1	K2	K3
# undetected of targets	3	5	3
Weapon 1	2	0	2
Weapon 2	1	0	1
Weapon 3	0	1	0
Weapon 4	0	1	0

5.   Conclusions

We have presented an approach to managing risk in stochastic environments, where distributions of stochastic parameters are uncertain. This approach is based on the methodology of risk management with Conditional Value-at-Risk risk measure developed in [7]. Although the presented approach has been used to solve one-stage and two-stage stochastic Weapon-Target Assignment problems, it is quite general and can be applied to wide class of problems with risks and uncertainties in distributions. Among the directions of future research we emphasize consideration of a stochastic WTA problem in NLP formulation, where the damage to the targets is maximized while constraining the risk of false target attack.

6.   Appendix. Formal Definition of CVaR

Consider a loss function $L(x, \xi)$ depending on a decision vector x and a stochastic vector ξ, and its cumulative distribution function (c.d.f.) $\Psi(x, \zeta)$:

$$\Psi(x, \zeta) = P[L(x, \xi) \leq \zeta].$$

Then the α-VaR (Value-at-Risk at confidence level α) function $\zeta_\alpha(x)$ corresponding to loss $L(x, \xi)$ is

$$\zeta_\alpha(x) = \min_{\zeta \in \mathcal{R}} \{\Psi(x, \zeta) \geq \alpha\}.$$

Approximately, Conditional Value-at-Risk with confidence level α (α-CVaR) is defined as conditional expectation of losses exceeding the α-VaR level. This definition is correct for continuously distributed loss functions. However, for loss functions with general non-continuous distributions (including discrete distributions) the α-CVaR function $\phi_\alpha(x)$ is defined as the expected value of random variable z_α ([8]):

$$\phi_\alpha(x) = \text{CVaR}_\alpha[L(x, \xi)] = \text{E}[z_\alpha],$$

where c.d.f. $\Psi_{z_\alpha}(x, \zeta)$ of z_α has the form

$$\Psi_{z_\alpha}(x, \zeta) = \begin{cases} 0, & \zeta < \zeta_\alpha(x), \\ [\Psi(x, \zeta) - \alpha]/[1 - \alpha], & \zeta \geq \zeta_\alpha(x). \end{cases}$$

In ([8]), it was shown that α-CVaR can be expressed as a convex combination of α-VaR and conditional expectation of losses strictly exceeding α-VaR:

$$\phi_\alpha(x) = \lambda_\alpha(x)\, \zeta_\alpha(x) + [1 - \lambda_\alpha(x)]\, \phi_\alpha^+(x), \tag{12}$$

where

$$\phi_\alpha^+(x) = E[L(x, \xi) \mid L(x, \xi) > \zeta_\alpha(x)], \tag{13}$$

which is also known as "upper CVaR" or Expected Shortfall, and

$$\lambda_\alpha(x) = [\Psi(x, \zeta_\alpha(x)) - \alpha]/[1 - \alpha], \quad 0 \leq \lambda_\alpha(x) \leq 1.$$

Similar to (13), another percentile risk measure, called "lower CVaR", or CVaR$^-$, can be defined:

$$\phi_\alpha^-(x) = E[L(x, \xi) \mid L(x, \xi) \geq \zeta_\alpha(x)].$$

Then, as it was shown in ([8]), the introduced risk functions satisfy the following inequality:

$$\zeta_\alpha(x) \leq \phi_\alpha^-(x) \leq \phi_\alpha(x) \leq \phi_\alpha^+(x).$$

We also note that in the case when behavior of the stochastic parameter ξ can be represented by a scenario model $\{\xi_s,\ s = 1, ..., S\}$ with equally probable scenarios ($\pi_s = 1/S$), the concept of CVaR has especially simple and transparent interpretation. Namely, if (for a fixed x) the scenarios $\xi_1, ..., \xi_S$ are indexed such that $L(x, \xi_1) \leq \cdots \leq L(x, \xi_S)$, then α-CVaR equals the weighted average of losses for the $\lceil (1 - \alpha)S \rceil$ worst scenarios:

$$\phi_\alpha(x) = \frac{1}{(1 - \alpha)\, S} \left[(s_\alpha - \alpha S)\, L(x, \xi_{s_\alpha}) + \sum_{s=s_\alpha+1}^{S} L(x, \xi_s) \right],$$

where number s_α is such that

$$S - s_\alpha \leq (1 - \alpha)S < S - s_\alpha + 1.$$

References

[1] C. Acerbi and D. Tasche, "On the coherence of expected shortfall", *Working paper, download from www.gloriamundi.org*, 2001.

[2] P. Artzner, F. Delbaen, J.-M. Eber, and D. Heath, "Coherent measures of risk", *Mathematical Finance*, 9: 203–228, 1999.

[3] G. G. denBroeger, R. E. Ellison and L. Emerling, "On optimum target assignments", *Operations Research*, 7: 322–326, 1959.

[4] A. Golodnikov, P. Knopov, P. Pardalos, and S. Uryasev, "Optimization in the space of distribution functions and applications in the bayesian analysis", In S. Uryasev, editor, *Probabilistic Constrained Optimization: Methodology and Applications*, pp. 102-131, Kluwer Academic Publishers, 2000.

[5] A. S. Manne, "A target assignment problem", *Operations Research*, 6: 346–351, 1958.

[6] R. Murphey, "An approximate algorithm for a stochastic weapon target assignment problem", In P. Pardalos, editor, *Approximation and Complexity in Numerical Optimization: Continuous and Discrete Problems*, 1999.

[7] R. T. Rockafellar and S. Uryasev, "Optimization of Conditional Value-at-Risk", *Journal of Risk*, 2: 21–41, 2000.

[8] R. T. Rockafellar and S. Uryasev, "Conditional Value-at-Risk for general loss distributions", *Research Report 2001-5. ISE Dept., Univ. of Florida.*, 2001.

Chapter 10

NETWORK-CENTRIC MFA TRACKING ARCHITECTURES

Suihua Lu
Department of Mathematics
Colorado State University
Fort Collins, CO 80523
lu@math.colostate.edu

Aubrey B. Poore
Numerica, Inc.
PO Box 271246
Fort Collins, CO 80527-1246
and
Department of Mathematics
Colorado State University
Fort Collins, 80523
poore@math.colostate.edu

Abstract   Multiple frame association (MFA) based on multiple dimensional assignment problems has been particularly successful approach to the central problem of data association for tracking multiple objects such as airplanes, ground vehicles, and missiles using multiple sensors on multiple platforms. These optimization methods have been developed for a centralized communication and tracking architecture, but such an architecture suffers from communication overloading, single-point failure, and data latency over the network of platforms. Thus, one must turn to decentralized architectures with the objectives being to affordably preserve the quality of MFA for centralized architecture while managing communication loading and achieving a consistent air picture across the multiple tracking platforms. Such objectives present some of the most challenging tracking research problems. Thus, the goal of this work is to develop and test some initial architectures for network centric (decentralized) multiple frame association (Network MFA), namely, MFA Centralized, Replicated MFA Centralized, Network MFA on Local Data and Network Tracks, and Network MFA on All

S. Butenko et al. (eds.), Cooperative Control: Models, Applications and Algorithms, 187-213.

Data and Network Tracks, are presented. Based on well-established performance
metrics, the relative merits of the architectures are also discussed.

Keywords: Multiple Frame Association (MFA), Data Associate Problem, MFA Centralized,
Replicated MFA Centralized, Network MFA on Local Data and Network Tracks,
Network MFA on All Data and Network Tracks

Introduction

Multiple target tracking methods divide into two broad classes, namely single frame and multiple frame methods. The single frame methods include nearest neighbor, global nearest neighbor and JPDA (joint probabilistic data association). The most successful of the multiple frame association methods are multiple hypothesis tracking (MHT) [4] and multiple frame assignment (MFA) [15, 16, 3], which is formulated as a multidimensional assignment problem. The performance advantage of the multiple frame methods over the single frame methods follows from the ability to hold difficult decisions in abeyance until more information is available and the opportunity to change past decisions to improve current decisions. In dense tracking environments the performance improvements of multiple frame methods over single frame methods is very significant, making it the preferred solution for many tracking problems.

The application of multiple frame tracking methods must consider an architecture in which the sensors are distributed across multiple platforms. Such geometric and sensor diversity has the potential to significantly enhance tracking and discrimination accuracy. A centralized architecture in which all measurements are sent to one location and processed with tracks being transmitted back to the different platforms is a simple one that is probably optimal in that it is capable of producing the best track quality (e.g., purity and accuracy) and a consistent air picture. The centralized tracker is, however, unacceptable for several reasons, notably the communication overloads, single-point-failure, and latency. Thus, one must turn to a distributed architecture for both estimation/fusion [5] and data association.

Thus, an important problem in the development of an advanced network-centric (decentralized) MFA that preserves the strengths of the MFA centralized architecture while managing communication loading and achieving a consistent air picture. It is important to note that the network centric architecture of CEC and JCTN provides a consistent or single integrated air picture (SIAP) across multiple platforms and limits the communications loads to within a practical limit [13]. They were, however, designed with single frame data association in mind. The multiple frame data association approaches of MFA offer substantially improved tracking and discrimination performance compared to current CEC and JCTN approaches while maintaining consistent threat pictures across platforms and limited communications loads characteristic of CEC and JCTN. Multiple frame data association methods are better able to handle closely spaced objects caused by dense threats with associated objects, false signals and clutter, radar multi-path, residual sensor registration biases, counter-measures, unresolved closely spaced objects, and data from 2-D sensors. Multiple frame data association methods offer improved performance in, for example, improved accuracy of the target tracks, discriminants, and covariance consistency and also reduced track switches, track breaks, and missed targets. What is more, local sensor corruption can be moderated in a network tracker that uses multiple frame data association. Metrics of concern for consistency of the threat pictures across platforms include common track numbers and also similar target state estimates, error covariance matrices and target discrimination decisions from platform to platform.

Development of a high performance near-optimal network-centric architecture (compared to centralized tracking) is a very difficult problem. It requires innovative multiple frame tracking architectures, new ideas about what measurement information contributes to near optimal performance and advances in establishing and maintaining a common track picture across multiple platforms. Many concepts for selecting the measurement information have been articulated; some (e.g., tracklets) have been analytically developed, but their quantitative performance is unknown and the inherent difficulty of developing a high performance near-optimal solution remains.

The essence of the current work can be summarized as follows. First, we assume that the communication network is fully connected. Then, one starts with a MFA centralized architecture as a baseline. To distribute this, the next architecture is that of replicated MFA centralized architecture in which a "centralized tracker" is placed on each platform and measurements from all sensors on the network are transmitted to the tracker on each platform. Track states are not transmitted back to the other platforms. To this architecture one next adds a rule to assist with the consistency of tracks across the different platforms. The rule that is used herein is similar to that used in the CEC and JCTN [13], namely that the tracker on each platform is in charge of assigning its measurements to the global set of tracks, i.e., network tracks, maintained on each platform. This assignment is then broadcast to the other platforms which are required to use this assignment. Given this rule, one next distinguishes two architectures. The Network MFA on All Data and Network Tracks in which both remote and local data are used in the local association problem and the Network MFA on Local Data and Network Tracks in which only local data is used in the association process. Thus, a consistent set of tracks are maintained some N frames back. To demonstrate the performances of these architectures, several metrics chosen form the JCTN Benchmark are utilized. (A copy of these metrics can be found at at the web site http://seal-www.gtri.gatech.edu/jctn/.)

While we do not address the problem of communication loading, we do measure it.

There are a number of additional issues and optimization problems that need to be addressed in the design of a multi-platform system [4]. These include (1) distributed data association and estimation (addressed herein); (2) consistent or single integrated air picture (addressed herein); (3) management of communication loading; (4) sensor biases as well as location and registration errors (sometimes called gridlock); (5) pedigree problems in which information may be counted more that once; and, (6) out-of-order, latent, and missing data due to both sensor and communication problems (addressed in the thesis by Lu [11]). These topics as well as others are discussed in the article by Moore and Blair [13], the books by Blackman and Popoli [4] and Blair and Bar-Shalom [3], and the thesis of Lu [11].

This chapter is organized as follows. The four architectures are presented in Section 1, while Section 2 explains some of the benefits found from extensive computational experience with the architectures [11]. Future directions are addressed in Section 3.

1.   Tracking and Communication Architecture

The four tracking architectures considered in this work are MFA Centralized, Replicated MFA Centralized, Network MFA on All Data and Network Tracks, and Network MFA on Local Data and Network Tracks. Each architecture is explained in the subsequent subsections.

1.1.   MFA Centralized Architecture

The basic centralized tracking architecture is explained in many references such as the book by Blackman and Popoli [4], the lecture notes by Drummond [8], and the article by Moore and Blair [13]. It is reviewed here for completeness. A centralized architecture (Figure 10.1) is one in which each platform sends its sensor measurements to a central processing unit where the tracking is performed. The tracker is called a network (i.e., global or composite) tracker. A key component of MFA tracking is the concept of a proper frame of data in which any particular target is seen at most once. However one defines a frame of data, the essential need is that of being able to determining how to score missed detections. (If a target is in the field of view of the sensor and should have been seen, it is scored as a missed detection. If the target is out of the field of view of the sensor, it should not have been seen and is not scored as a missed detection.) As measurements are received at the central processing unit, they are placed into (proper) frames of data which are then processed. This architecture is conceptually the simplest and will ideally produce the most accurate data associations and tracking. Furthermore, the tracks

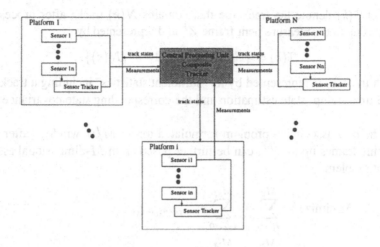

Figure 10.1. MFA Centralized Architecture

(global tracks) are broadcast back to all the platforms and in the absence of significant communication delays can produce a single integrated air picture (SIAP). The centralized architecture provides a baseline architecture against which other architectures can be compared. The centralized architecture does, however, suffer from the single-point-failure problem, communication over-loading due to track states and covariances being sent back to remote platforms, and system delays.

The multiple frame data association method used in this work for the centralized, as well as other architectures, is based on multi-dimensional assignment problems. In our approach to the MFA method, a (M/N)-double pane sliding window is used, where the M refers to the number of frames in the sliding window. All M frames within the window participate in track initiation, but only the most recent $N < M$ frames participate in track continuation. The reason for this framework is that it includes most methods commonly used in tracking but also allows considerably more flexibility in adjusting the windowing and assignment complexity using different length windows. What is more, this approach combines track initiation and maintenance in the same assignment formulation of the data association problem.

To explain this moving window further, let $Z(k)$ denote a frame of M_k reports $\{z_{i_k}^k\}_{i_k=1}^{M_k}$ and let Z^N denote the cumulative data set of N such sets:

$$Z(k) = \{z_{i_k}^k\}_{i_k=1}^{M_k} \quad \text{and} \quad Z^N = \{Z(1), \ldots, Z(N)\}. \tag{1}$$

Let $\mathcal{T}(k)$ denote the track file that contains $N(k)$ tracks after processing frames up to the measurement frame Z^k and represented by

$$\mathcal{T}(k) = \{T_j(k) : \quad j = 1, \cdots, N(k)\}. \tag{2}$$

Each track T_j is represented by its "sufficient statistics" including a track update time stamp, state estimation and the corresponding state covariance matrix.

The data association problem formulated for an M/N window, after processing frames up to Z^M, can be further posed as an M-dimensional assignment problem.

$$\text{Minimize} \quad \sum_{i_1=0}^{M_1} \cdots \sum_{i_M=0}^{M_M} c_{i_1 \cdots i_M} z_{i_1 \cdots i_M}$$

$$\text{Subject To:} \quad \sum_{i_2=0}^{M_2} \cdots \sum_{i_M=0}^{M_M} z_{i_1 \cdots i_M} = 1, \quad i_1 = 1, \ldots, M_1,$$

$$\sum_{i_1=0}^{M_1} \cdots \sum_{i_{k-1}=0}^{M_{k-1}} \sum_{i_{k+1}=0}^{M_{k+1}} \cdots \sum_{i_M=0}^{M_M} z_{i_1 \cdots i_M} = 1, \tag{3}$$

$$\text{for} \quad i_k = 1, \ldots, M_k \text{ and } k = 2, \ldots, M-1,$$

$$\sum_{i_1=0}^{M_1} \cdots \sum_{i_{M-1}=0}^{M_{M-1}} z_{i_1 \cdots i_M} = 1, \quad i_M = 1, \ldots, M_M,$$

$$z_{i_1 \cdots i_M} \in \{0,1\} \text{ for all } i_1, \ldots, i_M.$$

The notation for this multi-dimensional assignment problem is somewhat simplified in the following way. If the tracks with firm (hard) decisions is numbered k as in the above, then the first frame in the assignment problem is actually $k - M + N + 1$ and the last frame M is $k + N$. The track is indexed with the $T_j \in \mathcal{T}(k)$", since equation (5) uses it.

The score $c_{i_1 \cdots i_M} = -\ln L_{i_1 \cdots i_M}$ where $L_{i_1 \cdots i_M}$ is a likelihood ratio defined as:

$$L_{i_1 \cdots i_M} = \frac{\text{the probability that } z_{i_1}^1, \cdots, z_{i_M}^M}{\text{the probability that } z_{i_1}^1, \cdots, z_{i_M}^M \text{ are false alarms}}, \tag{4}$$

and

$$L_{0 \cdots 0 - T_j i_{M-N+1} \cdots i_M} = \frac{\text{the probability that } z_{k_{M-N+1}}^{M-N+1}, \cdots, z_{i_M}^M}{\text{emanates from the track } T_j}{\text{the probability that track } T_j \text{ terminates and}}. \tag{5}$$
$$z_{i_{M-N+1}}^{M-N+1}, \cdots, z_{i_M}^M \text{ are false alarms}$$

If one uses filtering with no smoothing in the estimation of the state across the M frames of data, then likelihood ratios and the scores [14] can be written recursively as:

$$L_{i_1 i_2 \cdots i_M} = \prod_{k=1}^{M} L_{i_k} = L_{i_1 i_2 \cdots i_{M-1}} L_{i_M} \tag{6}$$

$$c_{i_1 i_2 \cdots i_M} = \sum_{k=1}^{M} c_{i_k} = c_{i_1 i_2 \cdots i_{M-1}} + c_{i_M}. \tag{7}$$

It is important to note in this representation that the likelihood L_{i_M} and scores c_{i_M} are history dependent, i.e., depend on the particular sequence $\{i_1 i_2 \cdots i_{M-1}\}$. Otherwise, the multidimensional assignment problem would decompose into a sequence of two-dimensional ones. These likelihood ratios are developed in the work of Poore and Lu [14] and references therein and contain the sensor characteristics, probabilistic characterizations of target detection, false alarms, new and terminating targets.

Next, one develops a probabilistic graph that we call a *track tree* to represent all feasible combinations of existing tracks and measurements. Each branch in the tree has a likelihood ratio score associated with it. (The correct design of this *track tree* is fundamentally important in any tracking system in that it must be possible to enter and retrieve data as well as perform searches efficiently.) The data association problem which is formulated as an N dimensional assignment problem is solved using Lagrange relaxation method explained in [15]. Based on the solutions to the data association problem, the decisions between the track file $\mathcal{T}(k)$ and frame $Z(k+1)$ are fixed. Some tracks may be deleted; new tracks that initiate based on the observations in the first $(M - N + 1)$ frames that have not been associated with any of the existing tracks are added into the track file. Thus, the *track tree* is pruned and shifted, and the track file is updated to be $\mathcal{T}(k+1)$.

A 4/2 window is shown in Figure 10.2, where 4 frames of observations can be used to initiate new tracks. The existing tracks extend into the 3rd and 4th frames (extension window).

1.2.   Replicated MFA Centralized Architecture

Perhaps one of the simplest decentralized architectures is that of placing a "centralized tracker" on each platform. Each platform may or may not have local trackers, but measurements are assumed to be available from all local sensors. Thus, the centralized tracker processes all the data from all the platforms, but does not transmit track states back to the other platforms. As in real scenarios, all the trackers get local observations from the local sensors directly

Track initiation Window   **Track extension Window**

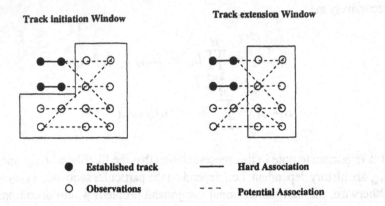

● **Established track**   ——— **Hard Association**
○ **Observations**   - - - **Potential Association**

Figure 10.2. Double Pane Sliding Window

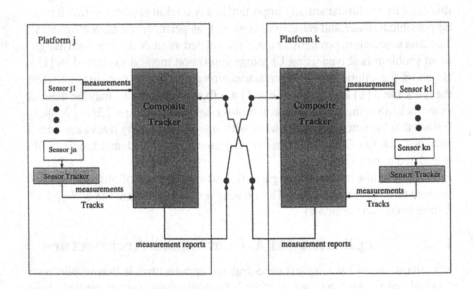

Figure 10.3. Replicated MFA Centralized Architecture

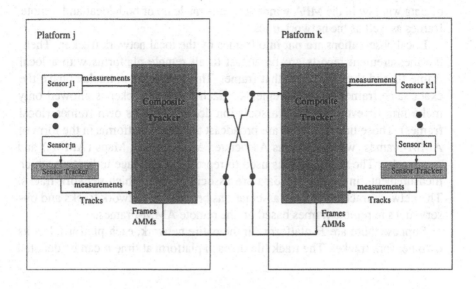

Figure 10.4. Network MFA on All Data and Network Tracks

and remote observations from remote platforms through the communication network.

This architecture removes the problem of single-point-failure. If one or more network trackers fail to perform correctly, or certain communication links are cut down, the network trackers on other platforms are still capable of the tracking. However, the network tracker may not have access to all remote measurement reports in which case the tracking performance can degrade gracefully.

However, due to communication delays in the network, the order of observations arriving at different platforms varies. Each network tracker is making its own tracking decisions based on the data it receives, regardless of the decisions of other platforms. Therefore, SIAP may not be achieved across the network.

1.3. Network MFA on All Data and Network Tracks

The Network MFA on All Data and Network Tracks evolves from the Replicated MFA Centralized architecture by adding the rule that each platform is in charge of assigning its own data to the network tracks. Figure 10.4 gives a further explanation of this architecture. Note that there is one network tracker on each platform that has access to all the data from all platforms. The frames of data window in the MFA window is thus made up of both local and remote frames as well as the network tracks.

Local observations are put into frames by the local network tracker. Then, the measurement reports are broadcast to all remote platforms with a local frame ID and the position in that frame. Thus, all network trackers have the exact same frames of measurements. Each network tracker is allowed only make firm (irrevocable) data association decisions on its own frames (local frames). Those firm decisions are broadcast to remote platforms in the form of AMM-Frames, which contains Associated Measurement Maps (AMMs) and new tracks. The term AMM is used to represent a message indicating which measurements in a frame of data are associated with which network tracks. The network tracker fixes the associations between its network tracks and observations in remote frames based on the remote AMM-Frames.

Suppose there are P platforms in the entire network, each platform has its own network tracker. The track file on each platform at time n can be denoted as

$$\mathcal{T}^p(n) = \{T_j^p(n) : \quad j = 1, \cdots, N^p(n)\} \quad \text{for} \quad p = 1, \cdots, P. \quad (8)$$

Let $Z^p(n)$ denote the nth frame that comes from platform p. Then the frames up to time n can be denoted as

$$Z^n = \{Z^{p_1}(1), \cdots, Z^{p_i}(i), \cdots, Z^{p_n}(n)\}.$$

If there are no communication and processing delays, then all the frames on different platforms are in exactly the same order. It is further assumed that all the trackers are well synchronized. Figure 10.5 shows three platforms (i, j, and k) in the network; each has a track extension window of size 2 ($N = 2$). For simplicity, only track extension is discussed in this paper.

Suppose at a certain time n (or t_n), the network track files on all platforms ($\mathcal{T}^i(n)$, $\mathcal{T}^j(n)$ and $\mathcal{T}^k(n)$) are the same. The network tracks are represented as black dots in Figure 10.5. The first frame in the extension window ($Z^i(n+1)$) belongs to platform i. Another frame $Z^j(n+2)$ is then ready to be processed. The network tracker on platform i sets up a data association problem based on the network tracks and frames in the window (Eqn. 9).

$$\text{Minimize} \quad \sum_{j=0}^{N^i(n)} \sum_{i_{n+1}=0}^{M_{n+1}} \sum_{i_{n+2}=0}^{M_{n+2}} c_{-ji_{n+1}i_{n+2}} z_{-ji_{n+1}i_{n+2}}$$

$$\text{Subject To:} \quad \sum_{i_{n+1}=0}^{M_{n+1}} \sum_{i_{n+2}=0}^{M_{n+2}} z_{-ji_{n+1}i_{n+2}} = 1, \quad j = 1, \ldots, N^i(n),$$

$$\sum_{j=0}^{N^i(n)} \sum_{i_{n+2}=0}^{M_{n+2}} z_{-ji_{n+1}i_{n+2}} = 1 \quad i_{n+1} = 1, \ldots, M_{n+1}, \tag{9}$$

$$\sum_{j=0}^{N^i(n)} \sum_{i_{n+1}=0}^{M_{n+1}} z_{-ji_{n+1}i_{n+2}} = 1 \quad i_{n+2} = 1, \ldots, M_{n+2},$$

$$z_{-ji_{n+1}i_{n+2}} \in \{0,1\}.$$

In equation 9, $z_{-ji_{n+1}i_{n+2}} = 1$ denotes that measurements $z_{i_{n+1}}^{n+1}$ and $z_{i_{n+2}}^{n+2}$ are associated with track $T_j^i(n) \in \mathcal{T}^i(n)$. $c_{-ji_{n+1}i_{n+2}}$ is the corresponding likelihood ratio score.

After solving the data association problem, it makes firm (irrevocable) decisions between the network tracks ($\mathcal{T}^i(n)$) and its local frame ($Z^i(n+1)$). The network tracker on platform i then updates its track file to be $\mathcal{T}^i(n+1)$. AMMs are broadcast as an AMM-Frame to platform j and k in the form of

$$\{(T_j^i(n), i_{n+1}) : j = 1, \cdots, N^i(n) \text{ and } i_{n+1} = 1 \cdots, M_{n+1}\}.$$

For the network trackers on platforms j and k, the first frame in their extension windows is a remote frame (from platform i) and thus these platforms wait for the AMM-Frame from platform i. After the AMM-Frame is received, the trackers use them to update the network tracks. For instance, the network tracker on platform k associate measurement $z_{i_{n+1}}^{n+1}$ to its corresponding network track $T_j^k(n)$ according to the AMM pair $(T_j^i(n), i_{n+1})$. The remote frame ($Z^i(n+1)$) is then removed from the sliding window.

Referring now to the third and fourth row in Figure 10.5, another frame (from platform k) is ready. Now the first frame in the extension window belongs to platform j, so the network tracker on platform j sets up the data association problem, solves it, fixes the decisions in that frame, broadcasts the AMM-Frame to platform i and k, and then moves the window forward. The network trackers on platform i and k wait for the AMM-Frame to update their network tracks.

Rows five and six can be explained similarly. Thus, Figure 10.5 illustrates three steps of the process. After those three steps, the network tracks on all platforms are still the same because they have the same observation history.

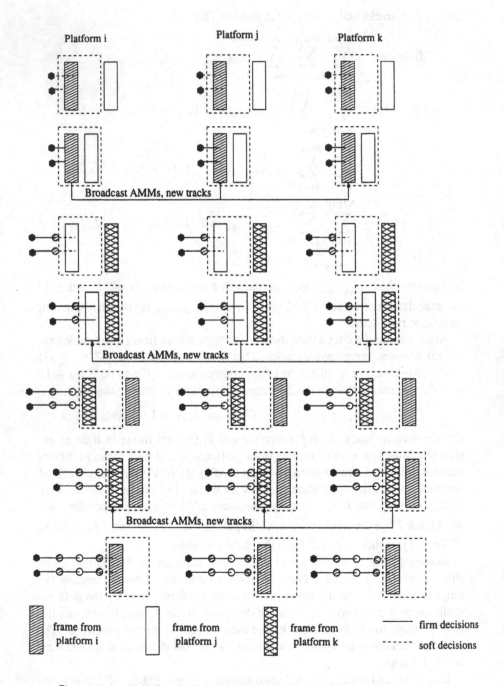

Figure 10.5.   Ideal Case for Network MFA on All Data and Network Tracks

In real systems, random delays in the transmission network are inevitable. Furthermore, processing time needed to put a remote frame together is different for each tracker. The communication delays and processing delays result in different order of frames on different platforms. Out-of-order data can create potentially serious problems such as conflicts in association and deadlock in processing the windows. The problems are illustrated here in this paper; the proposed solutions are addressed in detail in the thesis [11].

For the deadlock problem, suppose there are only two platforms in the network, denoted as platforms A and B. At time n, the order of the track lists and frames in the track extension window on platform A is $(\mathcal{T}^A(n), Z^A(n + 1), Z^B(n + 3), Z^B(n + 2), Z^A(n + 4))$, while the order on platform B is $(\mathcal{T}^B(n), Z^B(n+2), Z^A(n+4), Z^B(n+3), Z^A(n+1))$. According to the algorithm explained above, both platforms fix the decisions between $(\mathcal{T}^A(n), Z^A(n+ 1))$ and $(\mathcal{T}^B(n), Z^B(n + 2))$ and broadcast them as AMM-Frames. However, then they both going to wait for the other platforms for AMM-Frames to make the association between $(\mathcal{T}^A(n+1), Z^B(n+3))$ and $(\mathcal{T}^B(n+1), Z^A(n+4))$, and the deadlock problem occurs.

For conflicts in association, due to delays, the network track files on different platforms may be different and the orders of frames in the track extension window differs as well. Thus, whenever a network tracker receives a remote AMM pair, the corresponding association may not be regarded as feasible on the local platform.

1.4. Network MFA on Local Data and Network Tracks

The architecture Network MFA on Local Data and Network Tracks is similar to the Network MFA on All Data and Network Tracks discussed in the previous section except that only local data is used in the sliding window. Figure 10.6 illustrates this architecture. The network tracker on each platform only has access to its local data. On each platform, there are multiple sensors, some of which may have their own sensor trackers. It is assumed that measurements are available from all sensors, even those with sensor trackers. The data association decisions are based on scores of network tracks extending to the local frames in the window. Data association assignments are then broadcast to all platforms in the network in the form of AMR-Frames which contain Associated Measurement Reports(AMRs) and local new tracks. The term AMR is used to represent a message indicating a measurement and a network track ID to which it has been assigned by the own ship platform. AMR-Frames are used to update the track database on the remote platforms.

The algorithm of the Network MFA on Local Data and Network Tracks in the ideal case can be illustrated in Figure 10.7. There are three platforms in the

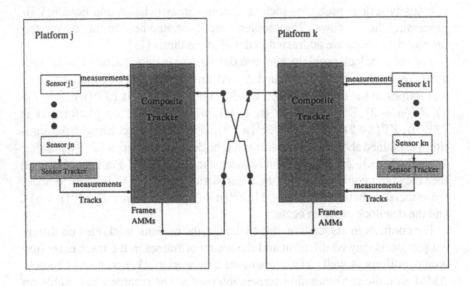

Figure 10.6.   Network MFA on Local Data and Network Tracks

network ($P = 3$). A track extension window of size 2 is chosen for all three platforms. It is further assumed that all network trackers are synchronized. For platform p, at time n, the network track file is denoted as $\mathcal{T}^p(n)$, and the frames in the track extension window is $Z^p(m \times P + p)$, where $m \times P < n < (m+1) \times P$.

Referring to the first and second rows in Figure 10.7, the window on each platform contains only local frames, and there is only one local frame in each of the track extension window. We start with the same set of network tracks on platform $\mathcal{T}^1(n)$, $\mathcal{T}^2(n)$ and $\mathcal{T}^3(n)$, which are represented by dark black dots in front of the track extension windows. The time n is chosen to be $n = m \times p$ for some integer m. The frame in the extension window is represented by $Z^1(n+1)$, $Z^2(n+2)$ and $Z^3(n+3)$ respectively. Suppose another local frame on platform 1 ($Z^1(n+4)$) is ready to be processed. Then the frame is inserted into the extension window and the *track tree* is extended to the frame.

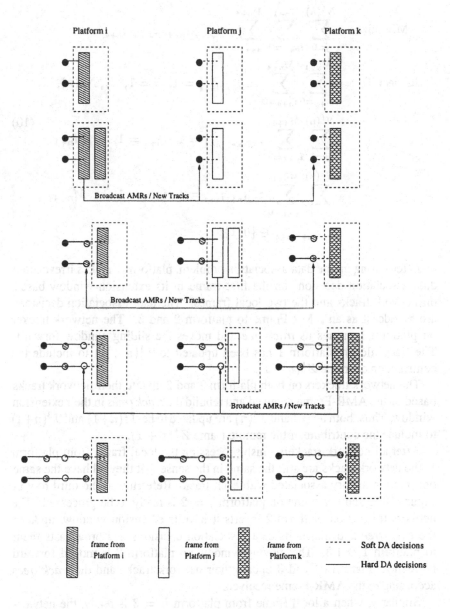

Figure 10.7. Ideal Case for Network MFA on Local Data and Network Tracks

The data association problem takes the following form:

$$\text{Minimize} \quad \sum_{j=0}^{N^1(n)} \sum_{i_{n+1}=0}^{M_{n+1}} \sum_{i_{n+4}=0}^{M_{n+4}} c_{-ji_{n+1}i_{n+4}} z_{-ji_{n+1}i_{n+4}}$$

$$\text{Subject To}: \quad \sum_{i_{n+1}=0}^{M_{n+1}} \sum_{i_{n+4}=0}^{M_{n+4}} z_{-ji_{n+1}i_{n+4}} = 1, \quad j = 1, \ldots, N^1(n),$$

$$\sum_{j=0}^{N^1(n)} \sum_{i_{n+4}=0}^{M_{n+4}} z_{-ji_{n+1}i_{n+4}} = 1 \quad i_{n+1} = 1, \ldots, M_{n+1}. \tag{10}$$

$$\sum_{j=0}^{N^1(n)} \sum_{i_{n+1}=0}^{M_{n+1}} z_{-ji_{n+1}i_{n+4}} = 1 \quad i_{n+4} = 1, \ldots, M_{n+4},$$

$$z_{-ji_{n+1}i_{n+4}} \in \{0,1\}.$$

After setting up the data association problem, platform 1 makes irrevocable data association decisions on the first frame in its extension window based on network tracks and the two local frames. The data association decisions are broadcast as an AMR-Frame to platform 2 and 3. The network tracker on platform 1 prunes its *track tree* and moves the sliding window forward. The track file on platform 1 has been updated to $\mathcal{T}^1(n+1)$ to include the contribution of frame $Z^1(n+1)$.

The network trackers on both platform 2 and 3 update their network tracks based on the AMR-Frame received and rebuild the *track trees* in their extension window. Thus, both $\mathcal{T}^2(n)$ and $\mathcal{T}^3(n)$ are updated to be $\mathcal{T}^2(n+1)$ and $\mathcal{T}^3(n+1)$ to include the contribute of the remote frame $Z^1(n+1)$.

After all network trackers finish processing the local frame from platform 1, the network tracks are still the same in the sense that they all have the same observation history associated with each track. Referring to the third row in Figure 10.7, another frame on platform $j = 2$ is ready to be processed. The network tracker on platform 2 inserts it into its extension window, updates the *track tree* and makes its data association decisions and broadcasts them to platform 1 and 3. The moving window on platform 2 is moved forward thereafter. Platform 1 and 3 update their network tracks and the *track trees* according to the AMR-Frame received.

Similarly, when a local frame from platform $k = 3$ is ready, the network tracker on platform 3 makes its firm data association decisions and broadcasts them to platform 1 and 2 as illustrated in the fourth row in Figure 10.7.

One can conclude that, ideally, when there are no communication delay and processing delays and the network trackers are well synchronized, the network tracks are exactly the same across the platforms in the sense that they have the same observation history.

The data association decisions on each platform are based on local frames in the extension window and network tracks which includes local and remote information. The remote information is included in the network tracks by processing the AMR-Frames. However, due to the contributions of remote AMRs and new tracks, the *track tree* needs to be updated after processing a remote AMR-Frame.

In a real system, it is impossible for all the network trackers to be well synchronized. However, the network tracker is event-driven, which means it is capable of processing events in the order they become ready and the algorithm works just the same [11].

2. Simulation Results: Comparisons and Discussions

This section contains a comparison of the four different tracking architectures: MFA Centralized, Replicated MFA Centralized, Network MFA on Local Data and Network Tracks, and Network MFA on All Data and Network Tracks.

The scenario we use contains four fighter aircraft and is 600 seconds in length. Two of the fighters start flying in close formation at around 120 seconds and all four fighters merge at about 460 seconds as illustrated in 10.8 The four surveillance platforms are ships with each having an S-band phased array radar and a UHF rotating radar. Figure 10.8 illustrates the target and platform trajectories and depicts the merging of the fighters. This problem is particularly difficult due to the closeness of flight as the fighters merge.

The network trackers in all architectures apply the MFA algorithm. A double pane sliding window of size $6/2$ is used along with a two-model interacting multiple model (IMM) filter based on two nearly constant velocity models with a high noise and low noise level. At least four observations are required to start a tracking filter when the tracker is turned on; however, thereafter, at least six observations are required to initiate a new track. Finally, a refiltering window of length 20 is used to avoid negative time updates.

2.1. Simulation Results Using a Perfect Communication Network

In this section, all simulations assume a perfect communication network in that there is no communication delay (when sending information from one platform to the next) and no lost data. However, the order of the observations arriving on different platforms may still be different. For observations taken at the same time, the local observations always arrive on local platforms faster than the remote ones. For each architecture, five Monte Carlo runs are per-

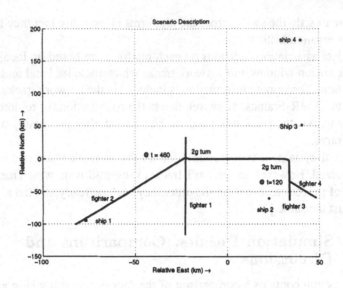

Figure 10.8.   Scenario Description

formed. For each metrics scoring request, the most recent "soft" decisions are used to give the state estimates.

For the case of perfect communication with no delays and dropped data, all four architectures have a perfect or near perfect metrics score on a large class of metrics that attempt to measure track continuity and a one-to-one match of computed tracks to truth, namely, spurious track mean ratio, redundant track mean ratio, track breakage, and track completeness. This is quite remarkable given that the local trackers and data arising from the local trackers are often very imperfect in these metrics. The reason for the success of the network tracker is that MFA at the network level can correct many if not all of the errors in the local tracker due to the more global information available at the network level.

The accuracy and consistency of all four architectures are reasonably close to each other with Network MFA on Local Data and Network Tracks being slightly worse than the other three architectures which use all data. Figure 10.9 shows the network track position accuracy differences between the centralized architecture and the Network MFA on All Data and Network Tracks architecture are negligible. Often it is the case that the local sensor tracker on a platform may produce erroneous tracks. Using local data only in the MFA may not be able to correct the problems as well as using all data in the MFA.

With respect to a consistent air picture, the centralized architecture naturally has an almost perfect cross-platform commonality history as measurement by non-common network tracker numbers and network state estimate differences.

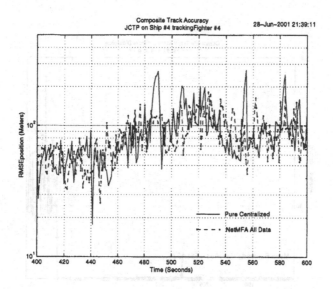

Figure 10.9. Comparison between MFA Centralized and Network MFA on All Data

Imperfections are due primarily to processing delays. The Replicated MFA Centralized architecture puts a centralized tracker on each platform and no track matching mechanism is applied. Each network tracker makes its own tracking decisions regardless of the other trackers in the network. Thus, the performance in cross-platform commonality history degrades somewhat. Figure 10.10 illustrates the network track state estimate difference for the Replicated MFA Centralized architecture.

The use of the rule that each platform is in charge of assigning its own measurements to the network tracks leads to considerable improvements found in the architectures of Network MFA on Local Data and Network MFA on All Data. We might note that the tracking accuracy is slightly better when using Network MFA on All Data and Network Tracks than with Network MFA on Local Data and Network Tracks. The Network MFA on All Data and Network Tracks architecture has an almost perfect cross-platform commonality history except at a discrete set of metrics scoring times due to processing delays. Figure 10.11 shows the network track state estimate difference for the Network MFA on All data and Network Tracks architecture.

Network MFA on Local Data and Network Tracks is worse in the cross-platform commonality history, primarily due to the use of soft data association decisions, but improves in these metrics if hard decisions are used. (This fact follows from the fact that there is a common track file N frames back.) Figure 10.12 shows the network track state estimate difference for the Network MFA on All Data and Network Tracks architecture.

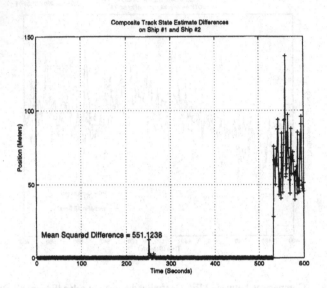

Figure 10.10.   Replicated MFA Centralized

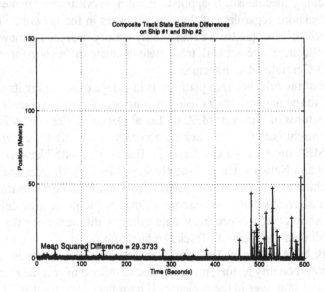

Figure 10.11.   Network MFA on All Data

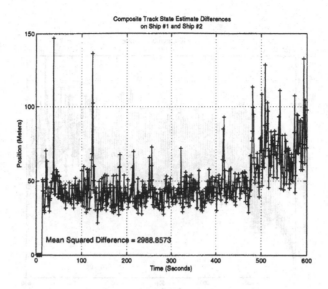

Figure 10.12. Network MFA on Local Data

With respect to communication loading, the MFA Centralized architecture is about three times that of the other architectures as illustrated in Figure 10.13, primarily due to the transmission of track states back to all platforms after the measurement update. The communication loading for Network MFA on All Data (Figure 10.14) is always a little heavier than the Replicated MFA Centralized architecture, which is caused by the extra messages sent in the network (end frame messages, AMMs, new tracks and end AMM Frame messages). In dense clusters, if the communication bandwidth is limited, the Network MFA on Local Data and Network Tracks may be the preferable architecture.

With respect to processor loadings, the network architectures are computationally more expensive than the centralized one. In the centralized architecture, there is only one network tracker in the network, whereas in all network architectures, each platform has its own network tracker. Also, the Network MFA on All Data and Network MFA on Local Data require ten to twenty percent more computing power than the Replicated MFA Centralized architecture.

2.2. Simulation Results Using an Imperfect Communication Network

When one introduces a moderate amount of communication delays and dropped data, the metrics that measure track continuity such as spurious track mean ratio, redundant track ratios, track breakage, track breakage remain nearly perfect; however, most other metrics degenerate somewhat.

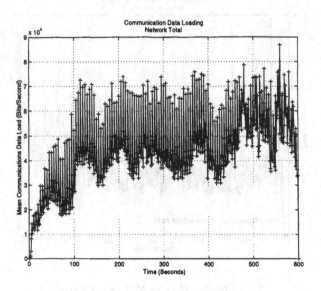

Figure 10.13.   MFA Centralized Architecture

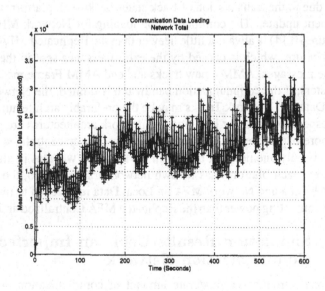

Figure 10.14.   Network MFA on All Data

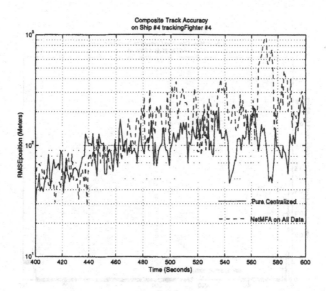

Figure 10.15.   Comparison between MFA Centralized and Network MFA on All Data

The MFA centralized tracking architecture is the best overall architecture except for communication loading (and the single-point-failure problem).

Figure 10.15 shows the network track position accuracy using MFA Centralized and Network MFA on All Data and Network Tracks tracking architecture. Compared with Figure 10.9, the tracking accuracy degenerate when delays and dropped data are present.

The delays cause non-common network track numbers for the Replicated MFA Centralized to degenerate significantly and this is primarily due to the lack of any mechanism to coordinate tracks or track numbers from platform to platform. All other metrics perform well for this architecture. In both Network MFA on Local Data and Network MFA on All Data, specific methods were developed to ensure cross-platform commonality history metrics. These include track numbering schemes [11] and track correlation for initiating tracks as well as data association rules that yield a consistent set of tracks N frames back, i.e., each platform is in charge of assigning its own measurements to the network tracks.

Figure 10.16 shows the non-common network track numbers for Network MFA on All Data and Network Tracks architecture, while Figure 10.17 shows the network track state estimate difference. Both cross-platform commonality history metrics degenerate a little compared to the perfect communication case.

The architecture Network MFA on Local data and Network Tracks demands the lowest communication bandwidth. However, the network trackers more

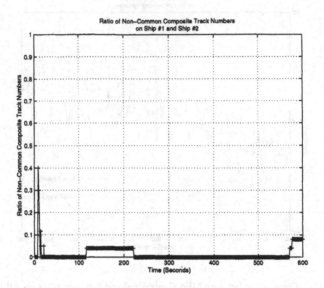

Figure 10.16.   Network MFA on All Data and Network Tracks

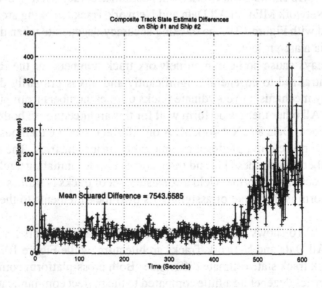

Figure 10.17.   Network MFA on All Data and Network Tracks

often initiate new tracks based on local data only, which may cause bigger estimation errors in the initialization step, and tracks may not be initiated sufficiently fast.

Network MFA on All Data and Network Tracks is more expensive in the sense of communication loadings and processing loadings; however, it generally provides a more consistent set of tracks across the network.

3. Future Directions of Research and Recommendations

This work combined with the computational results [11] demonstrate the exceptional performance of decentralized Network MFA in achieving a performance superior to single frame processing and close to that of a centralized processing across a network of platforms. These decentralized architectures are particularly superior in achieving a single integrated air picture (SIAP). Further work using dynamic data structures for the moving window can significantly moderate the problems of latent and missing data. Here are some future areas of investigation to improve the Network MFA architectures and algorithms.

Adaptive Push-Pull Scheme.   The network tracking system is a time-varying system. Instead of broadcasting all messages to remote platforms, the network tracker can push the messages that are of interest to a particular tracker. On the other hand, the tracker can request information from remote trackers based on the tracking performance.

Data Compression: Tracklets.   Now, information passed around the network are observations, AMRs, or track states. In the future, tracklets [12, 9, 8] can be used as a data compression technique for non-maneuvering targets. A tracklet is a track computed so that its errors are not cross-correlated with the errors of any other data in the system for the same apparent target. A tracklet is like a large (typically 6−dimensional) measurement and is equivalent to a track based on only the most recent measurements, typically 6 to 30 measurements.

Hybrid Architectures.   The network architectures proposed are Replicated MFA Centralized, Network MFA on Local Data and Network Tracks, and Network MFA on All Data and Network Tracks. In the future, instead of using one architecture in the network, the tracking system can be a hybrid system that contains some combinations of the architectures. For example, one might use Network MFA on All Data and Network Tracks for platforms with overlapping coverage and Network MFA on Local Data and Network Tracks when platforms do not have overlapping coverage.

Sparse Networks. In this work, we have assumed the communication network to be fully connected as in the CEC and JCTN networks; however, other networks, which we call sparse, are not and one must deal with routing information efficiently. Also, one must also deal with the pedigree problems that can occur when the information source is not know and is possibly used more than once.

Biases and Registration Errors. Any multi-platform or multi-sensor single platform tracking system must have a method for estimating and removing sensor biases and registration errors [4]. This problem is intimately tied to the data association problem in that these biases cannot be determined without knowing the association of measurements to tracks which cannot be determined properly without knowing the biases. If uncorrected, the registration errors can lead to large tracking errors and potentially to the formation of multiple tracks (ghosts) on the same target. Once estimated, the biases will be used to transform the measurement data.

Acknowledgments

This work was supported in part by the Office of Naval Research under Contract Number N00014-00-C-0264, by the Air Force Office of Scientific Research under Grant Number F49620-00-1-0108, and by the Boeing Corporation.

References

[1] Y. Bar-Shalom, *Multitarget-Multisensor Tracking: Advanced Applications*, Artech House, Norwood, MA, 1990.

[2] Y. Bar-Shalom and X.R. Li, *Multitarget-Mulitsensor Tracking Principles and Techniques*. YBS Publishing, Storrs, CT, 1995.

[3] Y. Bar-Shalom and W. D. Blair, *Multitarget/Multisensor Tracking: Applications and Advances III*, Artech House, Norwood, MA, 2000.

[4] S. Blackman and R. Popoli, *Design and Analysis of Modern Tracking Systems*, Artech House, Norwood, MA, 1999.

[5] C.-Y. Chong, S. Mori and K.-C. Chang, "Distributed multitarget multi-sensor tracking", *Multitarget-Multisensor Tracking: Advanced Applications*, 1990.

[6] F. E. Daum and R. J. Fitzgerald, "The importance of resolution in multiple target tracking", *Signal and Data Processing of Small Targets*, 329, 1994.

[7] O. E. Drummond, "Integration of features and attributes into target tracking", *Signal and Data Processing of Small Targets, Proc. SPIE*, 4048: 610-622, 2000.

[8] O. E. Drummond, "A hybrid fusion algorithm architecture and tracklets", *Signal and Data Processing of Small Targets, Proc. SPIE*, 3136: 485-502, 1997.

[9] O. E. Drummond, "Tracklets and a hybrid fusion with process noise", *Signal and Data Processing of Small Targets, Proc. SPIE*, 3163: 512-524, 1997.

[10] T. Kirubarajan, H. Wang and Y. Bar-Shalom, "Efficient multisensor fusion using multidimensional assignment for multitarget tracking", *SPIE Proceedings*, 3374: 14-25, 1998.

[11] Suihua Lu, *Network Multiple Frame Assignment Architectures*, Ph.D theis, Colorado State University, 2001.

[12] M. D. Miller and O. E. Drummond, "Trackets and covariance truncation options for theater millil tracking", *Proceedings of the Internationl Conference on Multisource-multisensor Information Fusion*, 1998.

[13] J. R. Moore and W. D. Blair, "Practical aspects of multisensor tracking", *Multitarget/Multisensor Tracking: Applications and Advances III*, 2000.

[14] A. B. Poore, S. Lu, and B. J. Suchomel, "Data association using multiple frame assignments", In David L. Hall and James Llinas, editors, *Handbook for Multi-sensor Data Fusion*, CRC Press, Inc., Boca Raton, FL, 2000.

[15] A. B. Poore and A. J. Robertson III, "A new class of Lagrangian relaxation based algorithms for a class of multidimensional assignment problems", *Computational Optimization and Applications*, 8: 129-150, 1997.

[16] A. B. Poore and X. Yan, "Some algorithmic improvements in multi-frame most probable hypothesis tracking", *Signal and Data Processing of Small Targets, Proc. SPIE*, 1999.

[17] R. Rothrock and O. E. Drummond, *Pilot JCTN algorithm Benchmarking Performance Metrics*, Office of Naval Research, 1999.

Chapter 11

TECHNOLOGIES LEADING TO UNIFIED MULTI-AGENT COLLECTION AND COORDINATION

Ronald Mahler

Lockheed Martin Tactical Systems, MS U2S25
3333 Pilot Knob Road, Eagan MN 55121
ronald.p.mahler@lmco.com

Ravi Prasanth

Scientific Systems Company, Inc., Suite 3000
500 West Cummings Park, Woburn MA 01801
prasanth@ssci.com

Abstract   The problem of managing swarms of UAVs consists of *multi-agent collection* (i.e., distributed robust data fusion and interpretation) and *multi-agent coordination* (i.e., distributed robust platform and sensor monitoring and control). These two processes should be feedback-connected in order to improve the over-all quality of data be collected on suitable targets. This paper summarizes work proposed by Lockheed Martin Tactical Systems (LMTS) of Eagan MN and its subcontractor Scientific Systems Co., Inc. (SSCI) of Woburn MA, under contract F49620-01-C-0031 of the AFOSR Cooperative Control Theme 2. LMTS and SSCI have proposed to (1) develop a mathematical programming framework for hybrid systems analysis and synthesis, (2) develop a computational hybrid control paradigm, (3) develop *transition-aware anytime algorithms* for time-bounded synthesis, and (4) develop suitable modeling and cooperative control of UAV swarms for a SEAD-type mission. Regarding multi-agent collection, LMTS and SSCI will (4) develop new theoretical approaches for integrating multiplatform, multisensor, multitarget sensor management into hybrid systems theory; (5) investigate real-time nonlinear filtering for detecting and tracking low-observable targets; (6) develop new approaches to distributed, robust data fusion; and (7) develop a language for Multi-Agent Coordination broad enough to encompass Bayesian, Dempster-Shafer, and fuzzy-logic inference. The basis of our approach is twofold: (a) a novel hybrid-systems control architecture that integrates the best of the current approaches; and (b) a new foundation for multisensor-multitarget problems called "finite-set statistics." Our approach integrates theoretically rigorous statistics (hybrid control, point process theory)

S. Butenko et al. (eds.), Cooperative Control: Models, Applications and Algorithms, 215-251.
© 2003 *Kluwer Academic Publishers.*

with potential practicability (computational hybrid control, computational non-linear filtering).

1.   Introduction

Unmanned Aerial Vehicles (UAV) are becoming an integral part of future military forces, and will be used for complex tasks such as reconnaissance, surveillance, Close Air Support (CAS), Suppression of Enemy Air Defenses (SEAD), aerial refueling, and precision strike missions. It is also envisioned that UAVs will be used for many tasks that are too demanding or hazardous for manned aircraft. Potential advantages of the UAVs over manned aircraft include: greater maneuverability, removal of the risk of casualties, significant weight savings, lower costs, potential for superior coordination, and opportunities for new operational paradigms. Even though the research in this area is in its initial stages, it has already been recognized that autonomy of individual vehicles and coordination among multiple UAVs are key capabilities for utilizing the full potential of UAV systems.

During a mission, the UAV formations will be required to achieve autonomous fault-tolerant and collision-free operation when carrying out a number of different tasks and missions. It is envisioned that "swarms" of such autonomous UAVs with an effective coordination strategy will lead to superior performance and efficient utilization of resources, and achieve effective force superiority in a future battlefield. Another important requirement in this context is to minimize communication between the vehicles and decrease the probability of failures, detection by enemy forces, or jamming of communication channels. Hence advanced autonomous fault-tolerant guidance and control algorithms for multiple UAVs in conjunction with effective decentralized multi-agent coordination strategies are of great interest to DoD.

The problem of managing swarms of autonomous, self-reconfiguring sensors and weapons divides into two general realms: *Multi-Agent Collection* (i.e., distributed robust data fusion and data interpretation) and *Multi-Agent Coordination* (i.e., distributed robust platform and sensor monitoring and control). The sensors on the various autonomous platforms must try to collect and interpret data from targets of interest, in the right places at the right times. In turn, the platforms must try to properly re-configure and position themselves to accomplish mission objectives based on this data, also in the right places at the right times. Ideally, these two processes should be connected by a feedback loop in which some platforms may be re-configured in order to improve the over-all quality of data be collected on certain targets. Consequently, data fusion/interpretation and platform configuration/control cannot be separated, since one function cannot be accomplished successfully without the other.

This paper summarizes a research program being conducted by Lockheed Martin Tactical Systems (LMTS) of Eagan MN and its subcontractor Scientific Systems Co., Inc. (SSCI) of Woburn MA, under contract F49620-01-C-0031 of the AFOSR Cooperative Control Theme 2. Our approach is ultimately aimed at unification of Multi-Agent Collection and Multi-Agent Coordination. Our approach is based on a potential unification of innovative distributed hybrid-systems control theory with a new multisensor-multitarget statistics called "finite-set statistics" (FISST).

1.1. Objectives: Multi-Agent Coordination

Existing models of hybrid systems are based on languages that are not sufficiently general enough to describe all of the functionalities of UAV swarms. For example, the controlled generalized hybrid dynamical system (CGHDS) [12] gives "nominal" models of a large class of hybrid systems, but provides no way to express uncertainties or ambiguities. Higher levels of swarm intelligence will require the incorporation of machine vision, target recognition and other complex uncertain tasks in the system model. Our research will aim at a general framework that extends beyond the usual propositional logic structures into the realm of Bayesian and Dempster-Shafer inference within which hybrid control problems can be stated as mathematical programming problems. A mathematical programming framework is expected to provide these capabilities as well as the ability to perform desktop verification and validation (V&V) efficiently.

Every successful control paradigm comes with efficient algorithms for computing controllers with guaranteed properties. This is true of LQG, $\mathcal{H}_\infty$ and robust control theories. Practicing engineers would estimate that about 5% of control system development effort goes towards the continuous-part (lowest level) of the hierarchical controller. The remaining 95% is spend on discrete decisions like exception handling, event-scheduling, safety, alarm detection, target classification and switching between operating modes. The aircraft industry routinely spends thousands of man-hours and millions of dollars for flight code verification alone. Controller design and verification cost is expected to rise significantly when the complexities of UAV swarms are considered. Our research in this area is akin to the LMI revolution of robust control theory and concerns two basic problems: (a) V&V and, (b) design of verifiable hybrid control systems.

The overall objective of this part of the proposed research is to develop efficient tools for the analysis and synthesis of hybrid control systems. Specifically, LMTS and SSCI will (1) develop a mathematical programming framework for hybrid systems analysis and synthesis, (2) develop a computational hybrid control paradigm, (3) develop *transition-aware anytime algorithms* for time-bounded synthesis, and (4) develop suitable modeling and cooperative control of UAV swarms for a SEAD-type mission. Our goal is to answer the following questions:

1. What is a language to formulate hybrid system problems involving heterogeneous components and functionalities such as GPS, machine vision and real-time signal processing?

2. What are the efficient computational tools for verification and validation (V&V) of hybrid systems? How to synthesize hybrid controllers with *verifiable properties*?

3. What are the new technical issues arising in real-time UAV swarms and MEMS applications? How to develop embedded software with real-time guarantees?

1.2. Objectives: Multi-Agent Collection

Progress in single-sensor, single-target problems has been greatly facilitated by the existence of a systematic, rigorous, and yet practical *engineering statistics*—e.g., random vectors, probability-mass and probability-density functions, differential and integral calculus, optimal state estimators, etc.—that supports the development of new concepts in this field. One would expect that multi-sensor, multi-object tracking would already rest upon a similarly systematic, rigorous, and yet practical engineering statistics. Surprisingly, until recently this has not been the case, even though a rigorous statistical foundation for multi-object problems—*point process theory*—has been in existence for decades. As a result, virtually all existing multisensor, multitarget algorithms are based on ingenious, but often convoluted, elaborations of single-sensor, single-target statistics. Finite-set statistics provides an "engineering-friendly" version of point process theory that is directly patterned after the "Statistics 101" formalism that most signal processing engineers already know and understand. The lack of such a statistics has probably hampered progress in multisensor-multitarget data fusion, detection, tracking, and target identification. This lack has also probably hampered the development of systematic, control-theoretic approaches to sensor management, distributed sensor management, and multiplatform coordination. It is expected that the research to be conducted during this project will address many, if not all, of these gaps.

The overall objective of the proposed research in Multi-Agent Collection is to develop effective new approaches for the collection, fusion, and interpretation of multitarget observations, using sensors onboard or offboard coordinated swarms of UAVs. As part of this effort, LMTS and SSCI will (1) develop a new theoretical approach for integrating multiplatform, multisensor, multitarget sensor management into Multi-Agent Coordination; (2) discover ways of incorporating this new sensor management approach into hybrid systems theory; (3) investigate the use of real-time Bayes nonlinear filtering in detecting, tracking, and identifying targets obscured by low signal-to-noise sensing conditions; (4) develop new approaches to distributed, robust data fusion; and (5) develop a new language for Multi-Agent Coordination broad enough to encompass more general inference techniques such as Bayesian statistics, the Dempster-Shafer theory of evidence, and fuzzy logic. These efforts will be aimed at answering the following questions:

1. Can sensor management be theoretically subsumed as a sub-discipline of distributed hybrid adaptive control; and if so, to what potential practical advantage?

2. Can computational nonlinear filtering techniques be developed that will effectively detect and extract low-observable targets from background noise and clutter; and if so, can this be done in real time?

3. Can data from multiple sensors/sources, on multiple platforms, be effectively fused in order to provide greater information about targets of interest? If so, is it possible to do so even if one or more of the sensors are ill-characterized statistically (i.e., their likelihood functions are not known with precision)?

1.3.   Overview: Multi-Agent Coordination

The existing approaches to Multi-Agent Coordination can be broadly divided into three categories: the *leader following, behavioral,* and *virtual structure* approaches. In *leader following,* one vehicle is designated as the leader with the rest of the vehicles designated as followers, and the followers track a function of the configuration of the leader [62], [25], [52]. The advantage of leader following is that the design of control laws for the followers is straightforward. Unfortunately, the leader-following framework is inherently centralized, inflexible, and complex. It also has a centralized single point of failure, requires heavy communication between the leader and the followers, and tends to be sensitive to failures and modeling errors of individual vehicles. In addition, the leader-following architecture only addresses the low-level control problem and not the group guidance problem.

The basic idea behind the *behavioral approach* is to prescribe several desired behaviors for each agent and to pre-specify the control action for each agent as a weighted average of the control algorithms corresponding to each of the behaviors. Possible behaviors include collision avoidance, obstacle avoidance, goal seeking, and formation keeping [43], [60]. One of the primary advantages of the behavioral approach is that it is inherently decentralized and requires limited computational resources. Its disadvantages are that it is difficult to guarantee convergence and robustness of the overall system, and that the guidance tasks become unwieldy. In fact, the so-called "synthesis" problem for emergent behavior architectures remains an open problem.

The *virtual structure (VS)* approach treats the group of vehicles as if they are elements of a single structure [5], [32], [4]. The control design is decomposed into three steps: (1) the desired dynamics of the virtual structure are defined;

(2) the motion of the virtual structure is transformed to the desired motion for each vehicle, and (3) tracking control laws for each vehicle are derived. One of the primary advantages of VS is that both guidance and control tasks are addressed within a unified framework as the guidance tasks can be specified by prescribing the appropriate dynamics to the virtual structure. The low level control tasks reduce to a tracking problem similar to that encountered in the leader-following approach.

LMTS and SSCI will investigate an architecture which is a novel integration all three approaches, leveraged off of SSCI's previous work [5], [30], [29]. Specifically, we will combine the decentralized nature of the *behavioral approach* with the *VS architecture*, using *leader following* techniques at the lowest level. We summarize our approach in the following subsections. Further detail can be found in Section 2.

1.3.1   Decentralized Formation Control Architecture.

The main challenge in designing a decentralized control scheme is that individual vehicles do not have the control and state information of all the vehicles in the formation, which makes it difficult to ensure that the control decisions made by individual vehicles will result in a well coordinated formation of vehicles. The key idea in the proposed decentralized scheme is to use the so-called *coordination variable* to coordinate the activities of all the vehicles in the formation. The coordination variable specifies the desired dynamics for each vehicle so that together they will achieve the desired goal. In a centralized scheme, the coordination variable is computed by a single controller and broadcast to all vehicles. In the proposed decentralized scheme, each vehicle will have its own coordination variable and controller. The problem is how to synchronize the coordination variables for all the vehicles, and how to assure that they converge to their desired values.

1.3.2   Autonomous Intelligent Controllers for Multiple UAVs.

Integral components of the autonomous formation control scheme include: (1) *path planning* algorithms to generate feasible paths for the entire formation; (2) *trajectory generation* algorithms to generate desired temporal trajectories that can be followed by all vehicles in the formation; and (3) a *formation hold* algorithm that initializes and maintains the formation throughout the mission. The design of each of the above components can be truly formidable. The path planner generates a path that avoids pop-up threats, and takes advantage of targets of opportunity whenever they arise. The trajectory generator optimizes the performance of the entire fleet while making sure that the commanded trajectories are within the performance limitations of all the vehicles. The formation hold autopilot maintains the formation in a robust and collision-free manner, even in the face of large maneuvers of the entire formation. Most importantly, to allow flexible and efficient coordinated control of multiple vehicles, all these three components need to be designed and implemented in a decentralized fashion. Our proposed autonomous intelligent control architecture is hierarchical. It is dictated by the requirements arising from autonomous Health Monitoring and Failure Detection and Identification (HM-FDI), control reconfiguration in the presence of different upsets, on-line decision making, situation awareness, real-time trajectory generation, flight path management, and mission replanning.

1.4. Overview: Multi-Agent Collection

Point process theory [14], [45], [53] is the statistical theory of systems that contain a finite but randomly-varying number of objects (such as sensor observations or target-tracks), each of which may in turn be randomly varying. *Finite-set statistics (FISST)* is an "engineering friendly" formulation of point process theory—which is to say, one that: (1) is geometric, in the sense that it treats multi-object systems as visualizable images; (2) preserves the "Statistics 101" formalism that most signal processing engineers already understand; and (3) is specifically oriented towards practical data fusion applications. Whereas conventional point process theory mathematically formulates a point process as the random measure $N_\Xi(S) = |\Xi \cap S|$ or random density $f_\Xi(y) = \sum_{x \in \Xi} \delta_x(y)$ of a random finite set Ξ, FISST represents the point process directly as Ξ. FISST provides a fully unified, scientifically defensible, probabilistic foundation for the following aspects of multisource, multitarget, multiplatform data fusion: (a) *Multi-Source Integration* (detection, identification, and tracking) based on Bayesian filtering and estimation theory; (2) *sensor management* using control theory; and (3) *performance evaluation* using information theory. One of its major purposes is to extend Bayesian (and other probabilistic) methodologies so that they are capable of dealing with *imperfectly characterized data and sensor models*; and *true sensor models and true target models for multisource-multitarget problems*. One consequence is that it encompasses certain expert-system approaches that are often described as "heuristic"—fuzzy logic, the Dempster-Shafer theory of evidence, and rule-based inference—as special cases of a single fully probabilistic (and in fact Bayesian) paradigm. We summarize our approach in the following subsections. Greater technical detail can be found in section 3.

1.4.1   Finite-Set Statistics (FISST).   FISST is based on a particular mathematical formulation of point process theory: stochastic geometry/random set theory [2], [49], [59], [42]. It derives from the following sequence of ideas:

(1) re-conceptualize the sensor suite as a single sensor;

(2) re-conceptualize the target set as a single target with multitarget state $X = \{\mathbf{x}_1, ..., \mathbf{x}_n\}$;

(3) re-conceptualize the set $Z = \{\mathbf{z}_1, ..., \mathbf{z}_m\}$ of observations, collected by the sensor suite at approximately the same time, as a single measurement of the target-set observed by the sensor suite;

(4) regard multisensor-multitarget states X and observations Z as simple point processes—i.e., as random variables on abstract spaces of finite sets, endowed with a specific topology called the Mathéron or "hit-or-miss" topology [42];

(5) just as single-sensor, single-target data can be modeled using a measurement model $\mathbf{Z} = h(\mathbf{x}, \mathbf{W})$, model multitarget multi-sensor data using a *multisensor-multitarget measurement model*—a randomly varying finite set $\Sigma = T(X) \cup C(X)$;

(6) just as single-target motion can be modeled using a motion model $\mathbf{X}_{k+1} = \Phi_k(\mathbf{x}_k, \mathbf{V}_k)$, model the motion of multitarget systems using a *multitarget motion model*—a randomly varying finite set $\Gamma_{k+1} = \Phi_k(X_k, \mathbf{V}_k)$;

Given this, we can mathematically reformulate multisensor, multitarget estimation problems as single-sensor, single-target problems. The basis of this reformulation is the concept of *belief-mass*. Every point S of the Mathéron topology (i.e., closed subset of Euclidean space) has a canonical fundamental closed neighborhood O_S such that $p_\Theta(O_S) = \Pr(\Theta \in O_S) = \Pr(\Theta \subseteq S)$ for any random closed subset Θ. The *non-additive* set function $\beta_\Theta(S) = p_\Theta(O_S)$ is the *belief-mass function* of Θ, and characterizes Θ in the same way that a probability-mass function $p_\mathbf{V}(S) = \Pr(\mathbf{V} \in S)$ characterizes a random vector $\mathbf{V}$. So:

(7) just as the probability-mass function $p(S|\mathbf{x}) = \Pr(\mathbf{Z} \in S)$ of a single-sensor, single-target measurement model $\mathbf{Z}$ is used to describe the statistics of ordinary data, use the belief-mass function $\beta(S|X) = \Pr(\Sigma \subseteq S)$ of a multisource-multitarget measurement model Σ to describe the statistics of multisource-multitarget data;

(8) just as the probability-mass function $p_{k+1|k}(S|\mathbf{x}_k) = \Pr(\mathbf{X}_k \in S)$ of a single-target motion model $\mathbf{X}_k$ is used to describe the statistics of single-target motion, use the belief-mass function $\beta_{k+1|k}(S|X_k) = \Pr(\Gamma_{k+1} \subseteq S)$ of a multitarget motion model Γ_{k+1} to describe the statistics of multitarget motion;

(9) just as the recursive Bayes filter [22]

$$f_{k+1|k}\left(\mathbf{x}|Z^k\right) = \int f_{k+1|k}\left(\mathbf{x}|\mathbf{y}\right) f_{k|k}\left(\mathbf{y}|Z^k\right) d\mathbf{y}$$

$$f_{k+1|k+1}\left(\mathbf{x}|Z^{k+1}\right) \propto f_k\left(\mathbf{z}_{k+1}|\mathbf{x}\right) f_{k+1|k}\left(\mathbf{x}|Z^k\right)$$

provides a theoretical basis for single-sensor, single-target detection, tracking, and identification, so the multitarget recursive Bayes filter

$$f_{k+1|k}\left(X|Z^{(k)}\right) \ = \ \int f_{k+1|k}\left(X|Y\right) f_{k|k}\left(Y|Z^{(k)}\right) \delta Y$$

$$f_{k+1|k+1}\left(X|Z^{(k+1)}\right) \ \propto \ f_k\left(Z_{k+1}|X\right) f_{k+1|k}\left(X|Z^{(k)}\right)$$

provides a theoretical basis for multisensor, multitarget detection, tracking, and identification. For more details, see Section 3.1 below.

(10) just as the probability-mass function $p(S|\mathbf{x}) = \Pr(\mathbf{Z} \in S)$ is used to describe the generation of conventional data $\mathbf{z}$, use *belief-mass functions* of the general form $\rho(\Theta|\mathbf{x}) = \Pr(\Theta \subseteq \Sigma|\mathbf{x})$ to describe the generation of ill-characterized data Θ such as features, rules, or natural-language statements (which can be expressed in Dempster-Shafer or fuzzy-set form, see Section 3.5 below).

1.4.2 Multi-Agent Sensor Management.

A basic capability that will eventually be required of swarms of smart UAVs is that they must be able to re-configure themselves and their on-board sensors to collect additional data from under-collected targets of interest. This, in turn, means that *sensor management* will be an inescapable aspect of multi-agent configuration. Finite-set statistics (FISST) may provide insight in how to incorporate sensor management into multi-agent coordination. This is because it potentially allows multiplatform-multisensor sensor management to be reformulated as a *direct generalization of (nonlinear) control theory*. We will directly generalize conventional single-sensor, single-target statistics and control theory with the aim of producing the elements of a control theory for multiplatform, multisensor, multitarget problems. We will determine how such an approach can be incorporated with hybrid-systems theory, with the aim of producing a practical approach to sensor management for Multi-Agent Coordination.

1.4.3 Real-Time Nonlinear Filtering for Low-Observable Application of UAVs.

Under many circumstances, UAV-based sensors will be subject to less-than-ideal sensing conditions. A typical example occurs when signal-to-noise ratio (SNR) is so low that an imaging sensor images a target that is not actually visible in any given image-frame. Provided that data is collected rapidly enough, a *recursive Bayes nonlinear filter* can be used to detect, extract, and track the target from frame to frame. In cooperation with the University of Alberta (Edmonton), LMTS is sponsoring the development of fundamental new real-time computational nonlinear filtering approaches such as infinite-dimensional exact filters and particle-system filters [1], [27]. LMTS and SSCI will investigate the application of nonlinear filters for UAV-based sensors encountering low-SNR conditions.

1.4.4   Systematic and Robust Multi-Agent Sensor Fusion.

It is expected that, even with the presence of multi-agent sensor management, targets will not be effectively detected, tracked, or identified without fusing data collected from more than one sensor and more than one UAV. Because it provides a rigorous and systematic approach for constructing multisensor-multitarget likelihood functions, finite-set statistics potentially provides a systematic foundation for multiplatform, multisensor, multitarget data fusion. The goal of this research will be to develop the foundations of such a distributed data fusion, and develop means of rendering them practicable. In addition, some sensors or sources will not be completely characterizable from a statistical point of view. Consequently, robust approaches to data fusion are required. We will use *generalized likelihood functions* $\rho(z|x)$ to model the *uncertainties* as well as the *certainties* in data z that are difficult to characterize either because its likelihood $f(z|x)$ is not known with sufficient fidelity (e.g. HRRR, SAR) or because z is inherently difficult to model in and of itself (e.g., datalink features, natural-language statements, rules).

1.5.   Organization of the Paper

In the remainder of the paper we summarize our approach in greater technical detail. Sections 2 and 2 describe our approaches to Multi-Agent Coordination and Multi-Agent Collection, respectively.

2.   Approach: Multi-Agent Coordination

In this section we summarize our approach in Multi-Agent Coordination in greater technical detail. In Section 2.1 we describe our approach for a mathematical programming framework for hybrid systems. Our approach to computational hybrid control is described in Section 2.2 and to phase-transitions and transition-aware anytime algorithms in Section 2.3. Finally, we describe our approach to cooperative control of UAV swarms in Section 2.4.

2.1.   Mathematical Programming for Hybrid Systems

Two recent strands of research have greatly influenced our approach. First, convex programming and linear matrix inequalities (LMIs) [11] have transformed the way control engineers think of and *solve* purely continuous control problems. These new techniques are popular because many hard problems in systems theory can be *relaxed* to convex/LMI problems leading to efficiently computable upper (or lower) bounds. Second, linear programming embeddings of predicate and partially interpreted logics have been developed by Borkar, Chandru and Mitter [10]. That is, checking the validity of certain types of logical statements can be reduced to mathematical programming problems. These two lines of research show that mathematical programming is a common theme in continuous-variable control and discrete-event systems. We shall combine these developments into a unified framework for computational hybrid control.

2.1.1 Framework for Hybrid Systems with Propositional Logic. Our work will be based on the distinction between *soluble* and *checkable* constraints introduced by Hooker [23]. Soluble constraints might be linear or convex inequalities like those arising from control specifications; whereas checkable constraints belong to NP and are of the type considered in the standard constraint satisfaction problem (CSP), for example dynamic resource allocation. Variables associated with soluble constraints are *solution variables* and those associated with checkable constraints are *search variables*. These variables are connected through *conditional constraints*. An example is the following mixed logical/convex program (MLCP):

$$\text{minimize } c^T x \text{ subject to } p_j(y,h) \rightarrow (x \in X_j), \quad j \in J | q_i(y,h), i \in I,$$

where X_j's are convex sets, p_j's and q_i's are logical formulas. The constraint set in the optimization problem above has a continuous part (on the left hand side of the bar) and a logical part (on the right hand side). The logical part consists of formulas q_i that involve atomic propositions y which are either true or false. An example is $q_i(y,h) = y_1 \vee y_2$ which says that y_1 or y_2 must be true. These formulas may also contain variables h that take on discrete values. The continuous part associates logical formulas p_j with convex constraints. That is, the constraint $x \in X_j$ is to be enforced when $p_j(y,h)$ is true. A triple (x, y, h) is feasible if (y, h) makes all the formula q_i's true and x belongs to all those X_j's for which $p_j(y,h)$ is true.

2.1.2 Random Set Theory Framework for Hybrid Systems.
The framework just described subsumes mixed integer/linear programs (MILP) arising in certain hybrid control problems [6] as well as other more general hybrid control problems [12]. However, it is not sufficiently general to accommodate Bayesian and other sophisticated forms of inference. However, finite-set statistics allows many expert-systems inference approaches—Bayesian, fuzzy logic, Dempster-Shafer, and perhaps also rule-based—to be unified under a single, fully probabilistic foundation called a *random closed subset* [21], [36], [28] (see Section 3.5). We will investigate the possible use of random set theory as a means of incorporating more general forms of inference (Dempster-Shafer, fuzzy logic, Bayesian) into hybrid control system theory.

2.2.   Computational Hybrid Control

In order to illustrate our philosophy, consider the design of *least restrictive controllers* that guarantee reachability specifications. A game theoretic approach [34][25] to this problem leads to a design procedure involving Hamilton-Jacobi (HJ) equations. Even in the relatively simple case of purely continuous states, HJ equations cannot be solved using a finite computation. Thus, it is necessary to allow some amount of design conservatism in order to develop techniques that are applicable to more practical and larger classes of hybrid systems than those for which least restrictive controllers are constructible. LMIs and mathematical programming techniques allow us to trade-off between computational complexity and conservatism via upper (or lower) bound problems (so called semi-definite programming (SDP) relaxations in robust control). A similar paradigm for hybrid systems is our objective. Of course, there are a number of new difficulties stemming from the continuous-discrete optimization-constraint satisfaction nature of hybrid problems that we need to address.

2.2.1   System-Theoretic Approach to V&V.   We will consider the following two approaches and develop software for hybrid system verification and validation (V&V). First, a *modular game-theoretic approach* based on the continuous dynamics [51] with discrete-event dynamics either being embedded in the continuous ones or discrete abstractions of the continuous dynamics being extracted. The overall system (with the controller) is viewed as an interconnection of various smaller agents/subsystems. Second, a *multiple Lyapunov function (MLF) approach* based on Lyapunov stability theory. The analysis procedure involves three (coupled) steps: (a) split the state space into a finite number of overlapping regions, (b) for each region, choose a (possibly quadratic) Lyapunov function that decreases along state trajectories in that region, and (c) piece them together to form a global Lyapunov function.

2.2.2   V&V using Abstractions based on Achievable Performance.   Sometimes the performance measure is continuous-time in nature, but the achievable performance depends on the particular discrete case. When this happens, it is possible to construct a formal logic verifier that works on the high level concept of achievable performance instead of trying to re-derive it from basic definition. As shown in [16], this can lead to sharp efficiently computable bounds. There are essentially three steps in this approach: (1) derive bounds on achievable continuous-time performance for each discrete state using standard convex/LMI programs; (2) characterize the worst-case feasible sets and associated continuous-time states; and (3) verify

the discrete-event system. Our work will adapt these steps to more general hybrid systems.

2.2.3 Mathematical Programming for Hybrid Systems.
There are two sources of computational complexity in hybrid systems. First, the logical part makes the problem at least NP-hard. Second, certain specifications on the continuous part of the hybrid system may lead to infinite dimensional convex constraints. Our research will focus on finite dimensional approximations of infinite dimensional convex constraints, primal-dual relaxations of hybrid mathematical programming problems and other computational issues. The following four approaches will be examined for V&V and the design of verifiable control systems: (1) *branch-and-cut algorithms; (2) semi-definite programming (SDP) relaxations; (3) resolution and inference algorithms;* and (4) *linear programming embedding of logic and other inferential schemes.* Embeddings of propositional, predicate and partially interpreted logic can be dealt with, as can embeddings of probabilistic and Dempster-Shafer inferences [23]. Work in this area will involve extension of Borkar-Chandru-Mitter embeddings, belief functions, higher-order logic and efficient implementations of Bayesian inference.

2.2.4 Heuristic Algorithms for Hybrid Systems.
There are many heuristic algorithms available for mixed integer/linear programming (MILP) problems. We will be concerned with developing similar procedures for hybrid systems. Research will focus on applications of simulated annealing (SA) and genetic algorithms (GAs), and extension of Bender's decomposition to MLCP. We will extend Bender's decomposition to the mixed logic/continuous variable case. GAs offer a combination of hill-climbing ability, stochastically and population-based search, and can be implemented in parallel.

2.3. Transition-Aware Anytime Algorithms

Many of the real-time decision-making problems that will be faced by a swarm are NP-hard. In fact, the computational cost, when viewed as a function of problem instance, exhibits a *phase transition* or *easy-hard-easy* pattern. Such patterns have been shown to exist in constraint satisfaction problems (CSP) of logical inference. However, they are expected to be present in more complicated hybrid problems. As the swarm evolves, it will encounter many problem instances that are easy to solve and some instances that are hard. So, a computational strategy for real-time applications must: (a) reliably classify problem instances, and (b) generate efficiently solvable approximations for hard instances. The transition-aware algorithms to be developed will use various problem structure parameters to determine if an instance is easy or hard. When a hard instance is encountered, these algorithms will need to make a decision to either *project to an easy problem* or *relax to an easy problem*. Projecting a hard problem to a nearby (but across the transition boundary) easy problem may be achieved by dropping less important constraints and objectives. Relaxing to an easy problem is achieved by obtaining a polynomial-time approximation via, for example, Lagrangian relaxation. A major part of this research will investigate when and how projection and relaxation methods should be used.

2.4.   Cooperative Control of UAV Swarms

The overall objective of this research is to model and design a decentralized cooperative controller for a scenario that includes flying a formation of UAVs to enemy territory, carrying out a SEAD type mission, and avoiding pop-up threats by replanning its path and re-configuring the formation.

2.4.1   Problem Formulation.   We will analytically define UAV models and performance requirements for the entire formation for tasks such as formation contraction and expansion, formation keeping, and threat and collision avoidance.

2.4.2   Development of decentralized hierarchical formation flying control architecture.   We will develop a decentralized feedback control architecture suitable for large numbers of independent, similar, and autonomous UAVs flying in a formation. The proposed architecture is a novel integration of SSCI's previous work on centralized coordinated control [4], [5] and decentralized control schemes based on the behavioral paradigm [30], [29]. Specifically, in the proposed architecture we plan to combine the decentralized nature of the behavioral approach with the Virtual Structure (VS) architecture, using leader following techniques at the lowest level. To fully explain our ideas, we will first describe a centralized version of the VS architecture in the context of close formation flight for UAVs. We will then indicate as to how the ideas from behavior-based control can be extended to a decentralized VS architecture.

The VS architecture which was introduced in [5] for spacecraft formation flying. Let the individual vehicles be represented by the block V_i and their local formation hold controllers by K_i. The commands for K_i are variables ξ (the *coordination variables*). A feedback signal from individual vehicles to the trajectory generation block is denoted by z_i (the *performance variable*). In VS the trajectory generation block outputs the signal w to the path planning block, and receives the waypoints of the desired path σ from it in return. For instance, σ can be a vector in $\mathbb{R}^3$ that specifies the next (inertial) waypoint for the formation, while the feedback signal w can be a binary signal that indicates to the path planner that the formation is within a certain distance of the current waypoint σ_{i-1}. Our fundamental assumption is that coordination requires *shared information*, encapsulated in the coordination variable. Let x_i be the configuration of the i'th vehicle and let x_i^d be the desired configuration. Then we require the existence of a mapping T_i such that $x_i^d = T_i(\xi)$. The interaction between the trajectory generation block (continuous dynamics) and the path planning block (discrete event dynamics) makes the overall system hybrid.

The VS architecture can be thought of as an extension of leader following, where the leader can be replaced by a virtual leader, with dynamics that can be arbitrarily defined, and can change depending on the commands from the path planning block. In addition, the VS architecture allows feedback from the individual vehicles to the trajectory generation block. One of the primary advantages of the VS architecture is that there is a clean separation between the guidance and control tasks. Low level control tasks are handled via tracking control laws, similar to leader following, while the guidance tasks are handled through the coordination variable ξ.

In this project we will explore techniques that allow decentralization of the VA architecture. Our initial idea is to create a local copy of the path planning and trajectory generation blocks on each vehicle. Each vehicle now maintains a local copy of the coordination variable ξ_i and the state of the path planner, i.e., σ_i. If $\xi_i(t) \equiv \xi_j(t)$ for each $i, j \in [1, N]$, then the action of the group will be identical to the centralized case. This architecture has the following advantages: (1) it is decentralized, accommodating as much communication bandwidth as is available; (2) both guidance and control tasks are addressed in a uniform framework; (3) guidance tasks are relatively easy to specify; (4) it includes implicit formation feedback, adding robustness to timing discrepancies, and the dynamics of poorly performing vehicles; and (5) it provides a decentralized extension of leader following.

The decentralized architecture introduces new complexity, caused by the fact that we must synchronize the coordination variables in order to ensure coordinated behavior. That is, the objective is defined as follows:

$$\xi_1(t) \to \xi_2(t) \to \cdots \to \xi_N(t) \to \xi_1(t)$$

asymptotically. The main difficulty in synchronizing the coordination variables comes from the fact that only local communication is possible, i.e., vehicle i can only communicate with vehicle $i + 1$ and can only receive communication from vehicle $i - 1$. This synchronization problem is very similar to the behavior-based schemes explored in [30], [29], [16]. That is, suppose that the dynamics of the coordination variable under the centralized scheme is given by $\dot{\xi} = f(\xi, \phi)$, where $\phi = \sum z_i$. Then under the decentralized scheme the dynamics for the ith coordination variable can be chosen as

$\dot{\xi}_i = f(\xi, z_i) + K(\xi_{i-1} - \xi_i)$, where K is a positive definite matrix. The second term in this expression has the effect of equalizing the coordination variables of all vehicles. Simple simulations have shown that synchronization is indeed achieved. We plan to investigate several theoretical aspects of the proposed scheme, including convergence, stability, and performance; and extend the techniques developed in [29] to mathematically guarantee synchronization under the proposed scheme.

2.4.3 Integration of Voronoi graph and Dynamic Programming-based path planning algorithm.

Our approach to real-time path planning will follow the techniques outlined in [44]. One of the main objectives of the path planning procedure is to avoid known threats and to quickly adapt the path when pop-up threats are detected. The output of the path planning block is the next waypoint for the virtual aircraft. In terms of avoiding pop-up threats, our approach will be to solve the path planning problem on-line again when a new threat is detected. Therefore, the path planning procedure needs to be computationally efficient. Given a target and a list of known threat locations, a threat-avoiding path to the target can be devised by constructing a Voronoi diagram [13] based on the locations of the threats. To generate an initial path, the formation starting point and the target point must be connected to the Voronoi graph in some way. We will simply connect the starting point and the target point to the three closest nodes of the Voronoi diagram. With the Voronoi diagram indicating numerous paths to the target, the final step in finding the initial path becomes determining which path to take. There are two costs associated with traveling along an edge of the Voronoi diagram: threat costs and fuel costs. Threat costs are determined based on the group exposure to enemy radar. An exact calculation of the threat cost for traveling along an edge would involve the integration of the cost along the edge. A simplified approach involves calculating the threat cost at several locations along an edge and taking into account the length of the edge. With the costs for traveling a particular edge of the Voronoi diagram determined, the Voronoi diagram is searched to determine the lowest cost path.

In SSCI's previous work with Draper Lab on the autonomous path planning demo using model helicopters, we have used the A^*, D^* and T^* algorithms for this purpose. For this project, we will use Dijkstra's algorithm, which requires that the graph be a directed graph. We will integrate the path planning algorithm described here into the proposed decentralized formation control scheme. We will also explore methods to refine this algorithm, including ways to smooth the final path and to include other costs in the algorithm.

2.4.4 Trajectory generation algorithms based on differential flatness.

Due to dynamical constraints and physical limits on control authority, not all trajectories connecting two waypoints can be followed by the UAV. It is the responsibility of the trajectory generation block to use the waypoints to generate a *feasible* state trajectory, as well as a set of nominal inputs that drive the UAV along this path. Recently, there has been great interest in the trajectory generation problem for a special class of systems called *differentially flat systems* [17], [50]. A nonlinear system $\dot{x} = f(x, u)$ is differentially flat if we can find *flat outputs* z of the form $z = \zeta(x, u, \dot{u}, \ldots, u^{(l)})$ such that and $x = x(z, \dot{z}, \ldots, z^{(l)})$ and $u = u(z, \dot{z}, \ldots, z^{(l)})$. Flat systems are ideally suited for trajectory generation tasks because trajectories planned in the flat output space can be lifted to the state and input space simply through an algebraic mapping.

We will use the differential flatness approach to design a trajectory generation algorithm for UAV formation. The basic idea in our design of the trajectory generation block is to generate a "virtual aircraft" that is to become a "virtual leader" for the entire formation of aircraft. In other words, the location of all aircraft in the formation will be referenced to the virtual aircraft, which in turn is assumed to have exactly the same dynamics as all the aircraft in the formation. Following [48] we will assume that each aircraft is equipped with standard autopilots for heading hold, Mach hold and altitude hold. The resulting aircraft/autopilot models are assumed to be first order for heading and Mach hold, and second order for altitude hold. From these assumptions it is possible to construct formulas for the state of the virtual aircraft, $\mathbf{x}_F = (X_F, Y_F, \psi_F, V_F, h_F, \dot{h}_F)^T$. Given this, a flat output is $z = (X_F, Y_F, h_F)^T$. Define a mapping from $t \to \tau(t) \in [0, 1]$ and let $\rho : [0, 1] \to \mathbb{R}^3$ be a parametrized path from the current configuration to the desired configuration at the next waypoint. To avoid an infinite dimensional problem, let each component of ρ be given by a third order polynomial with coefficients c_i. The trajectory in z is then obtained as $z(t) = \rho(\tau(t))$. The full state trajectory and nominal input for the virtual leader can then be lifted from z.

As a first step in this effort, we will use the algorithm described above in the trajectory generation block of the proposed decentralized control scheme. Then we will examine the possibility of using the differential flatness approach for trajectory generation of other UAV dynamics. We will also investigate various ways of generating trajectories that take into consideration actuation and state constraints.

2.4.5 Globally stable formation hold algorithms.

One of the main prerequisites for achieving the desired formation dynamics is tight control of pairs of individual vehicles—a highly nonlinear problem. Several authors have designed flight controllers based on nonlinear inversion. Recent attempts to extend such approaches to formation control either yield an unnecessarily complicated control system, or have a singularity in the state space.

In contrast, SSCI has developed a globally stable nonlinear formation control algorithm that assures that, for any pair of vehicles in the formation, the relative distance and velocity errors converge to zero exponentially from arbitrary initial conditions [33]. The main idea is to consider the dynamics of an error that is defined in terms of inertial as well as relative coordinates of the lead vehicle. More precisely, consider two aircraft, denoted by subscripts 1 and 2, with vehicle 1 being the leader and vehicle 2 the follower. Define

$$E = \left[\begin{array}{c} E_1 \\ E_2 \end{array} \right] = \left[\begin{array}{c} X_1 - X_2 \\ Y_1 - Y_2 \end{array} \right] - R(\psi_2) \left[\begin{array}{c} x_r \\ y_r \end{array} \right],$$

where $R(\psi_2)$ is a non-singular matrix parametrized by ψ_2, and x_r, y_r are the desired relative position of vehicle 1 from the body frame of vehicle 2. It can be easily shown that $E = 0$ implies that the actual relative position of vehicle 1 from vehicle 2 is equal to $[x_r \; y_r]^T$. Therefore, to achieve the desired relative position between the two vehicles, we can simply design a control law that stabilize E instead.

We have shown that the dynamics of E are feedback linearizable and can be stabilized by the inputs V_{2c} and $\dot{\psi}_{2c}$, provided x_r is non-zero. The existing algorithm has been developed for pairs of vehicles in two dimensions only. We plan to extend the algorithms to swarms of vehicles and to the three-dimensional case. Furthermore, we will include a collision avoidance component in the algorithm to ensure collision-free formation initialization from arbitrary initial configuration.

2.4.6   Integrate different layers in the autonomous intelligent control system architecture.

In the past several years there has been substantial progress in the area of fault-tolerant and reconfigurable control designs for both manned and unmanned aircraft. The results so far have demonstrated the potential of the reconfiguration techniques to maintain automatically the desired aircraft performance despite severe control effector failures and structural or battle damage. Several of those approaches have been extensively tested through simulations, and even flight tested. However, most of the proposed techniques are complex and it is not clear at all as to how to integrate them with the guidance and path-planning loops to achieve truly autonomous operation under different upsets, failures and unanticipated events. We plan to integrate different layers in our autonomous intelligent flight control system architecture. The Level 1 controller has already been developed by SSCI and tested through simulation of several advanced fighter aircraft. The overall system is based on the Multiple Models, Switching and Tuning

(MMST) methodology, and SSCI's extensions of it to the FDI problem, resulting in the Adaptive Interacting Multiple Observer (AIMO) technique.

2.4.7   Testing the model and control algorithms.   The decentralized formation control algorithms will be tested through extensive simulation studies.

3.   Approach: Multi-Agent Collection

In this section we summarize our approach in Multi-Agent Collection in greater technical detail. In Section 3.3 we describe our intended approach to multiplatform, multisensor, multitarget sensor management. Our approach to computational nonlinear filtering for detection and tracking in low-SNR environments is described in Section 3.4. Section 3.5 summarizes our approach to robust multisource data fusion.

3.1.   Elements of FISST

We summarized the overall FISST approach in Section 1.4.1. In this section we provide a little more detail. The FISST *multisensor-multitarget differential and integral calculus* is what transforms the mathematical abstractions of Section 1.4.1 into a form that can be used in practice [21], [40], [36].

If $Z = \{z_1, ..., z_m\}$ with $z_1, ..., z_m$ distinct, the *set derivative* of the (not necessarily additive) function $\beta(S)$ of a closed-set argument S is defined as an iterated constructive Radon-Nikodým derivative:

$$\frac{\delta\beta}{\delta z}(S) \triangleq \frac{\delta}{\delta z}\beta(S) \triangleq \lim_{\lambda(E_z)\searrow 0} \frac{\beta(S \cup E_z) - \beta(S)}{\lambda(E_z)}$$

$$\frac{\delta\beta}{\delta Z}(S) \triangleq \frac{\delta^m\beta}{\delta z_1 \cdots \delta z_m}(S) \triangleq \frac{\delta}{\delta z_1} \cdots \frac{\delta}{\delta z_m}\beta(S)$$

$$\frac{\delta\beta}{\delta\emptyset}(S) \triangleq \beta(S),$$

where E_z is a small closed neighborhood of z that "converges" to the singleton $\{z\}$. (*Caution:* The first of these three equations is simplified for the sake of clarity. A more complicated definition is required to encompass discrete variables and situations in which $S \cap E_z \neq \emptyset$, as well as to ensure that the limit is well-defined.) It is not possible to formulate the set derivative in a somewhat more conventional manner [37]. Given a random finite set (simple point process) Ξ, let

$$G_\Xi[h] = \prod_{x\in\Xi} h(x)$$

be its probability generating functional and let $1_S(x)$ denote the characteristic function of the closed subset S (of Euclidean space). Then it can be shown that

$$G_\Xi[1_S] = \beta_\Xi(S) = \Pr(\Xi \subseteq S),$$

and we could instead define

$$\frac{\delta\beta_\Xi}{\delta z}(S) = \frac{\partial G_\Xi}{\partial \delta_z}[1_S],$$

where $\frac{\partial G_\Xi}{\partial g}[h]$ denotes the functional derivative (Frechét derivative) of $G_\Xi[h]$ in the direction of g.

The inverse operation of the set derivative, the *set integral*, is defined as:

$$\int f(X)\delta X = \sum_{n=0}^{\infty} \frac{1}{n!} \int f(\{x_1,...,x_n\})dx_1\cdots dx_n.$$

The set derivative and set integral are inverse in the sense that

$$\beta(S) = \int_S \frac{\delta\beta}{\delta X}(\emptyset)\,\delta X, \qquad \left[\frac{\delta}{\delta X}\int_S f(Y)\,\delta Y\right]_{S=\emptyset} = f(X).$$

Then:

(1) Just as the likelihood function $L_z(x) = f(z|x)$ can be derived from the probability-mass function $p(S|x)$ of a single-sensor, single-target measurement model via differentiation, so the true multitarget likelihood function $L_Z(X) = f(Z|X)$ can be derived from the belief-mass function $\beta(S|X)$ of the multisensor-multitarget measurement model fusing the set derivative:

$$f(Z|X) = \frac{\delta\beta}{\delta Z}(\emptyset|X).$$

(2) Just as the Markov transition density $f_{k+1|k}(x_{k+1}|x_k)$ can be derived from the probability-mass function $p_{k+1|k}(S|x_k)$ of the single-target motion model via differentiation, so the true multitarget Markov transition density $f_{k+1|k}(X_{k+1}|X_k)$ can be derived from the belief-mass function $\beta_{k+1|k}(S|X_k)$ of the multitarget motion model via set-differentiation:

$$f_{k+1|k}(Y|X) = \frac{\delta\beta_{k+1|k}}{\delta Y}(\emptyset|X).$$

3.2.   Multitarget Recursive Bayes Filtering

Let $Z^{(k)} = \{Z_1,...,Z_k\}$ be a time-sequence of multisource-multitarget observations. Then one can construct *true multitarget posterior distributions* from the true multisource-multitarget likelihood and hence recursive multisensor-multitarget Bayes filtering equations:

$$f_{k+1|k}\left(X|Z^{(k)}\right) = \int f_{k+1|k}\left(X|Y\right) f_{k|k}\left(Y|Z^{(k)}\right) \delta Y, \quad (1)$$

$$f_{k+1|k+1}\left(X|Z^{(k+1)}\right) \propto f_k\left(Z_{k+1}|X\right) f_{k+1|k}\left(X|Z^{(k)}\right). \quad (2)$$

Here, $f_{k|k}(\emptyset|Z^{(k)})$ is the posterior likelihood that no targets are present and $f_{k+1|k}(\{x_1,...,x_n\}|Z^{(k)})$ is the posterior likelihood that n targets with states $x_1,...,x_n$ are present. This provides a basis for predicting and estimating the states of multitarget systems. From these distributions one can compute *simultaneous, provably optimal estimates of target number, kinematics, and identity without resort to the optimal report-to-track assignment characteristic of multi-hypothesis approaches.* (However, it turns out that multitarget Bayes-optimal estimation is not entirely straightforward, because *the classical Bayes-optimal estimators (the posterior expectation and the maximum a posteriori (MAP) estimates are not even defined!* Consequently, one must construct new multitarget estimators and prove that they are Bayes-optimal and otherwise well-behaved [21], [36].)

Single-target problems can be computationally simplified by propagating a statistical moment, e.g. the posterior expectation

$$\hat{x}_{k|k}^{EAP} = \int x \cdot f_{k|k}(x|Z^k) dx$$

in place of the posterior $f_{k|k}(x|Z^k)$. Likewise, multisensor-multitarget problems can be computationally simplified by computing a *multitarget* statistical moment—e.g., the *probability hypothesis density* (first-moment density)

$$\hat{D}_{k|k}(x|Z^{(k)}) = \int_{X \ni x} f_{k|k}(X|Z^{(k)}) \delta X$$

in place of the full multitarget posterior $f_{k|k}(X|Z^{(k)})$ [41], [35], [37].

3.2.1 A Short History of Bayes Multitarget Filtering.
The concept of multitarget Bayesian nonlinear filtering is a relatively new one. If one assumes that the number of targets is *not* known beforehand, the earliest exposition appears to be due to Miller, O'Sullivan, Srivastava, et. al. [56]. Their very sophisticated approach is also apparently the only approach to deal with continuous evolution of the multitarget state (all other approaches assume discrete state-evolution). Mahler [21], [36] was apparently the first to systematically deal with the general discrete state-evolution case (Bethel and Paras [8] assume discrete observation and state variables).

Portenko et. al. [47] use branching-process concepts to model changes in target number. Kastella's "joint multitarget probabilities (JMP)," introduced at LMTS in 1996, are a renaming of a number of early core FISST concepts (set integrals, multitarget Kullback Leibler metrics, multitarget posteriors, joint multitarget state estimators, the APWOP) devised two years earlier [46, pp. 27-28]. Indeed, a "JMP" itself is just a FISST multitarget posterior under a new name and notation with all state variables assumed discrete: $f_{\text{"JMP"}}(x_1, ..., x_n|Z) = \frac{1}{n!} f_{\text{FISST}}(x_1, ..., x_n|Z)$. The approach of Stone et. al. [58], which consists of citing the multitarget filtering equations 1 and 2, is best described as "heuristic". This is because (1) its theoretical basis is so imprecisely formulated that these authors have found it possible to both disparage and implicitly assume a random set framework at the same time; (2) its Bayes-optimality and "explicit procedures" are both frequently asserted but never actually justified or even spelled out with precision; and (3) its treatment of certain basic technical issues in Bayes multitarget filtering—specifically, their claim that "The [multitarget] posterior distribution...constitutes the Bayes estimate of the number and state of the targets...From this distribution we can compute other estimates when appropriate such as maximum a posteriori probability estimates or means" [58, p. 163]—is erroneous).

3.3.　　Multi-Agent Sensor Management

Sensor management can be usefully described as the process of "redirecting the right data-collection source at the right place or platform to the right target at the right time". A common approach to multisensor-multitarget sensor management is to assemble a patchwork of heuristic "bottom up" techniques—e.g., loosely integrating many distinct algorithms, each of which addresses some specific part of the problem (detection, tracking, sensor cueing, allocating and scheduling sensor dwells, allocating and scheduling platform flight paths, etc.). However, sensor management is inherently a *stochastic multi-object problem*: it involves groups of targets, groups of sensors/sources, and groups of platforms, whose states and numbers can and do vary randomly in space and time. Our approach addresses the Multi-Agent Collection and Coordination process as what it actually is: a problem in nonlinear adaptive control theory in which both the data sources being controlled and the targets being tracked by the control process are, mathematically speaking, multi-object systems.

However, no such approach is possible without the even more fundamental concept of *multi-object miss distance*. To see why this is the case, first consider a conventional optimal single-sensor, single-target control-system application such as a missile-tracking camera. A single target follows some trajectory. The camera automatically adjusts its elevation, azimuth, and focal length to keep the target centered as much as possible within its Field of View (FoV).

The problem naturally divides into two parts: (1) the observations periodically collected by the camera are used to compute estimates of the current and predicted target positions; and (2) voltages or other impulses are periodically chosen in order to direct the camera into a dynamic change of azimuth, elevation, and focal length so that, at the next sampling instant, the predicted target position will fall as closely as possible to the nominal center of the FoV. Specifically, one defines:

1 a "reference vector" $r_k = g_k(x_k)$ which is some function g_k of the target state x_k (e.g. r_k could be the predicted target position) and which is the quantity the sensor is expected to follow; and

2 a "controlled vector" $r_k^* = g_k^*(x_k^*)$ which is some function g_k^* of the state x_k^* of sensor s (e.g. r_k^* could be the nominal center of this sensor's field of view) and which is expected to follow the target.

We know that we are "following the target" successfully if a suitable distance metric $d(r_k, r_k^*)$ is kept as small as possible. (In linear control, this metric is typically a Mahalanobis distance $d(r, r^*)^2 = (r - r^*)^T R(r - r^*)$ where R is some symmetric, positive-definite matrix.) To avoid saturation of actuator inputs, the magnitudes $d(u_{k-1})$ of the control inputs u_{k-1} should be no larger than is necessary to effect the desired re-orientation of the sensor. A conventional (open-loop) *control law* is a sequence $u_0, u_1, ..., u_M$ of control inputs which jointly minimizes the expected values of the quantities $d(r_k, r_k^*)$ and $d(u_{k-1})$ throughout a time-period $k = 1, ..., M$ of interest.

In our approach, we use finite-set statistics (FISST) to directly generalize these ideas to multiplatform, multisensor, multitarget problems. The entire group of targets (of unknown number) can be reformulated as a single "meta-target" whose state is the set $X = \{x_1, ..., x_n\}$ of states x_i of all possible targets in the scene, detected or otherwise (including the possibility $X = \emptyset$, i.e. no targets present). The entire suite of sensors, each on their respective platforms, can be reformulated as a single "meta-sensor" whose collective state is the vector $\tilde{x}_k^*$ defined as the concatenation of the state-vectors x_k^* of all of the sensors (and more generally the state vectors x_k^{**} of all of the platforms). The "meta-control vector" $\tilde{u}_k$ is, likewise, the concatenation of the control vectors for all of the platforms, and all sensors on the platforms. Viewed in this manner, the sensor management problem can be reformulated as follows. The "meta-target" follows some trajectory in a generalized mathematical space. The "meta-sensor" attempts to predict the next state of the multitarget system and then to redirect itself so that its "meta-FoV" will overlap the predicted meta-state of the multitarget system at the next sampling instant. By analogy with the missile-tracking camera described above, one defines:

1 the "meta-reference" $R_k = g_k(X_k)$ to be some finite set-valued function of the multitarget state X_k (e.g. R_k could be the set of reference vectors of all of the targets) and which is the quantity that the meta-sensor is expected to follow; and

2 the "meta-control" $R_k^* = g_k^*(\tilde{\mathbf{x}}_k^*)$ to be some finite set-valued function of the meta-sensor state $\tilde{\mathbf{x}}_k^*$ (e.g. R_k^* could be the set of control vectors of all of the sensors) and which is expected to follow the meta-target.

Then, "following the meta-target" would have to mean that some suitable distance metric $d(R_k, R_k^*)$ is kept as small as possible. However, R_k and R_k^* are finite sets of objects which may have differing number. Consequently, a relatively conventional control-theoretic solution hinges upon our ability to define a theoretically justified but computationally tractable, multi-object miss-distance metric $d(R, R^*)$.

3.3.1   Multi-Object Distance Metrics.

One obvious choice for $d(R, R^*)$ is the *Hausdorff distance*, a common tool in image signal processing, where it is used to measure the similarity/dissimilarity of pixelated images. This is defined as

$$d_H(R, R^*) = \max\{\max_{\mathbf{r} \in R} \min_{\mathbf{r}^* \in R^*} \|\mathbf{r} - \mathbf{r}^*\|, \ \max_{\mathbf{r}^* \in R^*} \min_{\mathbf{r} \in R} \|\mathbf{r}^* - \mathbf{r}\|\}.$$

Although Hausdorff distance is "natural" for FISST in the sense that its metric topology on finite sets is the Mathéron topology, it has one peculiarity that limits its effectiveness as a multi-object miss-distance metric. *It is sensitive to differences in numbers of objects only when the objects being measured are close to each other; otherwise it tends to ignore object number.* Consequently, under this project we have made more systematic investigations of multitarget miss distance and this has led us to the *Wasserstein distances*.

Wasserstein distances are used in theoretical statistics to measure the distance between probability distributions [19], [9], and are also finding application in image processing [18], [31]. Suppose that $f(\mathbf{x})$ and $g(\mathbf{x})$ are two probability densities on some Euclidean space and that $h(\mathbf{x}, \mathbf{y})$ is a joint distribution whose marginals are just these two densities: $\int h(\mathbf{x}, \mathbf{y}) \, d\mathbf{x} = g(\mathbf{x})$ and $\int h(\mathbf{x}, \mathbf{y}) \, d\mathbf{y} = f(\mathbf{x})$. Then for any $p > 0$ the Wasserstein distance $d_{W,p}$ is defined as

$$d_{W,p}(f, g) = \left(\inf_h \int \|\mathbf{x} - \mathbf{y}\|^p \, h(\mathbf{x}, \mathbf{y}) \, d\mathbf{x} d\mathbf{y} \right)^{1/p}.$$

These are true distance metrics on the space of probability densities [20]. They are easily specialized to metrics on the space of non-empty finite subsets $X =$

$\{x_1, ..., x_n\}$ and $Y = \{y_1, ..., y_m\}$ by replacing X and Y by their empirical distributions:

$$f_X(u) = \frac{1}{n} \sum_{i=1}^{n} \delta_{x_i}(u), \qquad f_Y(u) = \frac{1}{m} \sum_{j=1}^{m} \delta_{y_j}(u).$$

In this case the Wasserstein distance $d_{W,p}(X, Y) = d_{W,p}(f_X, f_Y)$ is

$$d_{W,p}(X, Y) = \left(\inf_C \sum_{i=1}^{n} \sum_{j=1}^{m} C_{i,j} \|x_i - y_j\|^p \right)^{1/p}, \qquad (4)$$

where the infimum is taken over all $n \times m$ "transport matrices" [26] $C = \{C_{i,j}\}$ such that $C_{i,j} \geq 0$ for all $i = 1, ..., n; j = 1, ..., m$ and such that

$$\sum_{i=i}^{n} C_{i,j} = \frac{1}{m} \quad \text{(all } j = 1, ..., m\text{)}, \qquad \sum_{j=i}^{m} C_{i,j} = \frac{1}{n} \quad \text{(all } i = 1, ..., n\text{)}.$$

Although the metric topology for $d_{W,p}(X, Y)$ is not the Mathéron topology, its Borel-measurable subsets are the same as those for the Mathéron topology.

If $n = m$ equation 4 reduces to:

$$d_{W,p}(X, Y) = \min_\sigma \left(\frac{1}{n} \sum_{i=1}^{n} \|x_i - y_{\sigma i}\|^p \right)^{1/p}, \qquad (5)$$

where the minimum is computed over all possible permutations σ on the numbers $\{1, ..., n\}$. This means that $d_{W,p}(X, Y)$ is just the *minimal average l^p-distance between the two object-sets* and so it is the value of an *optimal assignment problem*. So, it can be tractably computed using standard optimal-assignment algorithms such as Munkres or Hungarian. Moreover, we have shown that the $n = m$ case can be generalized to $n \neq m$. Let γ be the greatest common divisor of m and n and let $n^* = n/\gamma$ and $m^* = m/\gamma$. Let X^* be the unordered list obtained from X by replacing each distinct x_i in X by n^* identical copies of x_i; and let Y^* be the unordered list obtained by replacing each distinct y_j in Y by m^* identical copies of y_j. Then both X^* and Y^* have the same number of elements $N = nn^* = mm^*$, which means that we can compute $d_{W,p}(X^*, Y^*)$ using the optimal-assignment formula of equation 5. It turns out that is the same thing as the Wasserstein distance, equation 4: $d_{W,p}(X, Y) = d_{W,p}(X^*, Y^*)$. That is, regardless of whether $n \neq m$ or $n = m$ the *Wasserstein distance can be interpreted as the solution of an optimal assignment problem and can be computed using standard optimal-assignment algorithms*. We have coded algorithms implementing the Wasserstein distances for $p < \infty$, using both the optimal assignment technique and the simplex method. Initial tests indicate that optimal assignment is much more efficient than the simplex method when $m = n$, but is of comparable efficiency otherwise.

3.3.2   Multi-Object Miss Distances for Sensor Management.

A good multi-object miss distance is not in itself a sufficient basis for multi-object control theory. To see why, suppose that $R = \{\mathbf{r}_1, ..., \mathbf{r}_n\}$ is a set of predicted track-locations and $R^* = \{\mathbf{r}_1^*, ..., \mathbf{r}_m^*\}$ the centers of the Fields of View (FoV's) of some collection of re-allocatable sensors. Then Euclidean distance cannot tell us anything about the uncertainty of $\mathbf{x}_i$ relative to all other tracks, nor tell us anything about the extent of the FoV of the j'th sensor (which is essential information if we are to know how to place the FoV of a sensor on top of a given target track). Because of this, Wasserstein distances based on Euclidean distance (or poorly chosen Mahalanobis distances) will *tend to follow the brightest targets, thereby losing the dimmest targets.* This behavior is the exact opposite of what one would want from a sensor management algorithm: find undetected targets and better resolve poorly detected ones, while at the same time losing as little information as possible about targets that have already been well resolved. Consequently, $\|\mathbf{x}_i - \mathbf{s}_j\|$ must be replaced by a different distance metric $d\left(\mathbf{r}_i, \mathbf{r}_j^*\right)$ that includes this missing information. We have begun experimentation with possible replacement metrics, especially of the Mahalanobis type.

3.4.   Nonlinear Filtering for Low-Observable Sensing

Under many circumstances, UAV-based sensors will be subject to less-than-ideal sensing conditions. A typical example occurs when signal-to-noise ratio (SNR) is so low that an imaging sensor images a target that is not actually visible in any given image-frame. Provided that data is collected rapidly enough, a *recursive Bayes nonlinear filter* can be used to detect, extract, and track the target from frame to frame. The theoretical foundations of the recursive Bayes filter are well-known [3, pp. 373-377], [22] and extensively investigated [55]. The basic idea is as follows. Suppose that $\mathbf{z}$ is data and $\mathbf{x}$ is the target state vector. Suppose that we have a measurement model $\mathbf{z} = F_k(\mathbf{x}) + \mathbf{W}_k$ where $\mathbf{W}_k$ is random noise. Also suppose that the between-measurements motion of the target is described by a motion model $\mathbf{x}_{k+1} = \Phi_k(\mathbf{x}_k) + \mathbf{V}_k$ where $\mathbf{V}_k$ is a random vector that models our uncertainty regarding target motion. From these quantities we can derive the sensor likelihood function $f_k(\mathbf{z}|\mathbf{x}) = f_{\mathbf{W}_k}(\mathbf{z} - F_k(\mathbf{x}))$ and Markov state-transition density $f_{k+1|k}(\mathbf{y}|\mathbf{x}) = f_{\mathbf{V}_k}(\mathbf{y} - \Phi_k(\mathbf{x}))$. Let $Z^k = \{\mathbf{z}_1, ..., \mathbf{z}_k\}$ be the observation-sequence. Then as noted earlier, the recursive Bayes filter is defined by the equations

$$f_{k+1|k}(\mathbf{x}|Z^k) = \int f_{k+1|k}(\mathbf{x}|\mathbf{y}) f_{k|k}(\mathbf{x}|Z^k) d\mathbf{y},$$

$$f_{k+1|k+1}(\mathbf{x}|Z^{k+1}) \propto f_{k+1}(\mathbf{z}_{k+1}|\mathbf{x}) f_{k+1|k}(\mathbf{x}|Z^k),$$

where the Bayes normalization constant in the second equation is

$$f_{k+1}(\mathbf{z}_{k+1}|Z^k) = \int f(\mathbf{z}_{k+1}|\mathbf{x}) f_{k+1|k}(\mathbf{x}|Z^k) d\mathbf{x}.$$

(*Note*: If one assumes that kinematics $\mathbf{x}$ evolves continuously rather than discretely in time, there is another way of constructing $f_{k+1|k}(\mathbf{x}|Z^k)$. In this case it has been well-known since the mid-1960s [24, p. 165] that $f_{k+1|k}(\mathbf{x}|Z^k)$ is a solution of a partial differential equation called the Fokker-Planck equation (FPE).) Despite the existence of good numerical techniques at that time— e.g., the Alternating Direction Implicit (ADI) technique for solving the FPE [57, pp. 142-150] dates from 1966 [61]—nonlinear filtering was dismissed as computationally intractable for real-time application [55, p. 465].

Despite the fact that computational power has increased greatly during the last thirty years, this assessment is still largely (but, fortunately, not entirely) valid—the development of real-time nonlinear filters remains an active and difficult basic research area. This is because an *infinite number* of parameters are required to represent the posteriors $f_{k|k}(\mathbf{x}|Z^k)$ in low-SNR applications (because of their tendency to have a large number of randomly-shifting modes). Even if these are *compressed* into n parameters $\vec{\pi} = (\pi_1(k), ..., \pi_n(k))$, this merely transforms the second nonlinear filtering equation (Bayes' rule) into a system of n nonlinear equations in n unknowns, $\vec{\pi}^{k+1|k+1} = G(\vec{\pi}^{k+1|k})$. In general, $n = O(n_0^d)$, where n_0 is the number of parameters in the *one-dimensional* filter and d is the dimensionality of the problem. Even if this system of equations were linear, computational efficiency would be order $O(n_0^{2d})$, with best-possible efficiency being $O(n_0^d)$. Consequently, new computational strategies are a fundamental necessity.

In cooperation with Prof. Michael Kouritzen of the University of Alberta (Edmonton), LMTS is sponsoring the development of fundamental new real-time computational nonlinear filtering approaches such as infinite-dimensional exact filters [27] and particle-system filters [1].

Infinite-Dimensional Exact Filter. An exact filter is one that achieves computational tractability by assuming that posterior distributions belong to some family of densities $f(\mathbf{x}; \mathbf{p})$—e.g., Gaussians or generalized exponentials— parametrized by some *finite-dimensional* vector $\mathbf{p}$. Exact filters will perform badly when the real-world posteriors deviate significantly from this form.

Also, restricting posteriors to the form $f(\mathbf{x}; \mathbf{p})$ restricts the generality of the likelihood function $f(\mathbf{z}|\mathbf{x})$. Well-known examples are the Kalman, Kalman-Bucy, Beneš [7], and Daum [15] filters. Kouritzin's *convolution filter* is apparently the only existing *infinite-dimensional* exact filtering technique [27]. Because it in effect allows the parameter $\mathbf{p}$ to be infinite-dimensional it can encompass more general posteriors and measurement models than the Beneš or Daum filters (which it subsumes as special cases). Assume that the old posterior $f_{k|k}(\mathbf{x}|Z^k)$ and its time-update $f_{k+1|k}(\mathbf{x}|Z^k)$ must be related by $f_{k+1|k}(\mathbf{x}|Z^k) = \int K(\mathbf{x}-\mathbf{y}) \, f_{k|k}(\mathbf{y}|Z^k) d\mathbf{y}$ for some (unknown) convolution kernel K. Once K has been determined off-line, one can compute $f_{k+1|k}$ in real time using Fast Fourier Transforms.

Branching Particle Systems Filter. The basic idea behind this filter is to time-propagate a large system of "particles" that behave like statistical samples drawn from the evolving posterior distribution. Each sample is propagated as though it were a moving particle, based on the assumed dynamics of the target being tracked. The number and distribution of the particles can be varied by allowing particles to vanish (if they appear to be low-probability) and "branch" (split) into multiple offspring (if they appear to be high-probability). Branching particle-system filters are known to have attractive properties:

(1) *very strong guaranteed-convergence properties*: for *every* sequence of observations a particle-systems filter will converge almost surely to the true posterior distribution.

(2) they can accommodate *arbitrary measurement models*.

(3) they can accommodate *very general motion models*. Specifically, such filters are based on much more general theory than approaches based on the Fokker-Planck equation (which assume differentiability and other properties). Consequently, they can accommodate much more general motion models such as discontinuous or near-discontinuous motion and motion with heavy-tailed statistics.

(4) they have *very good computational properties*. Specifically, if p is the number of particles, d is the number of state parameters, and $N = p^d$ is the number of unknown quantities, then such filters have computational order $O(N)$ in the detection phase and order $O(p)$ in the tracking phase (i.e., once detection has been accomplished).

LMTS and SSCI are currently applying some of these approaches to joint HRRR tracking and identification (for AFRL/SNAT) and to tracking ground targets using Air Borne Laser ladar data (for AFRL/DEBA). We will investigate the application of nonlinear filters for UAV-based sensors encountering low-SNR conditions.

3.5.   Robust Multi-Agent Data Fusion

The Bayes filter exploits to the best possible advantage the high-fidelity knowledge about the sensor contained in the likelihood function $f_k(\mathbf{z}|\mathbf{x})$. Many forms of data are well enough characterized that $f(\mathbf{z}|\mathbf{x})$ can indeed be constructed with sufficient fidelity. Other kinds of data—e.g., SAR or HRRR—are proving to be so difficult to simulate that it is unclear that sufficiently high fidelity will ever be achieved in real-time operation. If $f(\mathbf{z}|\mathbf{x})$ is too imperfectly understood, an algorithm will "waste" much data trying (and perhaps failing) to overcome the mismatch between reality and the deficient model $f(\mathbf{z}|\mathbf{x})$. Still other kinds of data—features extracted from signatures, English-language statements received over link, rules drawn from knowledge bases, etc.—are so "ambiguous" (poorly characterized from a statistical point of view) that probabilistic approaches are not even obviously applicable.

For example, consider the case of INTELL features extracted from data by operators and distributed via datalink. We cannot statistically characterize an operator, since his/her performance is affected by unknowable factors such as "fat-fingering," fatigue, excessive workload caused by high data rates, differences in training, differences in ability, and so on. How does one construct a likelihood $f_{INTELL}(\mathbf{z}_{INTELL}|\mathbf{x})$ for *this* kind of "sensor"? Because of such issues, it will typically be necessary to combine data from multiple sources, on-board or off-board, via *data fusion*. For example, a joint likelihood

$$f(\mathbf{z}_{HRRR}, \mathbf{z}_{INTELL}|\mathbf{x}) = f_{HRRR}(\mathbf{z}_{HRRR}|\mathbf{x}) \cdot f_{INTELL}(\mathbf{z}_{INTELL}|\mathbf{x})$$

for fusing HRRR signatures with INTELL datalink features can be constructed only if we know how to construct likelihoods for ill-characterized sensors, e.g. both HRRR and INTELL. In FISST we use *generalized likelihood functions* $\rho(\mathbf{z}|\mathbf{x})$ to model the *uncertainties* as well as the *certainties* in data $\mathbf{z}$ that is difficult to characterize either because its likelihood $f(\mathbf{z}|\mathbf{x})$ is not known with sufficient fidelity (e.g. HRRR, SAR) or because $\mathbf{z}$ is inherently difficult to model in and of itself (e.g., datalink features, natural-language statements, rules). We then fuse disparate kinds of data using a *generalized joint likelihood function*:

$$\rho(\mathbf{z}_1^{[1]}, ..., \mathbf{z}_s^{[s]}|\mathbf{x}, c) = \rho(\mathbf{z}_1^{[1]}|\mathbf{x}) \cdots \rho(\mathbf{z}_s^{[s]}|\mathbf{x}).$$

If data sources are not independent, unknowable statistical correlations can often be modeled using *correlation operators*. For example, one useful correlation operator is $x \wedge y \triangleq xy/(x + y - xy)$.

In addition, it is expected that, even with the presence of multi-agent sensor management, targets will not be effectively detected, tracked, or identified without fusing data collected from more than one sensor and more than one UAV. Because it provides a rigorous and systematic approach for constructing multisensor-multitarget likelihood functions, finite-set statistics potentially provides a systematic foundation for multiplatform, multisensor, multitarget data fusion. The goal of this research will be to develop the foundations of such a distributed data fusion, and develop means of rendering them practicable.

We will investigate the application of robust data fusion to Multi-Agent Coordination in the presence of ill-characterized onboard or offboard sensors. The goal will be to produce a unified, systematic, and practical approach to multiplatform, multisensor, multitarget data fusion.

In the subsections that follow, we briefly summarize the elements of this approach

3.5.1 Random Set Models of "Ambiguous" Data.

The FISST approach to data that is difficult to statistically characterize is based on the following idea: *Ambiguous data can be probabilistically represented as random closed subsets of (multisource) measurement space.*

Consider a simple example. Suppose that we know that we are given a measurement model $z = C\mathbf{x} + \mathbf{W}$, where $\mathbf{x}$ is target state, $\mathbf{W}$ is random noise, and C is an invertible matrix. Let B be an "ambiguous observation" in the sense that it is a subset of measurement space that constrains the possible values of z: $B \ni z$. Then the random variable Γ defined by $\Gamma = \{C^{-1}(z-\mathbf{W}) \mid z \in B\}$ is the randomly varying subset of all target states that are consistent with this ambiguous observation. That is, the ambiguous observation B also *indirectly* constrains the possible target *states*. Suppose, in addition, that we are not very certain about the validity of the constraint $z \in B$ but, rather, believe that there are many possible constraints—of varying plausibility–on z. Then we would model this kind of ambiguity as a randomly varying subset Θ of measurements, where the probability $\Pr(\Theta = B)$ represents our degree of belief in the plausibility of the specific constraint B. The random subset of all states that are consistent with Θ would then be $\Gamma = \{C^{-1}(z - \mathbf{W}) \mid z \in \Theta\}$. (*Caution*: The random closed subset Θ is a model of a *single observation collected by a single source*. It should not be confused with a *multisensor, multitarget* observation-set Σ, whose instantiations $\Sigma = Z$ are finite sets of the form $Z = \{z_1, ..., z_m, \Theta_1, ..., \Theta_{m'}\}$, where $z_1, ..., z_m$ are individual conventional observations and $\Theta_1, ..., \Theta_{m'}$ are random-set models of individual ambiguous observations.)

It is one thing to recognize that random sets provide a common probabilistic foundation for various kinds of statistically ill-characterized data. It is quite another to construct practical random set representations of such data. We indicate how three kinds of ambiguous data—*imprecise, vague*, and *contingent*—can be represented probabilistically by random sets.

Vague (fuzzy) data: A fuzzy membership function on some (finite or infinite) universe U is a function that assigns a number $f(u)$ between zero and one to each member u of U. The random subset $\Sigma_A(f)$, called the *canonical random set representation* of the fuzzy subset f, is defined by

$$\Sigma_A(f) \triangleq \{u \in U \mid A \leq f(u)\}.$$

Imprecise data: A Dempster-Shafer body of evidence B on some space U consists of non-empty subsets $B : B_1, ..., B_b$ of U and nonnegative weights $b_1, ..., b_b$ that sum to one. Define the random subset Σ of U by $p(\Sigma = B_i) = b_i$ for $i = 1, ..., b$. Then we say that Σ is the random set representation of B and write $B = B^\Sigma$. The Dempster-Shafer theory can be generalized to the case when the B_i are fuzzy membership functions. Such "fuzzy bodies of evidence" can also be represented in random set form.

Contingent data: Knowledge-base rules have the form $X \Rightarrow S =$ "*if X then S*", where S, X are subsets of a (finite) universe U. It has been shown [38], [39] that there is at least one way to represent knowledge-base rules in random set form. Specifically, let Φ be a *uniformly distributed* random subset of U—that is, one whose probability distribution is $p(\Phi = S) = 2^{-|U|}$ for all $S \subseteq U$. A random set representation $\Sigma_\Phi(X \Rightarrow S)$ of the rule $X \Rightarrow S$ is:

$$\Sigma_\Phi(X \Rightarrow S) \triangleq (S \cap X) \cup (X^c \cap \Phi).$$

3.5.2 Generalized Likelihood Functions.

The next step in a strict Bayesian formulation of the ambiguous-data problem would be to specify a *likelihood function* for ambiguous evidence that models our understanding of how likely it is that we will observe the specific ambiguous datum Θ, given that a target of state $\mathbf{x}$ is present. At this point we encounter practical problems. The required likelihood function must have the form

$$f(\Theta|\mathbf{x}) = \frac{\Pr(\mathcal{R} = \Theta, \mathbf{X} = \mathbf{x})}{\Pr(\mathbf{X} = \mathbf{x})},$$

where $\mathcal{R}$ is a random variable that ranges over all random closed subsets Θ of measurement space. However, $f(\Theta|\mathbf{x})$ cannot be a conventional likelihood function unless it satisfies a normality equation of the form $\int f(\Theta|\mathbf{x})d\Theta = 1$, where $\int f(\Theta|\mathbf{x})d\Theta$ is an integral that sums over all closed random subsets of measurement space. It is very unclear how one would go about constructing a $f(\Theta|\mathbf{x})$ that not only models a particular real-world situation but, also, provably integrates to unity. If we knew enough to specify $f(\Theta|\mathbf{x})$ with such exactitude, it would probably also be possible to construct a high-fidelity *conventional* likelihood $f(\mathbf{z}|\mathbf{x})$.

To address this problem, FISST employs an engineering compromise based on the following fact. *Bayes' rule is very general: It applies to all events and not just those having the specific Bayesian form $\mathbf{X} = \mathbf{x}$ or $\mathcal{R} = \Theta$.* That is, Bayes' rule states that $\Pr(E_1|E_2)\Pr(E_2) = \Pr(E_2|E_1)\Pr(E_1)$ for any events

E_1, E_2. Consequently let E_Θ be any event with some specified functional dependence on the ambiguous measurement Θ—for example, $E_\Theta : \Theta \supseteq \Sigma$ or $E_\Theta : \Theta \cap \Sigma \neq \emptyset$, where Θ, Σ are random closed subsets of observation space. Then

$$f(\mathbf{x}|E_\Theta) = \frac{\Pr(\mathbf{X} = \mathbf{x}, E_\Theta)}{\Pr(E_\Theta)} = \frac{\rho(\Theta|\mathbf{x}) \, f_0(\mathbf{x})}{\Pr(E_\Theta)},$$

where $f_0(\mathbf{x}) = \Pr(\mathbf{X} = \mathbf{x})$ is the prior distribution on $\mathbf{x}$ and where I call $\rho(\Theta|\mathbf{x}) \triangleq \Pr(E_\Theta|\mathbf{X} = \mathbf{x})$ a *generalized likelihood function*. Notice that $\rho(\Theta|\mathbf{x})$ will usually be *unnormalized* since events E_Θ are not mutually exclusive. Joint generalized likelihood functions can be defined in the same way.

Given this engineering compromise, we can use Bayes' rule to compute posterior distributions conditioned on ambiguous data modeled by closed random subsets $\Theta_1, ..., \Theta_m$:

$$f(\mathbf{x}|\Theta_1, ..., \Theta_m) \propto \rho(\Theta_1, ..., \Theta_m|\mathbf{x}) \, f_0(\mathbf{x})$$

with proportionality constant $p(\Theta_1, ..., \Theta_m) \triangleq \int \rho(\Theta_1, ..., \Theta_m|\mathbf{x}) \, f_0(\mathbf{x}) d\mathbf{x}$.

To produce generalized likelihood functions usable in application, we note that *generalized likelihood functions can be constructed using the concept of "model-matching" between observations and model signatures*. That is, let Θ be the random closed subset of measurement space U that models a particular piece of evidence, and let $\Sigma|_{\mathbf{X}=\mathbf{x}}$ be another random closed subset of U that models the generation of observations. We say that data Θ "geometrically matches" signature model $\Sigma|_{\mathbf{X}=\mathbf{x}}$ if $\Theta \supseteq \Sigma|_{\mathbf{X}=\mathbf{x}}$ (complete consistency between observation and signature), $\Theta \cap \Sigma|_{\mathbf{X}=\mathbf{z}} \neq \emptyset$ (non-contradiction between observation and signature), etc. Depending on the matching criterion we get a different generalized likelihood function:

$$\rho_1(\Theta|\mathbf{x}) = \Pr(\Sigma|_{\mathbf{X}=\mathbf{x}} \cap \Theta \neq \emptyset), \qquad \rho_2(\Theta|\mathbf{x}) = \Pr(\Sigma|_{\mathbf{X}=\mathbf{x}} \subseteq \Theta).$$

Practical generalized likelihood functions can be constructed through suitable choice of the random-set "model signatures" $\Sigma_{\mathbf{x}} \triangleq \Sigma|_{\mathbf{X}=\mathbf{x}}$.

For example, let g be a "fuzzy observation" and $f_{\mathbf{x}}$ a "fuzzy signature model" (where both g and $f_{\mathbf{x}}$ are fuzzy membership functions on measurement space). If $\Theta_g \triangleq \Sigma_A(g)$ and $\Sigma_{\mathbf{x}} \triangleq \Sigma_A(f_{\mathbf{x}})$ it can be shown that:

$$\rho(g|\mathbf{x}) = \Pr(\Sigma|_{\mathbf{X}=\mathbf{x}} \cap \Theta_g \neq \emptyset) = \max_{\mathbf{z}} \min\{g(\mathbf{z}), f_{\mathbf{x}}(\mathbf{z})\}.$$

4.   Conclusions

In this paper, we summarized a research program being conducted by Lockheed Martin Tactical Systems and its subcontractor Scientific Systems Co., Inc. under contract F49620-01-C-0031 of the AFOSR Cooperative Control Theme 2. We identified a range of technologies that we believe may ultimately lead to unification of Multi-Agent Collection (distributed robust data fusion and data interpretation) and Multi-Agent Coordination (distributed robust platform and sensor monitoring and control). Our approach combines theoretically rigorous statistics (hybrid control, point process theory) with potential practicability (computational hybrid control, computational nonlinear filtering).

Our approach is based on a potential unification of innovative distributed hybrid-systems control theory with an "engineering-friendly" version of point process theory called "finite-set statistics" (FISST). We summarized the manner in which we intend to integrate the behavioral approach with the VS architecture, using leader following techniques at the lowest level. We summarized the FISST approach, as well as its extension to "ambiguous" data types such as attributes, rules, and natural-language statements. We sketched the manner in which we intend to apply it as a theoretical basis for multiplatform, multisensor, multitarget sensor management. As the project progresses, we intend to integrate these parallel efforts to arrive at the desired unification.

References

[1] D.J. Ballantyne, H.Y. Chan, and M.A. Kouritzin, "A novel branching particle method for tracking", *SPIE Proc.*, 4048: 277-287, 2000.

[2] M. Bardin, "Multidimensional Point Processes and Random Closed Sets", *J. Applied Prob.*, 21: 173-178, 1984.

[3] Y. Bar-Shalom and X.-R. Li, *Estimation and Tracking: Principles, Techniques, and Software*, Artech House, 1993.

[4] R. W. Beard and F. Y. Hadaegh, "Constellation templates: An approach to autonomous formation flying", *World Automation Congress*, pages 177.1–177.6, Anchorage, Alaska, May 1998.

[5] R. W. Beard, J. Lawton, and F. Y. Hadaegh, "A feedback architecture for formation control", *Proc. Amer. Control Conf.*, pp. 4087–4091, Chicago, IL., June 2000.

[6] A. Bemporad and M. Morari, "Control of systems integrating logic, dynamics and constraints", *Automatica*, 35, 1999.

[7] V.E. Beneš, "Exact finite-dimensional filters for certain diffusions with nonlinear drift", *Stochastics*, 5: 65-92, 1981.

[8] R.E. Bethel and G.J. Paras, "A PDF multitarget-tracker", *IEEE Trans AES*, 30: 386-403, 1994.

[9] P.J. Bickel and D. A. Feedman, "Some asymptotic theory for the bootstrap", *Annals of Statistics*, 9: 1196-1217, 1981.

[10] V. Borkar, V. Chandru and S. Mitter, "Mathematical programming embeddings of logic", Preprint, 2000.

[11] S. Boyd, L. El Ghaoui, E. Feron and V. Balakrishnan, *Linear Matrix Inequalities in Systems and Control Theory*, SIAM, 1994.

[12] M. Branicky, *Studies in Hybrid Systems: Modeling, Analysis and Control*, Ph.D dissertation, MIT, June, 1995.

[13] T. H. Cormen, C. E. Leiserson and R. L. Rivest, *Introduction to Algorithms*, MIT Press, 1990.

[14] D.J. Daley and D. Vere-Jones, *An Introduction to the Theory of Point Processes*, Springer-Verlag, 1988.

[15] F. Daum, "Exact finite dimensional nonlinear filters for continuous time processes with discrete time measurements", in *Proc. IEEE Conf. Dec. and Contr.*, pp. 16-22, 1984.

[16] N. Elia and B. Brandin, "Verification of an automotive active leveler", *Proc. American Cont. Conf*, pp. 2476-2480, 1999.

[17] M. Fliess, J. Lévine, P. Martin, and P. Rouchon, "Linéarisation par bouclage dynamique et transformations de lie-bäcklun", *D.R. Acad. Sci. Paris, t. 317, Serie I*, pp. 981–986, 1993

[18] W. Gangbo and R.J. McCann, "Shape recognition via Wasserstein Distance", *Quarterly of Applied Math.*, Vol LVIII No. 4: 705-737, 2000.

[19] M. Gelbrich, "On a formula for the L^2 Wasserstein Metric between measures on Euclidean and Hilbert Spaces", *Math. Nachr.*, 147: 185-203, 1990.

[20] C.R. Givens and R.M.Shortt, "A class of Wasserstein Metrics for probability distributions", *Michigan Math. J.*,31: 231-240, 1984.

[21] I.R. Goodman, R.P.S. Mahler, and H.T. Nguyen, *Mathematics of Data Fusion*, Kluwer Academic Publishers, 1997.

[22] Y.C. Ho and R.C.K. Lee, "A Bayesian approach to problems in stochastic estimation and control", *IEEE Trans. AC*, AC-9: 333-339, 1964.

[23] H.J. Hooker, *Logic-based methods for optimization: Combining optimization and constraint satisfaction*, Wiley, 2000.

[24] A.H. Jazwinski, *Stochastic Processes and Filtering Theory*, Academic Press, 1970.

[25] V. Kapila, A. G. Sparks, J. M. Buffington, and Q. Yan, "Spacecraft formation flying: Dynamics and control", *J. Guidance, Control, and Dynamics*, 23: 561–564, 2000.

[26] V. Klee and C. Witzgall, "Facets and vertices of transportation polytopes", *Mathematics of the Decision Sciences, Part I*, American Mathematical Society, pp. 257-282, 1968.

[27] M.A. Kouritzin, "On exact filters for continuous signals with discrete observations", *IEEE Trans. Auto. Control*, 43: 709-71, 1998.

[28] R. Kruse, E. Schwencke, and J. Heinsohn, *Uncertainty and Vagueness in Knowledge-Based Systems*, Springer-Verlag, 1991.

[29] J. Lawton, *Multiple Spacecraft Elementary Formation Maneuvers*, PhD thesis, Brigham Young University, Provo, UT 84602, 2000.

[30] J. Lawton, B. Young, and R. Beard, "A decentralized approach to elementary formation maneuvers", *IEEE Trans. Robotics and Automation*, to appear.

[31] E. Levina and P. Bickel, "The Earth Mover's Distance is the Mallow's Distance: Some Insights From Statistics", *Proc. IEEE 8th Int'l Conf. on Computer Vision*, Vol. II: 251-256, July 9-12 2001.

[32] M.A. Lewis and K.-H. Tan, "High precision formation control of mobile robots using virtual structures", *Autonomous Robots*, 4: 387–403, 1997.

[33] S.-M. Li, J. D. Boskovic, and R. K. Mehra, "Globally stable automatic formation flight control in two dimensions", *2001 AIAA Guidance, Navigation, and Control Conf.*, 2001.

[34] J. Lygeros, C. Tomlin, S. Sastry, "Controller for reachability specifications for hybrid systems", *Automatica*, March 1999.

[35] R. Mahler, "Approximate multi-sensor, multi-target detection, tracking, and target identification using a multitarget first-order moment statistic", submitted to *IEEE Trans. AES*, 2001.

[36] R. Mahler, *An Introduction to Multisource-Multitarget Statistics and Its Applications*, Lockheed Martin Technical Monograph, 104 pages, 2000.

[37] R. Mahler (2001) "Multitarget moments and their application to multitarget tracking", *Proc. Workshop on Estimation, Tracking, and Fusion: A Tribute to Yaakov Bar-Shalom*, May 17, 2001, Naval Postgraduate School, Monterey CA, pp. 134-166, ISBN 0-9648-3124-4

[38] R. Mahler (1996) "Representing Rules as Random Sets, I: Statistical Correlations Between Rules", *Information Sciences*, Vol. 88, pp. 47-68

[39] R. Mahler (1996) "Representing Rules as Random Sets, II: Iterated Rules", *Int'l Jour. Intelligent Sys.*, Vol. 11, pp. 583-610

[40] R. Mahler, "Random set theory for target tracking and identification", in D.L. Hall and J. Llinas (eds.), *Handbook of Multisensor Data Fusion*, CRC Press, Boca Raton FL, pp. 14-1 to 14-133, 2001.

[41] R. Mahler, "A theoretical foundation for the Stein-Winter 'Probability Hypothesis Density (PHD)' multitarget tracking approach", *Proc. 2000 MSS Nat'l Symp. on Sensor and Data Fusion*, Vol. I (Unclassified), San Antonio TX, Infrared Information Analysis Center, pp. 99-118, 2000.

[42] G. Mathéron, *Random Sets and Integral Geometry*, J. Wiley, 1975.

[43] C.R. McInnes, "Autonomous ring formation for a planar constellation of satellites", *J. of Guidance, Control, and Dynamics*, 18: 1215–1217, 1995.

[44] T. McLain and R. Beard, "Cooperative rendezvous of multiple unmanned air vehicles", *Proc. AIAA Guidance, Navigation and Control Conf.*, Denver, CO, August 2000.

[45] J.E. Moyal, "The general theory of stochastic population processes", *Acta Mathematica*, 108: 1-31, 1962.

[46] S. Musick, K. Kastella, and R. Mahler, "A practical implementation of joint multitarget probabilities", *SPIE Proc.*, 3374: 26-37, 1998.

[47] N. Portenko, H. Salehi, and A. Skorokhod, "On optimal filtering of multitarget tracking systems based on point processes observations", *Random Operators and Stochastic Equations*, 1: 1-34, 1997.

[48] A. W. Proud, M. Pachter, and J. J. D'Azzo, "Close formation flight control", *AIAA Guidance, Navigation, and Control Conf.*, pp. 1231–1246, Portland, OR, August 1999.

[49] B.D. Ripley, "Locally finite random sets: foundations for point process theory", *Annals of Prob.*, 4: 983-994, 1976.

[50] P. Rouchon, M. Fliess, J. Lévine, and Ph. Martin, "Flatness, motion planning and trailer systems", *Proc. IEEE Control and Decision Conf.*, pp. 2700–2705, 1993.

[51] O. Shakernia, G. Pappas and S. Sastry, "Decidable controller synthesis for classes of linear systems", *Hybrid Systems*, 1999.

[52] S. Sheikholeslam and C. A. Desoer, "Control of interconnected nonlinear dynamical systems: The platoon problem", *IEEE Trans. Auto. Control*, 37: 806–810, 1992.

[53] D.L. Snyder and M.I. Miller, *Random Point Processes in Time and Space, Second Edition*, Springer, 1991.

[54] H.W. Sorenson, "Recursive estimation for nonlinear dynamic systems", in J.C. Spall, editor, *Bayesian Analysis of Statisical Time Series and Dynamic Models*, Marcel Dekker, 1988.

[55] H.W. Sorenson and D.L. Alspach, "Recursive Bayesian Estimation Using Gaussian Sums", *Automatica*, 7: 465-479, 1971.

[56] A. Srivastava, M.I. Miller, and U. Grenander, "Jump-diffusion processes for object tracking and direction finding", *Proc. 29th Allerton Conf. on Communication, Control, and Computing*, U. of Illinois Urbana, pp. 563-570, 1991.

[57] C. Strikwerda, *Finite Difference Schemes and Partial Differential Equations*, Chapman & Hall, 1989.

[58] L.D. Stone, C.A. Barlow, and T.L. Corwin, *Bayesian Multiple Target Tracking*, Artech House, 1999.

[59] D. Stoyan, W.S. Kendall, and J. Meche, *Stochastic Geometry and Its Applications*, Second Edition, John Wiley & Sons, 1995.

[60] X. Yun, G. Alptekin, and O. Albayrak, "Line and circle formation of distributed physical mobile robots", *J. Robotic Systems*, 14: 63–76, 1997.

[61] E.L. Wahspress, *Iterative Solution of Elliptic Systems and Application to the Neutron Diffusion Equations of Reactor Physics*, Prentice-Hall, 1966.

[62] P. K. C. Wang and F. Y. Hadaegh, "Coordination and control of multiple microspacecraft moving in formation", *J. Astronautical Sciences*, 44: 315–355, 1996.

Chapter 12

OPTIMAL TRAJECTORIES FOR COOPERATIVE CLASSIFICATION *

Meir Pachter

Professor, Dept. of Electrical and Computer Engineering
Air Force Institute of Technology, Wright-Patterson AFB OH, USA
meir.pachter@afit.edu

Jeffrey Hebert

Major, U.S. Air Force
jeffrey.hebert@afit.edu

Abstract   A projected role for autonomous uninhabited air vehicles is to classify and subsequently attack time critical targets, as well as perform battle damage assessment after an attack. Thus, the problem of determining optimal look angles for automatic target recognition/classification is addressed first. Next, minimum time trajectories for this mission, for a vehicle with a minimum turning radius, are constructed. Lastly, an algorithm for performing cooperative classification and/or battle damage assessment involving more than one air vehicle, is presented.

1.   Optimal Look Angles for ATR

It is widely recognized that the performance of Automatic Target Recognition (ATR) algorithms is strongly dependent on the target's aspect angle. Using the right look angle is conducive to good ATR.

The targets' universe of discourse is stipulated to consist of rectangles located in a plane and with an arbitrary orientation. We assume that the probability of successfully classifying a rectangular target with sides a and b, using

---

*The views expressed in this article are those of the authors and do not reflect the official policy of the U.S. Air Force, Department of Defense, or the U.S. Government.

S. Butenko et al. (eds.), Cooperative Control: Models, Applications and Algorithms, 253-281.
© 2003 *Kluwer Academic Publishers.*

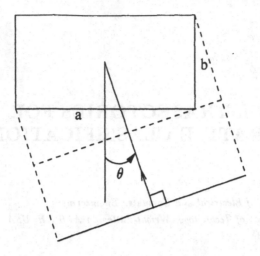

Figure 12.1. Geometry for the Optimal Look Angle ATR Problem

an Automatic Target Recognition (ATR) algorithm, is directly proportional to the projection of the visible sides of the rectangular object onto a line perpendicular to the aspect angle θ of the viewing sensor, as shown in Figure 12.1.

Without loss of generality assume $a \geq b$. The probability of classification $\rho(\theta)$ is then calculated as follows:

$$\rho(\theta) = \begin{cases} \frac{a\cos\theta + b\sin\theta}{a+b} & 0 \leq \theta \leq \frac{\pi}{2} \\ \frac{-a\cos\theta + b\sin\theta}{a+b} & \frac{\pi}{2} \leq \theta \leq \pi \\ \frac{-a\cos\theta - b\sin\theta}{a+b} & \pi \leq \theta \leq \frac{3\pi}{2} \\ \frac{a\cos\theta - b\sin\theta}{a+b} & \frac{3\pi}{2} \leq \theta \leq 2\pi. \end{cases} \tag{1}$$

A plot of $\rho(\theta)$ for $-\pi < \theta < \pi$ is provided in Cartesian coordinates in Figure 12.2(a). A polar plot is provided in Figure 12.2(b).

Consider the function $\rho(\theta)$ where $\theta \in (0, \pi/2)$. The first and second derivatives of $\rho(\theta)$ are

$$\rho'(\theta) = \frac{-a\sin\theta + b\cos\theta}{a+b} \tag{2}$$

$$\rho''(\theta) = \frac{-a\cos\theta - b\sin\theta}{a+b}. \tag{3}$$

Evidently, we have an extreme value of ρ at the aspect angle

$$\theta^* = \text{Arctan}\left(\frac{b}{a}\right). \tag{4}$$

Substituting Eq. (4) into Eq. (3) yields the result

$$\rho''(\theta^*) = -\frac{\sqrt{a^2 + b^2}}{a + b} < 0,$$

i.e., we have a relative maximum at the extreme point $\theta = \theta^*$. Substituting Eq. (4) into Eq. (1) finally yields

$$\rho(\theta^*) = \frac{\sqrt{a^2 + b^2}}{a + b}.$$

Due to symmetry, this maximum probability is repeated at the aspect angles

$$\theta^* = \left\{ \frac{\pi}{2} + \text{Arctan}\left(\frac{b}{a}\right), -\text{Arctan}\left(\frac{b}{a}\right), -\frac{\pi}{2} - \text{Arctan}\left(\frac{b}{a}\right). \right\}$$

At $\theta = \pi/2$, we have $\rho = \frac{b}{a+b}$. However, $\rho'(\pi/2)$ does not exist. Since

$$\rho'(\theta) < 0 \quad \text{for Arctan}\left(\frac{b}{a}\right) < \theta < \frac{\pi}{2}$$

and

$$\rho'(\theta) > 0 \quad \text{for } \frac{\pi}{2} < \theta < \frac{\pi}{2} + \text{Arctan}\left(\frac{b}{a}\right),$$

we conclude $\rho(\pi/2)$ is a relative minimum. Similarly, we find that $\rho(-\pi/2)$ is a relative minimum. Lastly, we note that

$$\rho(0) = \frac{a}{a + b}.$$

1.1. Multiple Look Classification

For multiple-look classification, where a predetermined probability of correct classification threshold needs to be achieved for a target to be classified, possibly employing more than one air vehicle, we consider the probability of correct classification, $\rho(\theta_i)$ for look i, where the aspect angle is θ_i. Thus, the probability ρ of correctly identifying a target after n independent snapshots have been taken is

$$\rho = 1 - \prod_{i=1}^{n} [1 - \rho(\theta_i)] \tag{5}$$

and for the special case of two looks, $n = 2$, we have the probability of correctly having classified the target

$$\rho = \rho(\theta_1) + \rho(\theta_2) - \rho(\theta_1)\rho(\theta_2). \tag{6}$$

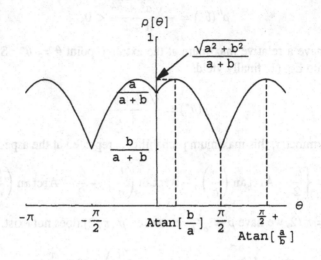

$$(a)$$

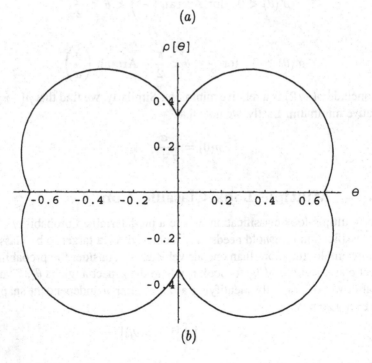

$$(b)$$

Figure 12.2. Parametric Plot of Probability of Classification for $-\pi \leq \theta \leq \pi$

The function $\rho(\theta)$ is the probability of being able to classify the target when a snapshot of the target is taken from an aspect angle of θ. Strictly speaking, a snapshot of the target is taken and is being reviewed, after which a binary (yes/no) decision is made:

- The target can be identified/classified

- The target cannot be identified/classified

Then, if n snapshots of the target are taken from aspect angles $\theta_1, \theta_2, \ldots, \theta_n$, the probability that the target has been correctly identified/classified, is given by Eq. (5). Here, the n snapshots are taken, and only thereafter, are they examined.

2. Optimal Angular Separation for Second Look

Consider the scenario where an initial snapshot of the target has been taken at an unknown aspect angle, θ. We wish to find the optimal change in aspect angle Δ that maximizes the average probability of identifying the target in two passes. Δ directly translates into a change in the vector of approach to the target. Without loss of generality, assume $\theta \in [0, \pi/2]$, $\Delta \in [0, \pi/2]$ and $0 \le |\Delta - \theta| \le \frac{\pi}{2}$. Thus, the optimization problem is posed

$$\max_{\Delta} \frac{1}{\pi/2} \left[\int_0^{\frac{\pi}{2}} \left(\rho(\theta) + \rho(\theta + \Delta) - \rho(\theta)\rho(\theta + \Delta) \right) \, d\theta \right]. \qquad (7)$$

Substituting Eq. (1) into Eq. (7), we obtain the following cost function

$$J = \int_0^{\frac{\pi}{2}} \left[\frac{a \cos \theta + b \sin \theta}{a + b} + \frac{a \cos(\Delta + \theta) + b \sin(\Delta + \theta)}{a + b} \right.$$
$$\left. - \frac{(a \cos \theta + b \sin \theta) \, (a \cos(\Delta + \theta) + b \sin(\Delta + \theta))}{(a + b)^2} \right] \, d\theta.$$

Integration yields

$$J(\Delta) = \frac{-(a^2 + b^2) \, \pi \, \cos \Delta}{4 \, (a + b)^2} - \frac{2 \cos \frac{\Delta}{2}}{a + b} \left((a + b) \sin \frac{\Delta}{2} + (a - b) \cos \frac{\Delta}{2} \right)$$
$$+ \frac{a^2 \sin \Delta - 2 \, a \, b \, \cos \Delta - b^2 \sin \Delta}{2 \, (a + b)^2}.$$

Equivalently,

$$J(\Delta) = \frac{\left[4 \, a \, b - (a^2 + b^2) \, (\pi - 4) \right] \cos \Delta + 4 \, (a + b) \left[(a + b) - \frac{1}{2}(a - b) \sin \Delta \right]}{4 \, (a + b)^2}.$$

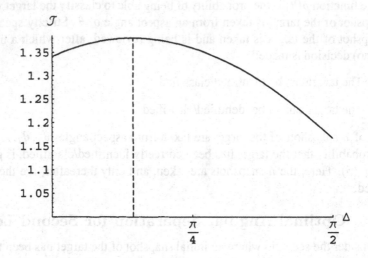

Figure 12.3. Example Cost Function for Second Target Identification Attempt

This function is plotted in Figure 12.3 for the target aspect ratio $a/b = 1/2$, and $0 \leq \Delta \leq \frac{\pi}{2}$.

We can find the maximum value for this function by setting

$$\frac{\partial J}{\partial \Delta} = 0$$

which yields

$$\frac{-2\left(a^2 - b^2\right)\cos\Delta + \left(a^2\left(\pi - 4\right) + b^2\left(\pi - 4\right) - 4\,a\,b\right)\sin\Delta}{4\left(a + b\right)^2} = 0.$$

We have an extreme value of J at

$$\Delta^* = \text{Arctan}\left(\frac{2\left(a^2 - b^2\right)}{\left(a^2 + b^2\right)\left(\pi - 4\right) - 4\,a\,b}\right). \tag{8}$$

Substituting Eq. (8) into the second derivative of J yields

$$\left.\frac{\partial^2 J}{\partial \Delta^2}\right|_{\Delta^*} = \frac{\sqrt{1 + \frac{4\left(a^2 - b^2\right)^2}{\left(a^2 + b^2\right)\left(\pi - 4\right) - 4\,a\,b^2}}\left[\left(a^2 + b^2\right)\left(\pi - 4\right) - 4\,a\,b\right]}{4\left(a + b\right)^2} < 0$$

which holds for all $a, b > 0$. Thus, the extreme value obtained at Δ^* is a relative maximum. As an example, for the case shown in Figure 12.3 where

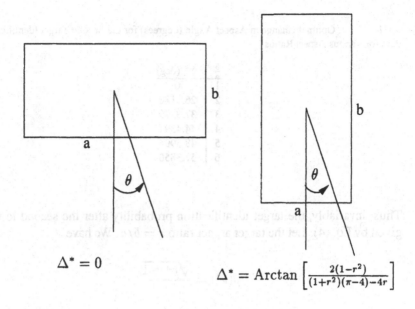

$$\Delta^* = 0$$

$$\Delta^* = \operatorname{Arctan}\left[\frac{2(1-r^2)}{(1+r^2)(\pi-4)-4r}\right]$$

Figure 12.4. Geometry for the Optimal Look Angle ATR Problem

the aspect ratio $a/b = 1/2$, the maximum value of J is achieved at

$$\Delta^* = \operatorname{Arctan}\left(\frac{6}{28 - 5\pi}\right)$$

$$= 26.018°.$$

It is important to note that Δ^* exists in the first quadrant, i.e., $0 \leq \Delta^* \leq \pi/2$, only if $b \geq a$. For all $a > b$, $J(\Delta)$ is maximized on the interval $[0, \pi/2]$ at the boundary $\Delta^* = 0$. Thus, for any rectangular object with $a > b$, the optimal change in aspect angle for the second look, given an unknown aspect angle θ, is zero - see, e.g., the illustration in Figure 12.4. Furthermore, for a square object, i.e., $a = b$, the optimal change in aspect angle is zero. Table 12.4 presents the optimal change in aspect angle for varying target aspect ratios for the case $b > a$.

2.1.   Feedback Control

If the ATR algorithm can provide an estimate of the aspect angle, θ, then the second pass should be flown such that the probability of classification of the second pass will be maximized, i.e., the second aspect angle

$$\theta^* = \operatorname{Arctan}\left(\frac{b}{a}\right).$$

Table 12.1. Optimal Change in Aspect Angle (degrees) for the Second Target Identification
Pass for Various Aspect Ratios

$\frac{b}{a}$	Δ^* (deg)
1	0
2	26.0180
3	37.8579
4	44.4394
5	48.5994
6	57.3850

Thus, invariably, the target identification probability after the second look is
given by Eq. (4). Let the target aspect ratio $r = b/a$. We have

$$\rho(\theta^*) = \frac{\sqrt{r^2+1}}{r+1}.$$

The probability of classification after two looks is

$$p(\theta) = \rho(\theta) + \frac{\sqrt{r^2+1}}{r+1} - \rho(\theta)\frac{\sqrt{r^2+1}}{r+1}$$

$$= \frac{\sqrt{r^2+1}}{r+1} + \left(1 - \frac{\sqrt{r^2+1}}{r+1}\right)\rho(\theta).$$

Thus, the average classification probability according to the feedback strategy
is

$$\bar{P}(r) = \frac{1}{\frac{\pi}{2}}\left\{\int_0^{\frac{\pi}{2}}\left[\frac{\sqrt{r^2+1}}{r+1} + \left(1 - \frac{\sqrt{r^2+1}}{r+1}\right)\rho(\theta)\right]d\theta\right\}$$

$$= \frac{\sqrt{r^2+1}}{r+1} + \frac{1}{\frac{\pi}{2}}\left(1 - \frac{\sqrt{r^2+1}}{r+1}\right)\int_0^{\frac{\pi}{2}}\frac{a\cos\theta + b\sin\theta}{a+b}\,d\theta.$$

Hence, the average probability of classification after two looks, as a function
of the target aspect ratio, r, is

$$\bar{P}(r) = \frac{2}{\pi} + \left(1 - \frac{2}{\pi}\right)\frac{\sqrt{r^2+1}}{r+1}.$$

A plot of $\bar{P}(r)$ is shown in Figure 12.5. Note that $\bar{P}(\infty) = 1$.

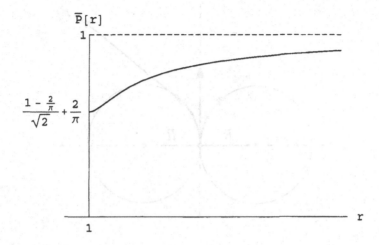

$$\overline{P}[r]$$

$$\frac{1-\frac{2}{\pi}}{\sqrt{2}} + \frac{2}{\pi}$$

Figure 12.5. Average Classification Probability after Two Looks when First Pass Aspect Angle is Known

3. Minimum Time Trajectories with a Minimum Turning Radius Constraint

In this section we consider minimum time optimal trajectories where a vehicle travelling in the plane with a constant velocity is constrained to have a minimum turning radius, R.

3.1. Specified Terminal Point

The first case considered entails a specified terminal point. The vehicle is initially at the origin, O with some initial heading angle and the terminal point, P_f, is outside the minimum turning radius circle. Without loss of generality, assume $P_f \in RH$ plane - see, e.g., Figure 12.6.

Proposition 1. *The minimum time trajectory, for the case where the specified terminal point P_f lies outside the minimum turning radius circle, consists of a hard turn into P_f until the bearing to P_f is 0°, followed by a straight line dash to P_f.*

Proof: This problem can be viewed as a special case of a two player differential game, where the second player is considered stationary and the first player is restricted to move at constant velocity with a limited maximum turning radius. The problem described is a special case of the Homicidal Chauffeur Game, and the minimum time trajectory of a hard turn into P_f until the bearing

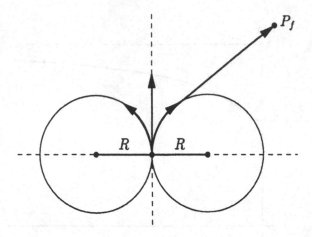

Figure 12.6. Minimum Time Trajectory Problem with Specified Terminal Point Outside the Minimum Turning Radius

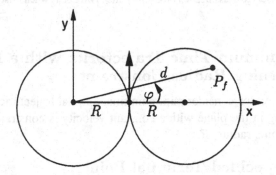

Figure 12.7. Minimum Time Trajectory Problem with Specified Terminal Point Inside the Minimum Turning Radius

to P_f is $0°$, followed by a straight line dash to P_f is the optimal strategy for the first player - see, e.g., [1]. ∎

The second case considered is the situation where the final point, P_f, is inside the minimum turning radius circle. The position of P_f is specified by the distance, d, from the center of the left minimum turning radius circle, and by the angle φ - see, e.g., Figure 12.7.

The parameters d and φ are constrained as follows. The equation of the right minimum turning radius circle is

$$(x - 2R)^2 + y^2 = R^2.$$

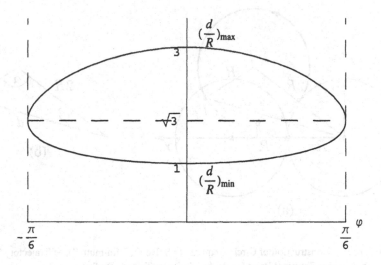

Figure 12.8. Plot of $\frac{d}{R}$ vs φ

The final point P_f has coordinates $(x, y) = (d \cos \varphi, d \sin \varphi)$. Since P_f is inside the right minimum turning radius circle, we have

$$(d \cos \varphi - 2R)^2 + d^2 \sin^2 \varphi \leq R^2$$
$$\Rightarrow \quad d^2 - 4Rd \cos \varphi + 3R^2 \leq 0.$$

Solving for the parameter $\frac{d}{R}$, yields the constraint

$$\max(0, 2\cos\varphi - \sqrt{4\cos^2\varphi - 3}) < \frac{d}{R} < 2\cos\varphi + \sqrt{4\cos^2\varphi - 3}. \quad (9)$$

Given P_f is contained within the right minimum turning radius circle, the angle φ is constrained by the two lines tangent to the right minimum radius turning circle which pass through the origin. By constructing right triangles with the following three points: the origin, the center of the right minimum turning radius circle, and each of the points of tangency, it is evident that

$$-\frac{\pi}{6} \leq \varphi \leq \frac{\pi}{6}. \quad (10)$$

In Figure 12.8 we plot the constraints (9) as a function of φ.

The solution of the optimal control problem entails the construction of a circle of radius R which is tangent to the left minimum turning radius circle and which passes through the point P_f. This requires construction of a triangle, given three sides: d, R and $2R$. The construction is illustrated in Figure 12.9(a).

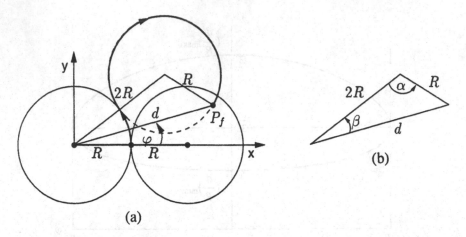

Figure 12.9. Construction of Circle Required to Solve the Minimum Time Trajectory Problem with Specified Terminal Point Inside the Minimum Turning Radius

The shown trajectory entails a hard turn to the left followed by a hard turn to the right, viz., a swerve maneuver. The length of the trajectory is determined by the angles of the constructed triangle, which is shown in Figure 12.9(b). Note: Since $d < 3R$, the solution triangle can always be constructed. From the law of cosines we have

$$d^2 = 4R^2 + R^2 - 4R^2 \cos \alpha$$

$$\Rightarrow \quad \alpha = \text{Arccos}\left(\frac{1}{4}\left[5 - \left(\frac{d}{R}\right)^2\right]\right).$$

Similarly

$$R^2 = 4R^2 + d^2 - 4Rd \cos \beta$$

$$\Rightarrow \quad \beta = \text{Arccos}\left(\frac{1}{4}\left[\frac{3}{\left(\frac{d}{R}\right)} + \left(\frac{d}{R}\right)\right]\right).$$

The path length, l, for the swerve maneuver is

$$\frac{l}{R} = \beta + \varphi + 2\pi - \alpha.$$

Hence,

$$\frac{l}{R} = 2\pi + \varphi + \text{Arccos}\left(\frac{1}{4}\left[\frac{3}{\left(\frac{d}{R}\right)} + \left(\frac{d}{R}\right)\right]\right) - \text{Arccos}\left(\frac{1}{4}\left[5 - \left(\frac{d}{R}\right)^2\right]\right) \quad (11)$$

The construction shown in Figure 12.9(b) will "fail", viz., the point at which

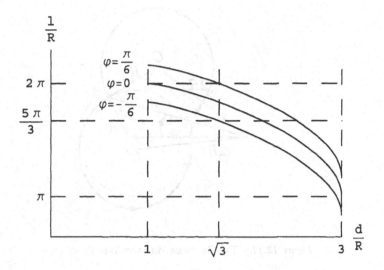

Figure 12.10. Plot of $\frac{1}{R}$

the right hand constructed circle is tangent to the left hand minimum radius circle will fall below the initial position, if

$$\varphi < 0.$$

Similarly, if

$$\varphi > 30°$$

the constructed circle cannot simultaneously contain points inside the right minimum turning radius circle and have a point tangent to the left minimum turning radius circle. However, these conditions need never occur.

Proposition 2. *All final points, P_f, in the interior of the right hand minimum radius circle can be reached by the constructed circle if the inequalities, $0 < \varphi < \pi/6$ and $\beta > 0$ are satisfied.*

Proof: Let $0 < \varphi < \pi/6$ such that A is any point in the interior of the right minimum turning radius circle, some distance, d, from the center of the left minimum turning radius circle, O - see, e.g., Figure 12.11. $\exists$ a pair of triangles, each with a common side OA a side of length $2R$ and a side of length R. Let the points of the triangles opposite OA be C and C'. Construct two circles of radius R at C and C'. Both circles will be tangent to the left minimum radius circle at one point. The angle between the point of tangency and center of the left turning radius circle is β. Associate the circle at C with $\beta > 0$ and the

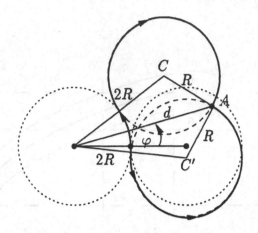

Figure 12.11.   Two Candidate Minimum Time Paths

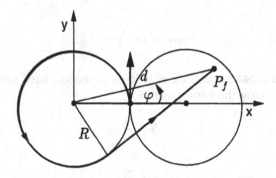

Figure 12.12.   Hard Turn to Left Followed By a Straight Line Dash

circle at C' with $\beta < 0$. The circle for which $\beta < 0$ must be disregarded since this requires movement in the opposite direction of the initial heading angle. Thus, the constructed circle with $\beta > 0$, is the only solution which intersects point A and permits movement in the direction of the initial heading vector.   ∎

We also must investigate the possibility of a hard turn to the left followed by a straight line dash, as shown in Figure 12.12. Now,

$$l = 2\pi R - \left(\text{Arccos}\left[\frac{1}{\left(\frac{d}{R}\right)}\right] - \varphi \right) R + \sqrt{d^2 + R^2} \quad \Rightarrow$$

$$\frac{l}{R} = 2\pi + \varphi - \text{Arccos}\left[\frac{1}{\left(\frac{d}{R}\right)}\right] + \sqrt{\left(\frac{d}{R}\right)^2 - 1}. \tag{12}$$

Equating Eqs. (11) and (12) yields the following transcendental equation, in terms of a single nondimensional parameter, $\frac{d}{R}$

$$\text{Arccos}\left(\frac{1}{4}\left[\frac{3}{\left(\frac{d}{R}\right)} + \left(\frac{d}{R}\right)\right]\right) - \text{Arccos}\left(\frac{1}{4}\left[5 - \left(\frac{d}{R}\right)^2\right]\right) =$$

$$\text{Arccos}\left[\frac{1}{\left(\frac{d}{R}\right)}\right] + \sqrt{\left(\frac{d}{R}\right)^2 - 1}. \tag{13}$$

A solution, $\frac{d}{R}$, of Eq. (13) must also satisfy the inequality (9). Obviously, $1 \leq \frac{d}{R} \leq 3$. Further, if P_f is restricted to the interior of the right minimum turning radius circle, then $1 < \frac{d}{R} < 3$ holds.

Proposition 3. *The transcendental equation (13) does not have a solution which satisfies*

$$1 < \frac{d}{R} < 3.$$

Proof: Suppose $\frac{d}{R}$ is a solution of the transcendental equation (13). Evidently, φ must satisfy $\varphi + \beta \leq 0$, which implies

$$\varphi \leq \bar{\varphi} = \text{Arccos}\left(\frac{1}{4}\left[\left(\frac{d}{R}\right) + \frac{3}{\left(\frac{d}{R}\right)}\right]\right).$$

Should a solution of Eq. (13) exist, there would be a boundary in the right minimum turning radius circle separating the optimal policies of two circular turns and one circular turn with a straight line dash - see, e.g., Figure 12.13. Consider that the final point, P_f, lies in the unshaded areas of the right minimum turning radius circle shown in Figure 12.13. The question arises, what would then be the optimal policy in those unshaded areas? If no unshaded areas exist, e.g., see Figure 12.14, then we have the case where

$$\frac{d}{R} = \sqrt{3}.$$

Inserting $\frac{d}{R} = \sqrt{3}$ into Eq. (13), we can check to see if this can be a solution. For the left hand side we have

$$\text{Arccos}\left(\frac{\sqrt{3}}{2}\right) + \text{Arccos}\left(\frac{1}{\sqrt{3}}\right).$$

One Circular Turn and Straight Line Dash

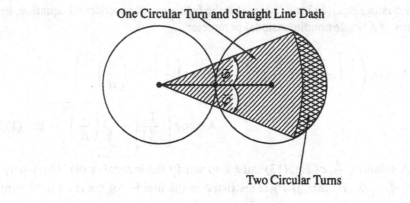

Two Circular Turns

Figure 12.13.   Hypothetical Boundary Separating Two Optimal Policies

One Circular Turn and Straight Line Dash

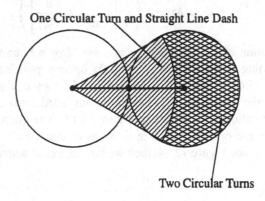

Two Circular Turns

Figure 12.14.   Hypothetical Boundary Separating Two Optimal Policies

For the right hand side we have

$$\text{Arccos}\left(\frac{1}{2}\right) + \sqrt{2}.$$

Since

$$\text{Arccos}\left(\frac{1}{2}\right) > \text{Arccos}\left(\frac{1}{\sqrt{3}}\right)$$

and

$$\sqrt{2} > \frac{\pi}{6} = \text{Arccos}\left(\frac{\sqrt{3}}{2}\right)$$

we see that $\frac{d}{R} = \sqrt{3}$ is *not* a solution of the transcendental Eq. (13). ∎

Theorem 1. *The minimum time trajectory, for the case where the final point, P_f, is inside the minimum turning radius circle, entails a swerve maneuver which consists of a hard turn to the left of*

$$\varphi + \text{Arccos}\left(\frac{1}{4}\left[\frac{3}{\left(\frac{d}{R}\right)} + \left(\frac{d}{R}\right)\right]\right)$$

followed by a hard turn to the right of

$$2\pi - \text{Arccos}\left(\frac{1}{4}\left[\frac{3}{\left(\frac{d}{R}\right)} + \left(\frac{d}{R}\right)\right]\right).$$

Proof: This result proceeds from the previous discussion. ∎

4. Minimum Time Trajectories for Target Classification

Consider a target classification problem where the sensor footprint is circular, of radius r, and it is offset from the sensor platform by a distance d. The sensor platform moves at a constant velocity and has a minimum turning radius of R. We define the distance

$$\rho = d + r$$

and stipulate that the target circle defined by a target at the center, P_f, with radius ρ, be approached orthogonally from the outside. See, e.g., Figure 12.15.

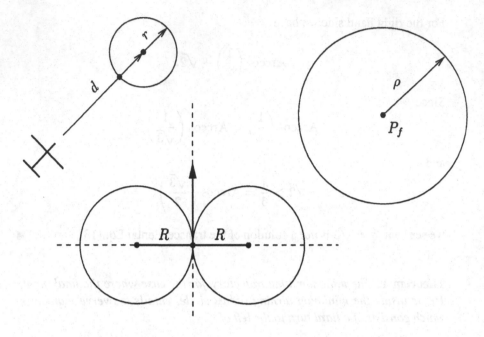

Figure 12.15. Minimum Time Trajectories for Search and Target Classification

4.1.   Type 1 Problems

Consider the case where P_f is outside the circle of radius ρ', which is concentric with the right minimum turning radius circle,

$$\rho' \doteq \sqrt{\rho^2 + R^2}.$$

The critical circle is shown in Figure 12.16.

Proposition 4. *When P_f is outside the critical circle the minimum time trajectory entails a hard turn into P_f until a bearing of $0°$ to P_f is established. Thereafter, a straight line path is followed until the target circle is met, as shown in Figure 12.17.*

Proof: See proof of Proposition 1.   ∎

4.2.   Type 2 Problems

Next we consider the case where P_f is inside the circle of radius ρ' which is concentric with the right minimum turning radius circle - see, e.g., Figure 12.18. This yields the following condition

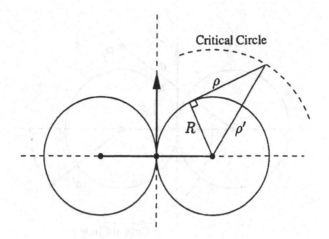

Figure 12.16. Critical Circle

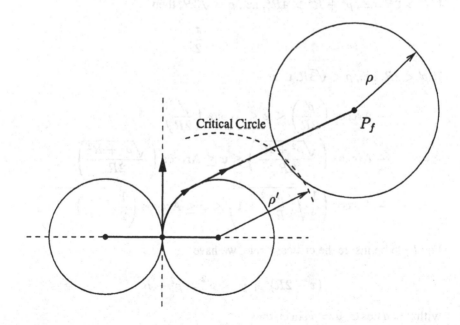

Figure 12.17. Minimum Time Trajectory for P_f Outside Critical Circle

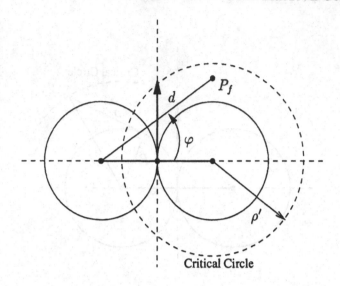

Figure 12.18. P_f Inside Critical Circle

If $\rho' > 2R$, i.e., $\rho^2 + R^2 > 4R^2$, i.e., $\rho > \sqrt{3}R$, then

$$-\frac{\pi}{2} < \varphi \leq \frac{\pi}{2}.$$

If $\rho' < 2R$, i.e., $\rho < \sqrt{3}R$, then

$$-\operatorname{Arccos}\left(\frac{\rho'}{2R}\right) \leq \varphi \leq \operatorname{Arccos}\left(\frac{\rho'}{2R}\right)$$

$$\Rightarrow \quad -\operatorname{Arccos}\left(\frac{\sqrt{\rho^2 + R^2}}{2R}\right) \leq \varphi \leq \operatorname{Arccos}\left(\frac{\sqrt{\rho^2 + R^2}}{2R}\right)$$

$$\Rightarrow \quad -\operatorname{Arccos}\left(\frac{1}{2}\sqrt{\left(\frac{\rho}{R}\right)^2 + 1}\right) \leq \varphi \leq \operatorname{Arccos}\left(\frac{1}{2}\sqrt{\left(\frac{\rho}{R}\right)^2 + 1}\right).$$

For P_f to be inside the critical circle, we have

$$(x - 2R)^2 + y^2 \leq \rho'^2 = \rho^2 + R^2$$

with $x = d\cos\varphi$, $y = d\sin\varphi$, thus

$$(d\cos\varphi - 2R)^2 + d^2\sin^2\varphi \leq \rho^2 + R^2$$

$$\Rightarrow \quad d^2 - 4Rd\cos\varphi + 3R^2 - \rho^2 \leq 0.$$

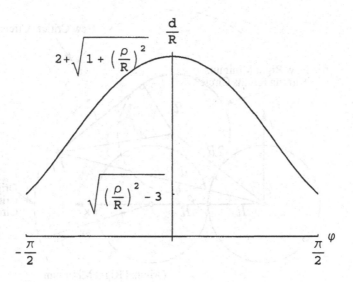

Figure 12.19. Plot of $\frac{d}{R}$ vs φ

We have the following inequality

$$\max\left(0, 2\cos\varphi - \sqrt{4\cos^2\varphi + \left(\tfrac{\rho}{R}\right)^2 - 3}\right) < \frac{d}{R}$$
$$< 2\cos\varphi + \sqrt{4\cos^2\varphi + \left(\tfrac{\rho}{R}\right)^2 - 3}. \tag{14}$$

Since

$$2\cos\varphi - \sqrt{4\cos^2\varphi + \left(\frac{\rho}{R}\right)^2 - 3} < 0, \qquad \forall\, \rho < \sqrt{3}R$$

and

$$2\cos\varphi - \sqrt{4\cos^2\varphi + \left(\frac{\rho}{R}\right)^2 - 3} > 0, \qquad \forall\, \rho > \sqrt{3}R,$$

Eq. (14) becomes

$$\begin{cases} 0 < \frac{d}{R} < 2\cos\varphi + \sqrt{4\cos^2\varphi + \left(\frac{\rho}{R}\right)^2 - 3}, & \rho < \sqrt{3}R \\ 2\cos\varphi - \sqrt{4\cos^2\varphi + \left(\frac{\rho}{R}\right)^2 - 3} < \frac{d}{R} < 2\cos\varphi + \sqrt{4\cos^2\varphi + \left(\frac{\rho}{R}\right)^2 - 3}, & \rho > \sqrt{3}R. \end{cases}$$

In Figure 12.19 we plot the constraint (14) as a function of φ.

Consider the situation where P_f is inside the circle of radius ρ', which is concentric with the right minimum turning radius circle. The minimum time trajectory involves a swerve maneuver. The swerve is just enough for P_f to be

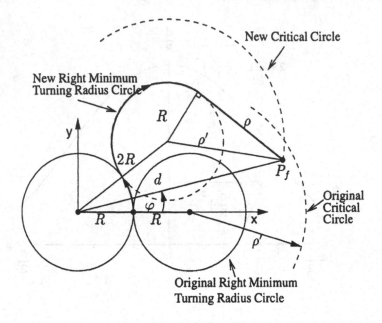

Figure 12.20.   Construction of New Critical Circle

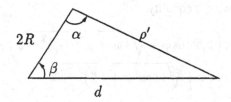

Figure 12.21.   Solution Triangle

reached by the boundary of a new critical circle of radius ρ', concentric with a new right minimum turning radius circle. This is depicted in Figure 12.20. We need to construct a solution triangle, e.g., see Figure 12.21, given its three sides d, $2R$ and ρ'. We calculate

$$\rho'^2 = d^2 + 4R^2 - 4Rd\cos\beta.$$

Thus,

$$\cos\beta = \frac{d^2 + 4R^2 - \rho'^2}{4Rd}$$
$$= \frac{3R^2 + d^2 - \rho^2}{4Rd} \quad \Rightarrow$$

$$\beta = \text{Arccos}\left(\frac{1}{4}\left[\frac{3}{\left(\frac{d}{R}\right)} + \left(\frac{d}{R}\right) - \frac{\left(\frac{\rho}{R}\right)^2}{\left(\frac{d}{R}\right)}\right]\right).$$

Similarly,

$$d^2 = \rho'^2 + 4R^2 - 4R\rho'\cos\beta.$$

Thus,

$$\cos\alpha = \frac{\rho'^2 + 4R^2 - d^2}{4R\rho'}$$

$$= \frac{5R^2 + \rho^2 - d^2}{4R\rho'}$$

$$= \frac{1}{4}\left[\frac{5 - \left(\frac{d}{R}\right)^2 + \left(\frac{\rho}{R}\right)^2}{\sqrt{1 + \left(\frac{\rho}{R}\right)^2}}\right] \quad \Rightarrow$$

$$\beta = \text{Arccos}\left(\frac{1}{4}\left[\frac{5 - \left(\frac{d}{R}\right)^2 + \left(\frac{\rho}{R}\right)^2}{\sqrt{1 + \left(\frac{\rho}{R}\right)^2}}\right]\right).$$

For the turn to the left, the change in heading is

$$\varphi + \beta.$$

For the turn to the right, the change in heading is

$$2\pi - \alpha - \text{Arctan}\left(\frac{\rho}{R}\right).$$

The path length is then expressed as

$$\frac{l}{R} = \beta + \varphi + 2\pi\left[\alpha + \text{Arctan}\left(\frac{\rho}{R}\right)\right].$$

Equivalently,

$$\frac{l}{R} = 2\pi + \varphi - \text{Arctan}\left(\frac{\rho}{R}\right) + \text{Arccos}\left(\frac{1}{4}\left[\frac{3}{\left(\frac{d}{R}\right)} + \left(\frac{d}{R}\right) - \frac{\left(\frac{\rho}{R}\right)^2}{\left(\frac{d}{R}\right)}\right]\right)$$

$$- \text{Arccos}\left(\frac{1}{4}\left[\frac{5 - \left(\frac{d}{R}\right)^2 + \left(\frac{\rho}{R}\right)^2}{\sqrt{1 + \left(\frac{\rho}{R}\right)^2}}\right]\right). \quad (15)$$

Remark 1. *The target circle is penetrated orthogonally by construction.*

Theorem 2. *When P_f is inside the circle of radius ρ', which is concentric with the right minimum turning radius circle, the minimum time trajectory involves*

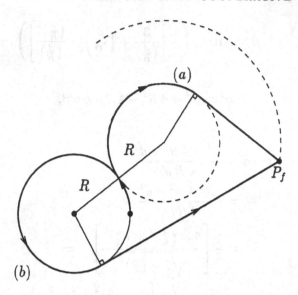

Figure 12.22.   Two Candidate "Optimal" Trajectories After An Initial Swerve

a swerve maneuver. The swerve is just enough for P_f to be reached by the boundary of a new critical circle of radius ρ', concentric with a new right minimum turning radius circle.

Proof: Consider the situation after an initial swerve which puts P_f on the boundary of a circle of radius ρ', centered at the new position of the right minimum turning radius circle.

From this time onward, the "optimal" trajectories, (a) and (b), are indicated in Figure 12.22. Similarly to the proof in Proposition 1, we realize that the optimal trajectory is (a).   ∎

5.   Cooperative Target Classification

Lastly, consider the problem of cooperative target classification. As in Section 4, the sensor footprint is circular, of radius r, and offset from the sensor platform by a distance d. Additionally, there is a commanded approach angle ξ, which corresponds, for instance, to an optimal second look at the target directed by the results of Sec. 2. Since our target is assumed rectangular, symmetry provides four candidate angles of approach, as shown in Figure 12.23.

The first step in developing the optimal trajectory to perform the cooperative target classification is to construct minimum turning radius circles such that the entry into the target circle is orthogonal. Eight circles are constructed, see, e.g., Figure 12.24.

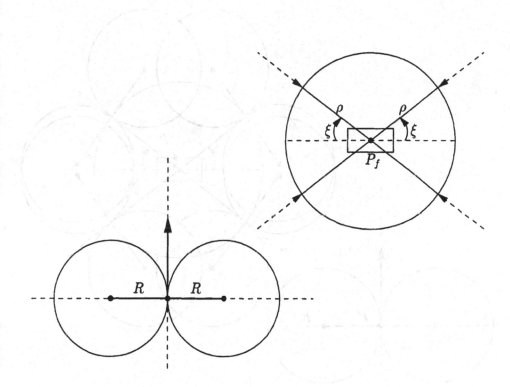

Figure 12.23. Cooperative Search and Target Classification Problem

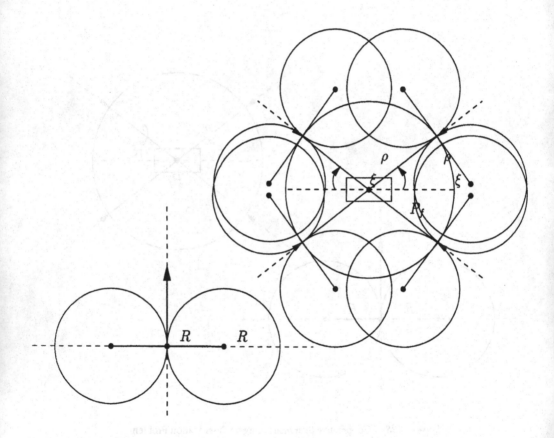

Figure 12.24.   Construction of the Minimum Turning Radius Circles for Orthogonal Entry into the Target Circle

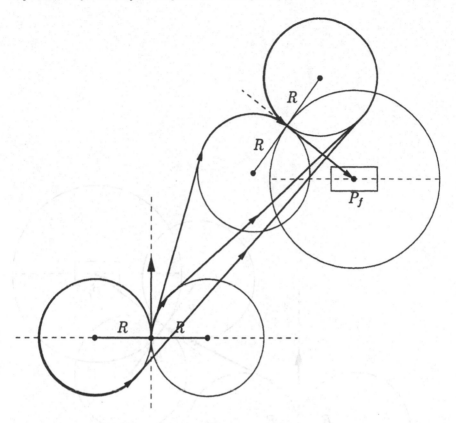

Figure 12.25. Construction of Some Candidate Minimum Time Trajectories for Cooperative Target Classification

All candidate trajectories consist of a hard turn, a straight line dash and a hard turn. The straight line dash is to be constructed tangent to the two minimum turning radius circles. Depending on the location of P_f and the sizes of the target circle and the minimum turning radii circles, a swerve maneuver may be considered as a portion of a candidate trajectory. Figure 12.25 depicts several candidate minimum time trajectories.

Lastly, the four candidate trajectories are ranked according to path length, and the minimum length trajectory is selected as optimal.

6. Conclusion

A cost function was developed for enhancing the probability of automatic target recognition and the corresponding optimization problem was solved. This requires approaching the target from specified directions and making sure

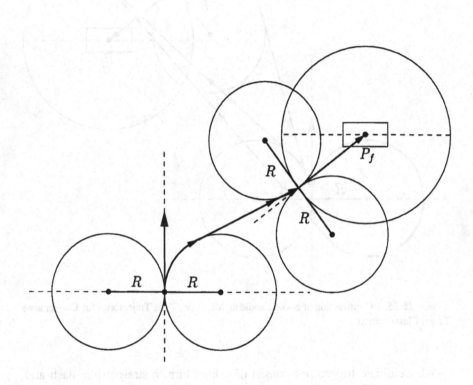

Figure 12.26.   Characterization of the Optimal Minimum Time Cooperative Classification Trajectory

that the air vehicle's sensor covers the target. Minimum time trajectories for air vehicles with a minimum turning radius were constructed for these tasks, and also for target attack and battle damage assessment. Lastly, the single vehicle results are encapsulated in a methodology addressing cooperative target classification and attack.

References

[1] R. Isaacs, *Differential Games*. John Wiley and Sons, New York, 1965.

Chapter 13

COOPERATIVE CONTROL DESIGN FOR UNINHABITED AIR VEHICLES

Marios Polycarpou
Dept. of Electrical and Computer Engineering and Computer Sciences
University of Cincinnati, Cincinnati, OH 45221-0030, USA
polycarpou@uc.edu

Yanli Yang
Dept. of Electrical and Computer Engineering and Computer Sciences
University of Cincinnati, Cincinnati, OH 45221-0030, USA
yangyanl@ececs.uc.edu

Yang Liu
Department of Electrical Engineering, The Ohio State University
2015 Neil Avenue, Columbus, OH 43210-1272, USA
liuya@ee.eng.ohio-state.edu

Kevin Passino
Department of Electrical Engineering, The Ohio State University
2015 Neil Avenue, Columbus, OH 43210-1272, USA
passino@ee.eng.ohio-state.edu

Abstract   The main objective of this research is to develop and evaluate the performance of strategies for cooperative control of autonomous air vehicles that seek to gather information about a dynamic target environment, evade threats, and coordinate strikes against targets. The chapter presents an approach for cooperative search by a team of uninhabited autonomous air vehicles, which are equipped with sensors to view a limited region of the environment, and are able to communicate with one another to enable cooperation. The developed cooperative search framework is based on two inter-dependent tasks: (i) on-line learning of the en-

S. Butenko et al. (eds.), Cooperative Control: Models, Applications and Algorithms, 283-321.
© 2003 *Kluwer Academic Publishers.*

vironment and storing of the information in the form of a "search map"; and (ii) utilization of the search map and other information to compute on-line a guidance trajectory for the vehicle to follow. We develop a real-time approach for on-line cooperation between air vehicles, which is based on treating the paths of other vehicles as "soft obstacles" to be avoided. Based on artificial potential field methods, we develop the concept of "rivaling force" between vehicles as a way of enhancing cooperation. We study the stability of vehicular swarms in a multi-dimensional framework to try to understand what types of communications are needed to achieve cooperative search and engagement, and characteristics that affect swarm aggregation and disintegration. Simulation results are presented to illustrated the concepts developed in the chapter.

1.   Introduction

This chapter presents an approach for cooperative search among a team of distributed agents and some stability results for multi-dimensional swarms. Although the presented framework is quite general, the main motivation for this work is to develop and evaluate the performance of strategies for cooperative control of autonomous air vehicles that seek to gather information about a dynamic target environment, evade threats, and possibly coordinate strikes against targets. Recent advances in computing, wireless communications and vehicular technologies are making it possible to deploy multiple uninhabited air vehicles (UAVs) that operate in an autonomous manner and cooperate with each other to achieve a global objective ([29, 58, 41, 30, 39]). A large literature of relevant ideas and methods can also be found in the area of "swarm robotics" (e.g., see [22, 33, 8]) and, more generally, coordination and control of robotic systems (e.g., see [44, 33, 52, 3, 10, 40, 11]). Related work also includes the techniques developed using the "social potential field" method ([62, 9, 54]) and multi-resolution analysis ([2]).

We consider a team of vehicles moving in an environment of known dimension, searching for targets of interest. The vehicles are assumed to be equipped with: 1) target sensing capabilities for obtaining a limited view of the environment; 2) wireless communication capabilities for exchanging information and cooperating with one another; and 3) computing capabilities for processing the incoming information and making on-line guidance decisions. It is also assumed that each vehicle has a tandem of actuation/sensing hardware and an inner-loop control scheme for path following. We focus solely on the design of the guidance controller (outer-loop control), and for convenience we largely ignore the vehicle dynamics.

The vehicles are assumed to have some maneuverability limitations, which constrain the maximum turning radius of the vehicle. The maneuverability constraint is an issue that is typically not encountered in some of the literature on "collective robotics," which describes swarms of robots moving in a terrain ([21]). The developed cooperative search framework is based on two inter-dependent tasks: (i) on-line learning of the environment and storing of the information in the form of a "search map"; and (ii) utilization of the search map and other information for computing on-line a guidance trajectory for the vehicle. We develop a real-time approach for on-line cooperation between agents based on treating the paths of other vehicles as "soft obstacles" to be avoided. Using artificial potential field methods we develop the concept of "rivaling force" between agents as a way of enhancing cooperation. The distributed learning and planning approach for cooperative search is illustrated by computer simulations. In the rest of the chapter, we will be using the general term "agent" to represent a UAV or other type of appropriate vehicle.

1.1. Related Research Work on Search Methods

Search problems occur in a number of military and civilian applications, such as search-and-rescue operations in open-sea or sparsely populated areas, search missions for previously spotted enemy targets, seek-destroy missions for land mines, and search for mineral deposits. A number of approaches have been proposed for addressing such search problems. These include, among other, optimal search theory ([73, 45]), exhaustive geographic search ([71]), obstacle avoidance ([12, 70]) and derivative-free optimization methods ([14]).

Search theory deals with the problem of distribution of search effort in a way that maximizes the probability of finding the object of interest. Typically, it is assumed that some prior knowledge about the target distribution is available, as well as the "payoff" function that relates the time spent searching to the probability of actually finding the target, given that the target is indeed in a specific cell ([73, 45]). Search theory was initially developed during World War II with the work of Koopmam and his colleagues at the Anti-Submarine Warfare Operations Research Group (ASWORG). Later on, the principles of search theory were applied successfully in a number of applications, including the search for and rescue of a lost party in a mountain or a missing boat on the ocean, the surveillance of frontiers or territorial seas, the search for mineral deposits, medical diagnosis, and the search for a malfunction in an industrial process. Detailed reviews of the current status of search theory have been given by [74], [65], and [6].

The optimal search problem can be naturally divided according to two criteria that depend on the target's behavior. The first division depends on whether the target is evading or not; that is, whether there is a two-sided optimization by both the searcher and the target, or whether the target's behavior is independent of the searcher's action. The second division deals with whether the target is stationary or moving. The two divisions and their combinations form four different categories. A great deal of progress in solving stationary target problems in the optimal search framework has been made, and solutions have been derived for most of the standard cases ([73]). For the moving target problem, the emphasis in search theory has shifted from mathematical and analytical solutions to algorithmic solutions ([6]). A typical type of search problem, called the path constraint search problem (PCSP), that takes into account the movement of the searcher, was investigated by several researchers ([23, 72, 37, 36]). Because of the NP-complete nature of this problem, most authors proposed a number of heuristic approaches that result in "approximately optimal" solutions. The two-sided search problem can be treated as a game problem for both the searcher and target strategies. This has been the topic of a number of research works ([18, 38, 83]). So far, search theory has paid little attention to the problem of having a team of cooperating searchers. A number of heuristic methods for solving this problem have been proposed by [19].

The Exhaustive Geographic Search problem deals with developing a complete map of all phenomena of interest within a defined geographic area, subject to the usual engineering constraints of efficiency, robustness and accuracy ([71]). This problem received much attention recently, and algorithms have been developed that are cost-effective and practical. Application examples of Exhaustive Geographic Search include mapping mine fields, extraterrestrial and under-sea exploration, exploring volcanoes, locating chemical and biological weapons and locating explosive devices ([71, 31, 35, 13]).

The obstacle avoidance literature deals with computing optimal paths given some kind of obstacle map. The intent is to construct a physically realizable path that connects the initial point to the destination in a way that minimizes some energy function while avoiding all the obstacles along the route ([12, 70]). Obstacle avoidance is normally closely geared to the methods used to sense the obstacles, as time-to-react is of the essence. The efficiency of obstacle avoidance systems is largely limited by the reliability of the sensors used. A popular way to solve the obstacle avoidance problem is the potential field technique ([43]). According to the potential field method, the potential gradient that the robot follows is made up of two components: the repulsive effect of the obstacles and the attractive effect of the goal position. Although it is straightforward to use potential field techniques for obstacle avoidance, there are still several difficulties in using this method in practical vehicle planning.

Derivative-Free Optimization methods deal with the problem of minimizing a nonlinear objective function of several variables when the derivatives of the objective function are not available ([14]). The interest and motivation for examining possible algorithmic solutions to this problem is the high demand from practitioners for such tools. The derivatives of objective function are usually not available either because the objective function results from some physical, chemical or economical measurements, or, more commonly, because it is the result of a possibly very large and complex computer simulation. The occurrence of problems of this nature appear to be surprisingly frequent in the industrial setting. There are several conventional deterministic and stochastic approaches to perform optimization without the use of analytical gradient information or measures of the gradient. These include, for example, the pattern and coordinate search ([81, 51]), the Nelder and Mead Simplex Method ([57]), the Parallel Direct Search Algorithm ([20]), and the Multi-directional Search Method ([80]). In one way or another, most derivative free optimization methods use measurements of the cost function and form approximations to the gradient to decide which direction to move. [59] provides some ideas on how to extend non-gradient methods to team foraging.

2. Cooperative Control Formulation

We consider N agents deployed in some search region $\mathcal{X}$ of known dimension. As each agent moves around in the search region, it obtains sensory information about the environment, which helps to reduce the uncertainty about the environment. This sensory information can be in the form of an image, which can be processed on-line to determine the presence of a certain entity or target. Alternatively, it can be in the form of a sensor coupled with automatic target recognition (ATR) software. In addition to the information received from its own sensors, each agent also receives information from other agents via a wireless communication channel. The information received from other agents can be in raw form or it may be pre-processed, and it may be coming at a different rate (usually at a slower rate) or with a delay, as compared to the sensor information received by the agent from its own sensors.

Depending on the specific application, the global objective pursued by the team of agents may be different. In this chapter, we focus mainly on the problem of cooperative search, where the team of agents seeks to follow a trajectory that would result in maximum gain in information about the environment; i.e., the objective is to minimize the uncertainty about the environment. However, the ideas and approach presented here can be extended to other missions such as cooperative engagement and classification, evading threats, etc.

Each agent has two basic control loops that are used in guidance and control, as shown in Figure 13.1. The "outer-loop" controller for agent $\mathcal{A}_i$ utilizes sensor information from $\mathcal{A}_i$, as well as sensor information from $\mathcal{A}_j$, $j \neq i$, to compute on-line a desired trajectory (path) to follow, which is denoted by $P_i(k)$. The sensor information utilized in the feedback loop is denoted by v_i and may include information from standard vehicle sensors (e.g. pitch, yaw, etc.) and information from on-board sensors that has been pre-processed by resident ATR software. The sensor information coming from other agents is represented by the vector

$$V_i = [v_1, \ldots, v_{i-1}, v_{i+1}, \ldots, v_N]^\mathsf{T},$$

where v_j represents the information received from agent $\mathcal{A}_j$. Although in the above formulation it appears that all agents are in range and can communicate with each other, this is not a required assumption—the same framework can be used for the case where some of the information from other agents is missing, or the information from different agents is received at different sampling rates, or with a communication delay. The desired trajectory $P_i(k)$ is generated as a digitized look-ahead path of the form

$$P_i(k) = \{p_i(k), p_i(k+1), \ldots, p_i(k+q)\},$$

where $p_i(k+j)$ is the desired location of agent $\mathcal{A}_i$ at time $k+j$, and q is the number of look-ahead steps in the path planning procedure.

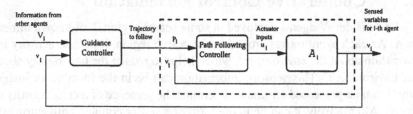

Figure 13.1.   Inner- and outer-loop controllers for guidance and control of air vehicles.

The inner-loop controller uses sensed information v_i from $\mathcal{A}_i$ to generate inputs u_i to the actuators of $\mathcal{A}_i$ so that the agent will track the desired trajectory $P_i(k)$. We largely ignore the agent dynamics, and hence concentrate on the outer-loop control problem. In this way, our focus is solidly on the development of the controller for guidance, where the key is to show how resident information of agent $\mathcal{A}_i$ can be combined with information from other agents so that the team of agents can work together to minimize the uncertainty in the search region $\mathcal{X}$.

The design of the outer-loop control scheme is broken down into two basic functions, as shown in Figure 13.2. First, it uses the sensor information received to update its "search map", which is a representation of the environment—this will be referred to as the agent's *learning* function, and for convenience it will be denoted by $\mathcal{L}_i$. Based on its search map, as well as other information (such as its location and direction, the location and direction of the other agents, remaining fuel, etc.), the second function is to compute a desired path for the agent to follow—this is referred to as the agent's *guidance decision* function, and is denoted by $\mathcal{D}_i$. In this setting we assume that the guidance control decisions made by each agent are autonomous, in the sense that no agent tells another what to do in a hierarchical type of structure, nor is there any negotiation between agents. Each agent simply receives information about the environment from the remaining agents (or a subset of the remaining agents) and makes its decisions, which are typically based on enhancing a global goal, not only its own goal. Therefore, the presented framework can be thought of as a *passive cooperation* framework, as opposed to *active cooperation* where the agents may be actively coordinating their decisions and actions.

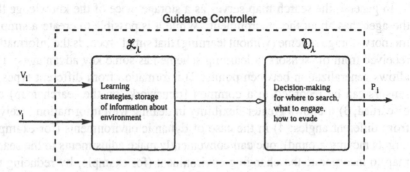

Figure 13.2. Learning and decision-making components of the outer-loop controller for trajectory generation of air vehicles.

2.1. Distributed Learning

Each agent has a three dimensional map, which we will refer to as "search map," that serves as the agent's knowledge base of the environment. The x and y coordinates of the map specify the location in the target environment (i.e., $(x, y) \in \mathcal{X}$), while the z coordinate specifies the certainty that the agent "knows" the environment at that point. The search map will be represented mathematically by an on-line approximation function as

$$z = \mathcal{S}(x, y; \theta),$$

where (x, y) is a point in the search region $\mathcal{X}$, and the output $z \in [0, 1]$ corresponds to the certainty about knowing the environment at the point (x, y) in the search region. If $\mathcal{S}(x, y; \theta) = 0$ then the agent knows nothing (is totally uncertain) about the nature of the environment at (x, y). On the other hand, if $\mathcal{S}(x, y; \theta) = 1$ then the agent knows everything (or equivalently, the agent is totally certain) about the environment at (x, y). As the agent moves around in the search region it gathers new information about the environment which is incorporated into its search map. Also incorporated into its search map is the information received by communication with other agents. Therefore, the search map of each agent is continuously evolving as new information about the environment is collected and processed.

We define $\mathcal{S} : \mathcal{X} \times \mathfrak{R}^q \mapsto [0, 1]$ to be an on-line approximator (for example, a neural network), with a fixed structure whose input/output response is updated on-line by adapting a set of adjustable parameters, or weights, denoted by the vector $\theta \in \Re^q$. According to the standard neural network notation, (x, y) is the input to the network and z is the output of the network. The weight vector $\theta(k)$ is updated based on an on-line learning scheme, as is common for example in training algorithms of neural networks.

In general, the search map serves as a storage place of the knowledge that the agent has about the environment. While it is possible to create a simpler memory/storage scheme (without learning) that simply records the information received from the sensors, a learning scheme has some key advantages: 1) it allows generalization between points; 2) information from different types of sensors can be recorded in a common framework (on the search map) and discarded; 3) it allows greater flexibility in dealing with information received from different angles; 4) in the case of dynamic environments (for example, targets moving around), one can conveniently make adjustments to the search map to incorporate the changing environment (for example, by reducing the output value z over time using a decay factor).

In this general framework, the tuning of the search map can be viewed as "learning" the environment. Mathematically, $\mathcal{S}$ tries to approximate an unknown function $\mathcal{S}^*(x, y, k)$, where for each (x, y), the function $\mathcal{S}^*$ characterizes the presence (or not) of a target; the time variation indicated by the time step k is due to (possible) changes in the environment (such as having moving targets). Hence, the learning problem is defined as using sensor information from agent $\mathcal{A}_i$ and information coming from other agents $\mathcal{A}_j$, $j \neq i$ at each sampled time k, to adjust the weights $\hat{\theta}(k)$ such that

$$\left\| \mathcal{S}(x, y; \hat{\theta}(k)) - \mathcal{S}^*(x, y, k) \right\|_{(x,y) \in \mathcal{X}}$$

is minimized.

Due to the nature of the learning problem, it is convenient to use spatially localized approximation models so that learning in one region of the search space does not cause any "unlearning" at a different region ([84]). The dimension of the input space (x, y) is two, and therefore there are no problems related to the "curse of dimensionality" that are usually associated with spatially localized networks. In general, the learning problem in this application is straightforward, and the use of simple approximation functions and learning schemes is sufficient; e.g., the use of piecewise constant maps or radial basis function networks, with distributed gradient methods to adjust the parameters, provides sufficient learning capability. However, complexity issues do arise and are crucial since the distributed nature of the architecture imposes limits not only on the amount of memory and computations needed to store and update the maps but also in the transmission of information from one agent to another.

At the time of deployment, it is assumed that each agent has a copy of an initial search map estimate, which reflects the current knowledge about the environment $\mathcal{X}$. In the special case that no a priori information is available, then each point on the search map is initialized as "completely uncertain." In general, each agent is initialized with the same search map. However, in some applications it may be useful to have agents be "specialized" to search in certain regions, in which case the search environment for each agent, as well as the initial search map, may be different.

2.2.   Distributed Path Planning

One of the key objectives of each agent is to on-line select a suitable path in the search environment $\mathcal{X}$. To be consistent with the motion dynamics of physical vehicles (and, in particular, air vehicles), it is assumed that each agent has limited maneuverability, which is represented by a maximum angle θ_m that the agent can turn from its current direction. For simplicity we assume that all agents move at a constant velocity μ (this assumption can be easily relaxed).

To describe the movement path of agent $\mathcal{A}_i$ between samples, we define the *movement sampling time T_m* as the time interval in the movement of the agent. In this framework, we let $p_i(k)$ be the position (in terms of (x, y) coordinates) of i-th agent at time $t = kT_m$, with the agent following a straight line in moving from $p_i(k)$ to its new position $p_i(k+1)$. Since the velocity μ of the air vehicle is assumed to be constant, the new position $p_i(k + 1)$ is at a distance μT_m from $p_i(k)$, and based on the maneuverability constraint, it is within an angle $\pm\theta_m$ from the current direction, as shown in Figure 13.3. To formulate the optimization problem as an integer programming problem, we discretize the arc of possible positions for $p_i(k + 1)$ into m points, denoted by the set

$$\overline{\mathcal{P}}_i(k + 1) = \left\{ \bar{p}_i^1(k + 1),\ \bar{p}_i^2(k + 1),\ \dots\ \bar{p}_i^j(k + 1),\ \dots \bar{p}_i^m(k + 1) \right\}.$$

Therefore, the next new position for the i-th agent belongs to one of the elements of the above set; i.e., $p_i(k + 1) \in \overline{\mathcal{P}}_i(k + 1)$.

The agent selects a path by choosing among a possible set of future position points. In our formulation we allow for a recursive q-step ahead planning, which can be described as follows:

- When agent $\mathcal{A}_i$ is at position $p_i(k)$ at time k, it has already decided the next q positions: $p_i(k + 1), p_i(k + 2), \dots, p_i(k + q)$.

- While the agent is moving from $p_i(k)$ to $p_i(k + 1)$ it selects the position $p_i(k + q + 1)$, which it will visit at time $t = k + q + 1$.

To get the recursion started, the first q positions, $p_i(1), p_i(2), \dots, p_i(q)$ for each agent need to be selected a priori. Clearly, $q = 1$ corresponds to the special

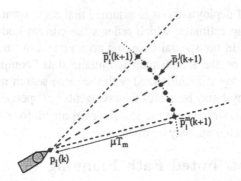

Figure 13.3. Selection of the next point in the path of the vehicle.

case of no planning ahead. The main advantage of a planning ahead algorithm is that it creates a buffer for path planning. From a practical perspective this can be quite useful if the agent is an air vehicle that requires (at least) some trajectory planning. Planning ahead is also useful for cooperation between agents since it may be communicated to other vehicles as a guide of intended plan selection. This can be especially important if there are communication delays or gaps, or if the sampling rate for communication is slow. On the other hand, if the integer q is too large then, based on the recursive procedure, the position $p_i(k)$ was selected q samples earlier at time $k - q$; hence the decision may be outdated, in the sense that it may have been an optimal decision at time $k - q$, but based on the new information received since then, it may not be the best decision anymore. The recursive q-step ahead planning procedure is illustrated in Figure 13.4 for the case where $q = 6$.

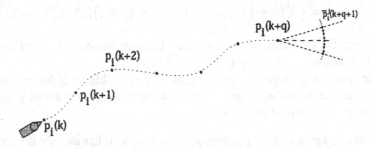

Figure 13.4. Illustration of the recursive q-step ahead planning algorithm.

Given the current information available via the search map, and the location/direction of the team of agents (and possibly other useful information, such as fuel remaining, etc.), each agent uses a multi-objective cost function J to select and update its search path. At decision sampling time T_d, the agent evaluates the cost function associated with each path and selects the optimal path. The decision sampling time T_d is typically equal to the movement sampling time T_m. The approach can be thought of as an "adaptive model predictive control" approach where we learn the model that we use to predict ahead in time, and we use on-line optimization in the formation of that model, and in evaluating the candidate paths to move the agent along.

A key issue in the performance of the cooperative search approach is the selection of the multi-objective cost function associated with each possible path. Our approach is quite flexible in that it allows the characterization of various mission-level objectives, and trade-offs between these. In general, the cost function comprises of a number of sub-goals, which are sometimes competing. Therefore the cost criterion J can be written as:

$$J = \omega_1 J_1 + \omega_2 J_2 + \cdots + \omega_s J_s,$$

where J_i represents the cost criterion associated with the i-th subgoal, and ω_i is the corresponding weight. The weights are normalized such that $0 \leq \omega_i \leq 1$ and the sum of all the weights is equal to one; i.e., $\sum_{i=1}^{s} \omega_i = 1$. Priorities to specific sub-goals are achieved by adjusting the values of weights ω_i associated with each subgoal.

The following is a list (not exhaustive) of possible sub-goals that a search agent may include in its cost criterion. Corresponding to each sub-goal is a cost-criterion component that need to be designed. For a more clear characterization, these sub-goals are categorized according to three mission objectives: Search (S), Cooperation (C), and Engagement (E). In addition to sub-goals that belong purely to one of these classes, there are some that are a combination of two or more missions. For example, SE1 (see below) corresponds to a search and engage mission.

S1 *Follow the path where there is maximum uncertainty in the search map.* This cost criterion simply considers the uncertainty reduction associated with the sweep region between the current position $p_i(k)$ and each of the possible candidate positions $\bar{p}_i^j(k+1)$ for the next sampling time (see the rectangular regions between $p_i(k)$ and $\bar{p}_i^j(k+1)$ in Figure 13.5). The cost criterion can be derived by computing a measure of uncertainty (or potential "gain" in knowledge) in the path between $p_i(k)$ and each candidate future position $\bar{p}_i^j(k+1)$.

S2 *Follow the path that leads to the region with the maximum uncertainty (on the average) in the search map.* The first cost criterion pushes the agent towards the path with the maximum uncertainty. However, this may not be the best path over a longer period of time if it leads to a region where the average uncertainty is low. Therefore, it's important for the search

agent to seek not only the instantaneous minimizing path, but also a
path that will cause the agent to visit (in the future) regions with large
uncertainty. The cost criterion can be derived by computing the average
uncertainty of a triangular type of region associated with the heading
direction of the agent (see the triangular regions ahead of $\bar{p}_i^j(k+1)$ in
Figure 13.5).

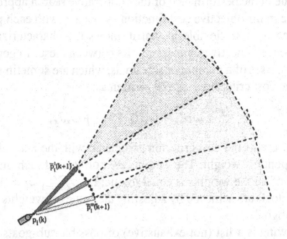

Figure 13.5. Illustration of the regions that are used in the cost function for finding the optimal search path.

C1 *Follow the path where there is the minimum overlap with other agents.*
Since the agents are able to share their new information about the search
region, it is natural that they may select the same search path as other
agents (especially since in general they will be utilizing the same search
algorithm). This will be more pronounced if two agents happen to be
close to each other. However, in order to minimize the global uncertainty
associated with the emergent knowledge of all agents, it is crucial that
there is minimum overlap in their search efforts. This can be achieved
by including a cost function component that penalizes agents being close
to each other and heading in the same direction. This component of the
cost function can be derived based on the relative locations and heading
direction (angle) between pairs of agents. This component of the cost
function is investigated more thoroughly in Section 3.

SE1 *Follow the path that maximizes coverage of the highest priority targets.*
In mission applications where the agents have a target search map with
priorities assigned to detected targets, it is possible to combine the search
of new targets with coverage of discovered targets by including a cost
component that steers the agent towards covering high priority targets.

Therefore, this leads to a coordinated search where both coverage and priorities are objectives.

E1 *Follow the path toward highest priority targets with most certainty if fuel is low.* In some applications, the energy of the agent is limited. In such cases it is important to monitor the remaining fuel and possibly switch goals if the fuel becomes too low. For example, in search-and-engage operations, the agent may decide to abort search objectives and head towards engaging high priority targets if the remaining fuel is low.

EC1 *Follow the path toward targets where there will be minimum overlap with other agents.* Cooperation between agents is a key issue not only in search patterns but also—and even more so—in engagement patterns. If an agent decides to engage a target, there needs to be some cooperation such that no other agent tries to go after the same target; i.e., a coordinated dispersed engagement is desirable.

The above list of sub-goals and their corresponding cost criteria provide a flavor of the type of issues associated with the construction of the overall cost function for a general mission. In addition to incorporating the desired sub-goals into the cost criterion (i.e., maximize benefit), it is also possible to include cost components that reduce undesirable sub-goals (minimize cost). For example, in order to generate a smooth trajectory for a UAV such that it avoids—as much as possible—the loss of sensing capabilities during turns, it may be desirable to assign an extra cost for possible future positions on the periphery (large angles) of the set $\overline{\mathcal{P}}_i$.

3. On-Line Cooperation by Distributed Agents

The framework developed in this chapter is based on distributed agents working together to enhance the global performance of a multi-agent system — in contrast to a framework where distributed agents may be competing with each other for resources. Therefore, one of the key issues in cooperative control is the ability of distributed agents to coordinate their actions and avoid overlap. In a decentralized environment, cooperation between agents may not come natural since every agent tries to optimize its own behavior. In typical complex scenarios it may not be clear to an individual agent how its own behavior is related to the global performance of the multi-agent system.

To illustrate this, consider the following "Easter egg hunt" scenario: each agent is asked to pick up Easter eggs from a field. For simplicity, we assume that the location of the eggs is known (no search is necessary). Each agent is initialized at some location in the field and its goal is to decide which direction to go. The velocity of each agent is fixed and once an agent is at the location of an egg then that egg is considered as having been picked. The global performance criterion is *for all the eggs to be pick up in the minimum possible time*. This simple scenario provides a nice framework for illustrating some of the key concepts of cooperative behavior. For example, an agent A_i may be tempted to head towards the direction of the closest Easter egg even though this may not enhance the global performance criterion if another agent A_j is closer to that egg and will get there before agent A_i. On the other hand, just because agent A_j is closest to that particular egg it does not necessarily imply that it will pick it up before agent A_i (it may go after some other eggs). If the Easter egg hunt problem was to be solved in a centralized framework then it would be rather easier to assign different eggs to different agents. However, in a distributed decision making setting, each agent is required to make decisions for enhancing the global performance criterion without having a clear association between its own action and the global cost function. In uncertain environments (for example, if the location of a certain Easter egg is not known unless the agent is within a certain distance and possibly within a certain heading angle from the egg) decisions need to be made on-line and therefore the cooperation issue becomes more challenging.

Cooperation between agents can be considered at different levels. For example, if each agent can perform several tasks (such as search for targets, classification, engagement and evaluation of attack) then cooperation between agents may involve coordinating their behavior while making decisions on which task to perform at what time. In this chapter, we are primarily focusing on the *cooperative search* problem. Therefore, the global objective of the team of agents is to update the overall search map (which represents the knowledge of the environment) in the minimum amount of time. To achieve this, each agent has a responsibility to select its path to benefit the team by selecting a path with minimum overlap with other agents' paths, as described earlier (sub-goal C1). Next we develop a real-time approach to realize the cooperative search activities among a team of distributed agents.

Before going into the details we present the main idea of the cooperative search scheme. Each agent possesses information about past paths of other agents via inter-agent communication. As discussed before, this information is used for updating the search map of each agent. Therefore, an agent is able to avoid going over paths previously searched by other agents simply by evaluating its search map and following a path that would result in maximum gain. However, this does not prevent an agent from following a path that another agent is about to follow, or has followed since the last communication contact. Therefore, the main idea of the proposed on-line cooperation scheme is for each agent to try to avoid selecting a path that may be followed by another agent in the near future. In this framework, paths of other agents are treated as "soft obstacles" to be avoided in path selection. However, special consideration is given to scenarios where path overlap may occur at approximately right angles, since in this case the overlap time is quite minimum, thereby not worth causing an interception in an agent's path planning. In other words, the scenario that should be avoided is two agents close to each other and heading in approximately the same direction. By treating paths of other vehicles as "soft obstacles" we employ a type of *artificial potential field* method ([43]) to derive an algorithm for generating the "rivaling force" that neighboring agents' paths may exert on a certain vehicle. The overall rivaling force exerted on an agent is taken into consideration in deciding which direction the vehicle will follow. Next we discuss the details of this approach.

3.1. Rivaling Force Between Agents

According to the proposed cooperative search framework, at time k, agent $\mathcal{A}_i$ uses the q-step ahead planning to select the position $p_i(k+q+1) \in \overline{\mathcal{P}}_i(k+q+1)$, which it will visit at time $t = k + q + 1$. By communicating with other vehicles at time $t = k - d$ (where d is the communication delay), agent $\mathcal{A}_i$ knows their q-step ahead positions $p_j(k + q - d)$ and heading directions $h_j(k + q - d)$ (measured in degrees from a reference direction). The rivaling force $F_{ij}(k)$ exerted by agent $\mathcal{A}_j$ onto agent $\mathcal{A}_i$ at time k is non-zero if both of the following conditions hold:

1 The location $p_j(k + q - d)$ of agent $\mathcal{A}_j$ is within a maximum distance $\bar{\mu}$ and maximum angle $\pm\bar{\varphi}$ from the location of agent $\mathcal{A}_i$ (see the shaded region in Figure 13.6).

2 The difference in heading angle $\chi_{ij}(k)$ between agent $\mathcal{A}_j$ and agent $\mathcal{A}_i$ lies within either $[-\bar{\chi}, \ \bar{\chi}]$ or $[180^0 - \bar{\chi}, \ 180^0 + \bar{\chi}]$, where $\bar{\chi}$ is the maximum allowed difference in heading angle.

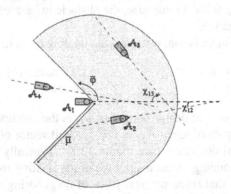

Figure 13.6. Illustration of conditions that generate non-zero rivaling forces between agents.

The first condition imposes a requirement that agent $\mathcal{A}_j$ needs to be sufficiently close to agent $\mathcal{A}_i$ before it exerts any rivaling force on $\mathcal{A}_i$. In addition to the distance, the angle between the two locations needs to be within $\pm \bar{\varphi}$. This requirement prevents a vehicle A_j which is behind A_i from exerting any rivaling force on A_i. In such a situation, there will be a rivaling force in the opposite direction from A_i to A_j. In the scenario shown in Figure 13.6, agents A_2 and A_3 satisfy Condition 1 with respect to their position to agent A_1, while agent A_4 does not satisfy Condition 1.

The second condition imposes the requirement that in order for agent A_j to exert a rivaling force on agent A_i it must either be heading in approximately the same direction, or be coming from approximately the opposite direction. This condition prevents the generation of any rivaling force if the two vehicles are heading in approximately perpendicular directions. Due to maneuverability constraints on the vehicles, the possible overlap in the paths of two agents is significant only if the heading angles are close to each other. At the same time, it is not desirable to impede the path of a vehicle if there is another vehicle coming at approximately right angles. In the scenario shown in Figure 13.6, agents A_2 and A_4 satisfy Condition 2 with respect to their heading direction in relation to agent A_1 (because both angles χ_{12}, χ_{14} are small), while agent A_3 does not satisfy Condition 2. Therefore, only agent A_2 satisfies both Conditions 1 and 2, and therefore it is the only one that exerts any rivaling force on agent A_1.

For vehicles satisfying both Conditions 1 and 2, the next step is to compute the magnitude and direction of the rivaling force exerted on agent A_i. The main objective here is that the magnitude of the rivaling force $F_{ij}(k)$ exerted by agent A_j onto agent A_i at time k should be "large" if agent A_i is close to the path of agent A_j, and should get smaller as agent A_i is further away from the path of agent A_j. This approach is similar to artificial potential field methods, which are used in many applications, including the problem of obstacle avoidance of robotic systems. In our case, the obstacle to be avoided is actually the path of another vehicle.

Based on this formulation, we select the rivaling force to be of the form

$$F_{ij}(k) = \begin{cases} k_1 e^{-\alpha \rho_{ij}} \vec{p}_{ij} & \text{if Conditions 1 and 2 hold} \\ 0 & \text{otherwise,} \end{cases} \tag{1}$$

where k_1, α are positive design constants, ρ_{ij} is the shortest distance between agent A_i and the path of agent A_j, and $\vec{p}_{ij}$ is a unit vector of the corresponding normalized partial derivative (see Figure 13.7). Typically k_1 will be a large constant, corresponding to the magnitude of the rivaling force if the distance ρ_{ij} is zero. Note that since we treat paths of neighboring agents as "soft obstacles" there is no need to set the magnitude of the rivaling force to ∞ as is sometimes done in the case of obstacle avoidance problems. The design parameter $\alpha > 0$ corresponds to the rate at which the rivaling force is decreasing as the distance ρ_{ij} is increasing. The rivaling force is not necessarily symmetric (i.e., $F_{ij}(k) \neq F_{ji}(k)$ since it depends on the relative position and heading direction of the two agents. In fact, as we saw earlier, it is possible for $F_{ij}(k)$ to be zero while $F_{ji}(k)$ is quite large (this would occur if agent $\mathcal{A}_j$ is behind $\mathcal{A}_i$ and heading in approximately the same direction). Figure 13.7 illustrates the potential field lines associated with the path of agent $\mathcal{A}_2$, and the resulting rivaling force exerted by $\mathcal{A}_2$ onto $\mathcal{A}_1$.

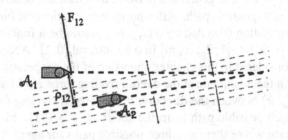

Figure 13.7. Illustration of the potential field lines associated with the path of agent $\mathcal{A}_2$, and the resulting rivaling force exerted by $\mathcal{A}_2$ onto $\mathcal{A}_1$.

As seen from Figure 13.7, the path of $\mathcal{A}_2$ that generates a rivaling force onto $\mathcal{A}_1$ includes not only the forward path but also some of the backward (previous) path. The reason for this is that communication delays may cause $\mathcal{A}_1$ to have incomplete (outdated) information about the path followed by $\mathcal{A}_2$. It is also noted that the actual path of an agent may not be a straight line as assumed in Figure 13.7. However, due to maneuverability constraints, this is a reasonable and simple approximation of the actual path for cooperation purposes.

The overall rivaling force exerted by the entire team of agents upon an agent $\mathcal{A}_i$ at time k is given by

$$F_i(k) = \sum_{j \neq i} F_{ij}(k). \tag{2}$$

Intuitively, according to the overall rivaling force $F_i(k)$ exerted on it, agent $\mathcal{A}_i$ is impelled to select a path $p_i(k + q + 1)$, among the possible set of paths $\overline{P}_i(k + q + 1)$, that is more in line with avoiding the paths of other vehicles. Therefore, in addition to the magnitude of the rivaling force, a key parameter is the angle difference between the direction of the overall rivaling force $F_i(k)$ and the direction of each possible path from the set $\overline{P}_i(k + q + 1)$, which we denote by $\theta_i(j, k)$. From a cooperative viewpoint, the objective is to select the path with the minimum $\theta_i(j, k)$ among $j \in [1, 2, \ldots m]$.

3.2. Cooperation Cost Function

Using the algorithm described in Section 3.1, each agent can compute the rivaling force exerted on it by other agents that are located in close proximity and, based on the overall rivaling force, select an optimal path that would minimize the overlap with paths of other vehicles. However, avoidance of path overlap is only one of an agent's objectives. Indeed, its main objective is to search for (and possibly engage) targets. Therefore, the goal of cooperation needs to be quantified as a cost function component and integrated with the remaining components of the cost criterion.

To integrate the cooperative sub-goal with other objectives, the cooperation cost function is required to generate a performance measure of cooperation associated with each possible path. After normalization, the cost function component for cooperation (denoted by $J(i, j, k)$) should be a function mapping each possible path $j \in [1, 2, \ldots, m]$ into an interval $[0, 1]$. According to the formulation considered in this chapter, the value of the cooperation cost function depends on the magnitude of the overall rivaling force $F_i(k)$ and the angle difference $\theta_i(j, k)$ between the direction of the overall rivaling force and the direction of each possible path from the set $\overline{P}_i(k + q + 1)$. Figure 13.8 illustrates the case where there are three possible paths for agent $\mathcal{A}_1$ to follow. The corresponding angles $\theta_1(1, k)$, $\theta_1(2, k)$, $\theta_1(3, k)$ are denoted by θ_1, θ_2, θ_3 respectively for diagrammatic simplicity. Hence, we consider a general

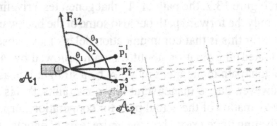

Figure 13.8.   Illustration of computing the cooperation cost function.

function

$$J(i, j, k) = f(|F_i(k)|, \ \theta_i(j, k)),$$

where $f : \Re^+ \times [-\pi, \pi] \mapsto [0, 1]$ is required to have the following attributes:

- As the magnitude of the rivaling force $F_i(k)$ becomes larger, the differences in the normalized cost function values between alternative paths should become larger. In other words, if $|F_i(k)|$ is large then cooperation is a crucial issue and therefore there should be a significant difference in the cooperation cost function to steer the agent into selecting the path of maximal cooperation. On the other hand, if $|F_i(k)|$ is small then cooperation is not a crucial issue, therefore the cooperation cost function component should be approximately equal for each alternative path plan, thereby allowing the agent to make its path decision based on the cost function associated with the other sub-goals.

- As the magnitude of the angle difference $\theta_i(j, k)$ becomes larger, the differences in the normalized cost function values between alternative paths should become larger. Again, if $|\theta_i(j, k))|$ is small then cooperation is not a crucial issue, therefore, the cooperation cost function component is approximately equal for each alternative path plan. If $|\theta_i(j, k))|$ is large then cooperation is a crucial issue and therefore there should be a significant difference in the cooperation cost function to steer the agent into selecting the path of maximal cooperation.

Deriving an appropriate function f with these attributes is rather straightforward. In the simulations presented in the next section, we use the following cooperative cost function

$$J(i, j, k) = \exp^{\gamma_0 |F_i(k)| \cos(\frac{\theta_i(j,k)}{2})}, \tag{3}$$

where γ_0 is a positive design constant.

It is important to note that the specific functions selected in Equation (1) for the rivaling force and in Equation (3) for the cooperative cost function, are not as important as the attributes of these functions. Specifically, other functions with the same attributes can be utilized to obtain similar results.

4.   Simulation Results

The approach described in this paper has been implemented and evaluated by several simulation studies. A representative sample of these studies is presented in this section. First, we describe the details of the cost function criterion and then present two simulations studies. In the first simulation study, a team of UAVs is searching in a mostly unknown environment. In the second simulation, we consider a scenario where the environment consists of three targets whose location belongs to a certain probability distribution.

4.1.   Design of Simulation Experiment

According to the proposed cooperative path planning approach, each agent uses a multi-objective cost function J to select and update its search path. This approach is quite flexible in that it allows the characterization of various mission-level objectives and facilitates possible trade-offs. The simulation examples presented in this section consider only the first three of the sub-goals (S1, S2, C1) described earlier. These sub-goals correspond to the main issues associated with the cooperative search problem.

The cost functions associated with each sub-goal are computed as follows:

- The first cost function $J_{S1}(i, j, k)$ is the gain of agent $\mathcal{A}_i$ on sub-goal S1 if it selects path $j \in [1, 2, \ldots, m]$ at time k. It is a positive value denoting the gain on the certainty of the search map by following path

j at time k. The following function is used to evaluate the actual gain obtained by selecting the jth path:

$$J_{S1}(i, j, k) = \sum_{(x,y)\in R_{i,j}} [S(x, y; \theta(k)) - S(x, y; \theta(k-1))], \quad (4)$$

where (x, y) denotes any point in the search area $R_{i,j}$ that will be encountered if agent A_i follows path j, and $S(x, y; \theta(k))$ is the certainty value of point (x, y) at time k.

- The second cost function $J_{S2}(i, j, k)$ is used to evaluate the potential gain based on the average uncertainty of a triangular region $R'_{i,j}$ associated with the heading direction j. The cost function $J_{S2}(i, j, k)$ is generated by

$$J_{S2}(i, j, k) = \sum_{(x,y)\in R'_{i,j}} (1 - S(x, y; \theta(k))), \quad (5)$$

where (x, y) denotes all the points in the region $R'_{i,j}$.

- The third cost-function is used to evaluate the sub-goal C1, which was formulated as

$$J_{C1}(i, j, k) = \exp^{\gamma_0 |F_i(k)| \cos(\frac{\theta_i(j,k)}{2})}. \quad (6)$$

After normalizing the three cost-functions and selecting appropriate weight coefficients, the overall multi-objective cost function is described by

$$J(i, j, k) = w_1 \cdot \overline{J}_{S1}(i, j, k) + w_2 \cdot \overline{J}_{S2}(i, j, k) + w_3 \cdot \overline{J}_{C1}(i, j, k), \quad (7)$$

where $\overline{J}_q$ for $q \in \{S1, S2, C1\}$ denote the normalized cost functions and w_i are the weights, which satisfy $w_1 + w_2 + w_3 = 1$. In the simulation examples different weight values were used to illustrate various aspects of cooperation. The normalized cost functions $\overline{J}_q$ are computed by

$$\overline{J}_q = \frac{J_q(i, j, k)}{\max_j \{J_q(i, j, k)\}}.$$

Therefore, each cost function $\overline{J}_q \in [0, 1]$. An agent A_i selects a path based on which $j \in [1, \cdots, m]$ gives the largest value, as computed by (7).

4.2. High Uncertainty Environment

The first simulation study considers a scenario of high uncertainty in the environment. The search region is a 200 by 200 area. It is assumed that there is some a-priori information about the search region: the green (light) polygons indicate complete certainty about the environment (for example, these can represent regions where it is known for sure—due to the terrain—that there are no targets); the blue (dark) polygons represent partial certainty about the environment. The remaining search region is assumed initially to be completely uncertain. First we consider the case of two agents, and then we use a team of five agents.

In both simulations we are using the recursive q-step ahead planning algorithm with $q = 3$. The weights of the cost function are set to: $w_1 = 0.3125$, $w_2 = 0.375$, $w_3 = 0.3125$, which gives approximately equal importance to each of the three sub-goals. The parameters of the potential field function used for sub-goal C1 are set to: $k_1 = 50$, $\alpha = 1$, $\gamma_0 = 1$. The results for the case of two agents are shown in Figure 13.9. The upper-left plot shows a standard

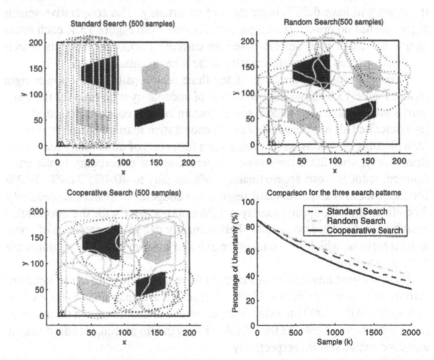

Figure 13.9. Comparison of the cooperative search pattern with a "standard" search pattern and a random search pattern for the case of two moving agents. The upper-left plot shows a standard search pattern for the first 500 time samples; the upper-right plot shows the corresponding search pattern in the case of a random search, subject to some bounds to restrict the agent from deserting the search region; The lower-left plot shows the cooperative search pattern based on the recursive q-step ahead planning algorithm; the lower-right plot shows a comparison of the performance of the three search patterns in terms of reducing uncertainty in the environment.

search pattern for the first 500 time samples, while the upper-right plot shows the corresponding result for a random search, which is subject to the maneuverability constraints. The standard search pattern utilized here is based on the so-called zamboni coverage pattern ([1]). The lower-left plot shows the result of the cooperative search method based on the recursive q-step ahead planning algorithm.

The search map used in this simulation study is based on piecewise constant basis functions, and the learning algorithm is a simple update algorithm of the form $\hat{\theta}(k + 1) = 0.5\hat{\theta}(k) + 0.5$, where the first encounter of a search block results in the maximum reduction in uncertainty. Further encounters result in reduced benefit. For example, if a block on the search map starts from certainty value of zero (completely uncertain) then after four visits from (possibly different) agents, the certainty value changes to $0 \mapsto 0.5 \mapsto 0.75 \mapsto 0.875 \mapsto 0.9375$. The percentage of uncertainty is defined as the distance of the certainty value from one. In the above example, after four encounters the block will have 6.25% percentage of uncertainty. The cooperative search algorithm has no pre-set search pattern. As seen from Figure 13.9, each agent adapts its search path on-line based on current information from its search results, as well as from search results of the other agents.

To compare the performance of the three search patterns, the lower-right plot of Figure 13.9 shows the percentage of uncertainty with time for the standard search pattern, the random search pattern and the cooperative search pattern described above. The ability of the cooperative search algorithm to make path planning decisions on-line results in a faster rate of uncertainty reduction. Specifically, after 2000 time steps the percentage of uncertainty in the environment reduces from approximately 85% initially to 40.4%, 34.4%, 29.2% for the random search, standard search, and cooperative search, respectively. Therefore, there is approximately a 15% improvement with the cooperative search over the standard search. This is mainly due to the presence of some known regions, which the standard search and random search algorithms are not trying to avoid.

The corresponding results in the case of five agents moving in the same environment is shown in Figure 13.10. The results are analogous to the case of two agents. After 2000 time steps the percentage of uncertainty in the environment reduces to 13.9%, 12.0%, 7.1% for the random search, standard search, and cooperative search, respectively.

In these simulation studies, we assume that the sampling time $T_m = 1$ corresponds to the rate at which each agent receives information from its own sensors, updates its search map and makes path planning decisions. Information from other agents is received at a slower rate. Specifically, we assume that the communication sampling time T_c between agents is five times the movement sampling time; i.e., $T_c = 5T_m$. For fairness in comparison, it assumed

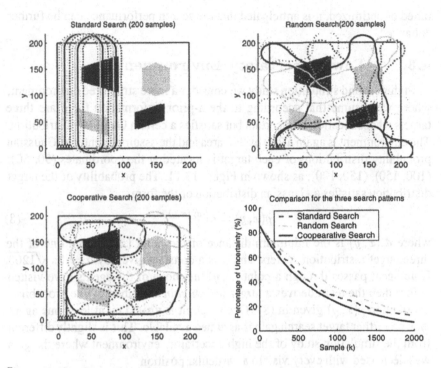

Figure 13.10. Comparison of the cooperative search pattern with a "standard" search pattern and a random search pattern for the case of five moving agents. The upper-left plot shows a standard search pattern for the first 200 time samples; the upper-right plot shows the corresponding search pattern in the case of a random search, subject to some bounds to restrict the agent from deserting the search region; The lower-left plot shows the cooperative search pattern based on the recursive q-step ahead planning algorithm; the lower-right plot shows a comparison of the performance of the three search patterns in terms of reducing uncertainty in the environment.

that for the standard and random search patterns the agents exchange information and update their search maps in the same way as in the cooperative search pattern, but they do not use the received information to make on-line decisions on where to go.

It is noted that in these simulations the path planning of the cooperative search algorithm is rather limited since at every sampled time each agent is allowed to either go straight, left, or right (the search direction is discretized into only three possible points; i.e., $m = 3$). The left and right directions are at angles of -15^0 and $+15^0$ respectively from the heading direction, which reflects the maneuverability constraints of the vehicles. As the complexity of the cooperative search algorithm is increased and the design parameters (such as the weights associated with the multi-objective cost function) are fine-tuned or optimized, it is anticipated that the search performance can be further enhanced.

4.3.   Low Uncertainty Environment

In this second simulation study we consider a more structured environment, where we assume that according to the a-priori information there are three targets whose location is uncertain but satisfies a certain Gaussian distribution. The environment is again a 200 by 200 area and the assumed center of Gaussian probability distributions of each target is located at the coordinates (50, 50), (100, 150), (150, 100), as shown in Figure 13.11. The probability of the target distribution satisfies a Gaussian distribution of the form

$$p(x, y) = e^{\frac{1}{\sigma} d_c^2 (x,y)}, \tag{8}$$

where $d_c(x, y)$ is the minimum distance of the point (x, y) from one of the three target distribution centers, and σ is a constant given by $\sigma = 2\pi \sqrt{1200}$. If an agent passes through a point (x, y) that none of the agents have visited before then the team derives a *target search gain* described by the probability distribution $p(x, y)$ given in (8). Once a point is visited by at least one agent then no further target search gain is assumed available. This is slightly different from the simulation study of the high uncertainty environment where the gain was decreased with every visit to a particular position.

In the simulation shown in Figure 13.11 we compare the performance of three different runs, all based on the search procedure developed in this paper using the recursive q-step ahead planning algorithm. The team of agents consists of five vehicles with the same maneuverability constraints as in the first simulation study. The only difference between the three runs is the amount of cooperation included, as defined by the third cost function component J_{C1}. The upper left plot shows the trajectories of the team of agents using the cooperative search algorithm with the weights selected as $w_1 = 1/8$, $w_2 = 2/8$, $w_3 = 5/8$. In the second simulation run, shown in the upper right plot, we show the trajectories selected by the five vehicles for a weakly cooperative system with the weights selected as $w_1 = 1/4$, $w_2 = 2/4$, $w_3 = 1/4$. Finally, in the third simulation run there is no cooperation between the five agents, in the sense that the weights are set to: $w_1 = 1/3$, $w_2 = 2/3$, $w_3 = 0$.

As seen from Figure 13.11, in the case of the cooperative search algorithm (upper left) the five vehicles split up between the two nearest targets and soon they also cover the distant target. In the case of the weakly cooperative search algorithm (upper right) the five agents first go to the nearest target on the lower left, and from there, some agents go to the other two targets. In the case of non-cooperation (lower left plot) all five vehicles head for the nearest target on the lower left and spend considerable time there before they move on to the other targets (in fact, the simulation shows 200 time steps—as compared

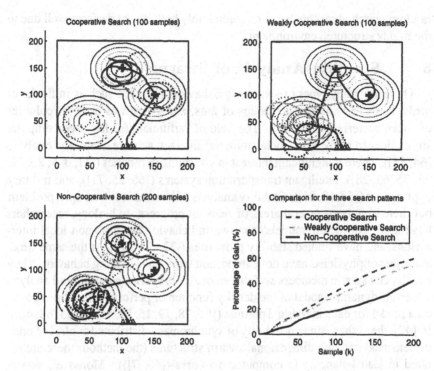

Figure 13.11. Comparison of the cooperative search pattern with a "weakly cooperative" search pattern and a non-cooperative search pattern for the case of five moving agents searching for three targets located according to a Gaussian distribution function around three center points. The upper-left plot shows a cooperative search pattern for the first 100 time samples; the upper-right plot shows the corresponding search pattern in the case of a weakly cooperative search algorithm; The lower-left plot shows the non-cooperative search pattern for the first 200 time samples; the lower-right plot shows a comparison of the performance of the three search patterns in terms of the percentage of target search gain over time for each of the three search patterns.

to 100 samples for the other two simulation runs—because during the first 100 steps all five vehicles remained at the first target). With no cooperation there is significant overlap of the paths of vehicles.

The performance of the three search patterns for the first 200 time steps is shown in the lower right plot of Figure 13.11 in terms of the percentage of target search gain over time. The percentage of target search gain is computed as the total gain of all five vehicles at time k divided by the initial total target search gain in the environment. After 200 time steps the target search gain for the cooperative search is 59.3%, for the weakly cooperative search it is 54.1% and for the non-cooperative search it is 42.8%. It is noted that in this simulation study we do not show the performance of a "standard search pattern" and the random search algorithm because comparably both do not perform well due to the highly structured environment.

5.   Stability Analysis of Swarms

There are many types of swarming behaviors in nature such as in flocks of birds, herds of wildebeests, groups of ants, and swarms of bees, or colonies of social bacteria ([8, 59, 67]). The field of "artificial life" has used computer simulations to study how "social animals" interact, achieve goals, and evolve ([64, 55]). There is significant interest in swarming in robotics ([11, 3, 5, 26, 52, 22, 75, 63, 28]), intelligent transportation systems ([66, 25, 77]), and military applications ([58, 69, 85]). Stability analysis of swarms is still an open problem but there have been several areas of relevant progress. In biology, researchers have used "continuum models" for swarm behavior based on non-local inter-actions, and have studied stability properties ([32, 50, 56]). At the same time, a number of physicists have done important work on swarming behavior. They usually call swarm members *self-driven* or *self-propelled particles* and analyze either the dynamic model of the density function or perform simulations based on a model for each individual particle ([61, 78, 79, 15, 82, 16, 17, 53, 68, 46]). In [42], the authors studied stability of synchronized distributed control of one-dimensional and two-dimensional swarm structures (the methods there are re-lated to load balancing in computer networks ([60, 7])). Moreover, swarm "cohesiveness"was characterized as a stability property and a one-dimensional asynchronous swarm model was constructed by putting many identical single finite-size vehicular swarm members together, which have proximity sensors and neighbor position sensors that only provide delayed position information in [48, 49]. For this model, [48, 49] showed that for a one-dimensional station-ary edge-member swarm, total asynchronism leads to asymptotic collision-free convergence and partial asynchronism leads to finite time collision-free con-vergence even with sensing delays. Furthermore, conditions were given in [49, 47], under which an asynchronous mobile swarm following (pushed by) an "edge-leader" can maintain cohesion during movements even in the pres-ence of sensing delays and asynchronism. Also, this work was expanded upon in [27], with the study of other local decision-making mechanisms and by ex-ploiting the analysis approaches from the theory of parallel and distributed computation ([7]). Stability of inter-vehicle distances in "platoons" and traffic in intelligent transportation systems have been studied ([4, 76, 24, 34, 66, 77]) and stable robust formation control for aircraft and microsatellites is relevant ([69, 85]).

5.1. Modeling and Analysis Via Nonlinear Asynchronous Difference Equations

Swarm stability for the $M \geq 2$ dimensional case with a fixed communication topology will be discussed in this section by extending the results in [48, 49, 47]. Our approach uses a discrete time discrete event dynamical system approach ([60]), and unlike the studies of platoon stability in intelligent transportation systems and flight formation control we avoid detailed characteristics of low level "inner-loop control" and vehicle dynamics in favor of focusing on high level mechanisms underlying qualitative swarm behavior when there are imperfect communications.

5.1.1 Single Swarm Member Model. An M-dimensional ($M \geq 2$) swarm is a set of N swarm members that moves in the M-dimensional space. Assume each swarm member has a finite physical size (radius) $w > 0$ and its position is the center of it. It has a "proximity sensor," which has a sensing range with a radius $\varepsilon > w$ around each member. In the $M = 2$ case, it is a circular-shaped area with a radius $\varepsilon > w$ around each member as shown in Figure 13.12. Once another swarm member reaches a distance of ε from it, the sensor *instantaneously* indicates the position of the other member. However, if its neighbors are not in its sensing range, the proximity sensor will return ∞ (or, practically, some large number). The proximity sensor is used to help avoid swarm member collisions and ensures that our framework allows for finite-size vehicles, not just points. Each swarm member also has a "neighbor position sensor" which can sense the positions of neighbors around it if they are present. We assume that there is no restriction on how close a neighbor must be for the neighbor position sensor to provide a sensed value of its position. The sensed position information may be subjected to random delays (i.e., each swarm member's knowledge about its neighbors' positions may be outdated). We assume that each swarm member knows its own position with no delay. Notice that we define the position, distance and sensor sensing range of the finite-size swarm member with respect to its center, not its edge.

Swarm members like to be close to each other, but not too close. Suppose $d > \varepsilon$ is the desired "comfortable distance" between two adjacent swarm neighbors, which is known by every swarm member. Each swarm member senses the inter-swarm member distance via both neighbor position and proximity sensors and makes decisions for movements via some position updating algorithms, which is according to the error between the sensed distance and the comfortable distance d. And then, the decisions are inputted to its "driving device," which provides locomotion for it. Each swarm member will try to move to maintain a comfortable distance to its neighbors. This will tend to make the group move together in a cohesive "swarm."

5.1.2 M-Dimensional Asynchronous Swarm Model with a Fixed Communication Topology.

An M-dimensional swarm is formed by putting many of the above single swarm members together on the M-dimensional space. A example of $M = 2$ dimensional swarm is shown in Figure 13.12. Let $x^i(t)$ denote the position vector of swarm member i at time t. We have $x^i(t) = [x_1^i(t), x_2^i(t), \ldots, x_M^i(t)]^\top \in R^M$, $i = 1, 2, \ldots, N$, where $x_m^i(t)$, $m = 1, 2, \ldots, M$, is the mth position coordinate of member i. We assume that there is a set of times $T = \{0, 1, 2, \ldots\}$ at which one or more swarm members update their positions. Let $T^i \subseteq T, i = 1, 2, \ldots, N$, be a set of times at which the ith member's position $x^i(t), t \in T^i$, is updated. Notice that the elements of T^i should be viewed as the indices of the sequence of physical times at which updates take place, not the real times. These time indices are non-negative integers and can be mapped into physical times. The T^i, $i = 1, 2, \ldots, N$, are independent of each other for different i. However, they may have intersections (i.e., it could be that $T^i \cap T^j \neq \emptyset$ for $i \neq j$), so two or more swarm members may move simultaneously. Here, our model assumes that swarm member i senses its neighbor positions and update its position only at time indices $t \in T^i$ and at all times $t \notin T^i$, $x^i(t)$ is left unchanged. A variable $\tau_j^i(t) \in T$ is used to denote the time index of the real time where position information about its neighbor j was obtained by member i at $t \in T^i$ and it satisfies $0 \leq \tau_j^i(t) \leq t$ for $t \in T^i$. Of course, while we model the times at which neighbor position information is obtained as being the same times at which one or more swarm members decide where to move and actually move, it could be that the *real time* at which such neighbor position information is obtained is earlier than the real time where swarm members moved. The difference $t - \tau_j^i(t)$ between current time t and the time $\tau_j^i(t)$ is a form of communication delay (of course the actual length of the delay depends on what real times correspond to the indices $t, \tau_j^i(t)$). Moreover, it is important to note that we assume that $\tau_j^i(t) \geq \tau_j^i(t')$ if $t > t'$ for $t, t' \in T^i$. This ensures that member i will use the most recently obtained neighbor position information. Furthermore, we assume swarm member i will use the real-time neighbor position information $x^j(t)$ provided by its proximity sensors instead of information from its neighbor position sensors $x^j(\tau_j^i(t))$ if its neighbor j is connected to i on its communication topology. This information will be used for position updating until member i gets more recent information, for example, from its neighbor position sensor. Next, based on [7], we specify two assumptions that we use to characterize asynchronism for swarms.

Assumption 1. (Total Asynchronism): *Assume the sets* T^i, $i = 1, 2, \ldots, N$, *are infinite, and if for each k, $t_k \in T^i$ and $t_k \to \infty$ as $k \to \infty$, then* $\lim_{k \to \infty} \tau_j^i(t_k) = \infty$, $j = 1, 2, \ldots, N$ *and* $j \neq i$.

This assumption guarantees that each swarm member moves infinitely often and the old position information of neighbors of each swarm member is eventually purged from it. On the other hand, the delays $t - \tau_j^i(t)$ in obtaining position information of neighbors of member i can become unbounded as t increases. Next, we specify a more restrictive type of asynchronism, but one which can accurately model timing in actual swarms.

Assumption 2. (Partial Asynchronism): *There exists a positive integer B (i.e., $B \in Z^+$, where Z^+ represents the set of positive integers) such that: (a) For every i and $t \geq 0$, $t \in T$, at least one of the elements of the set $\{t, t+1, ..., t+B-1\}$ belongs to T^i. (b) There holds $t - B < \tau_j^i(t) \leq t$ for all $i, j = 1, 2, ..., N$ and $j \neq i$, and all $t \geq 0$ belonging to T^i.*

Notice that for the partial asynchronism assumption, each member moves at least once within B time indices and the delays $t - \tau_j^i(t)$ in obtaining position information of neighbors of member i is bounded by B, i.e., $0 \leq t - \tau_j^i(t) < B$.

Assume that initially a set of N swarm members is randomly distributed in the M-dimensional space. We assume that $|x| = \sqrt{x^T x}$ and $|x^i(0) - x^j(0)| > d$, for $i, j = 1, 2, ..., N, i \neq j$ initially. We are studying several ways to establish communication topologies, including an initial distributed construction of the communication topology, and distributed dynamic reconfiguration of the communication topology (e.g., based on locality measures). To keep things simple here suppose that the communication topology is fixed based on the initial conditions (in an appropriate way so that evolving dynamics will not result in certain types of deadlock and collisions). An example of such a communication topology in the $M = 2$ case is shown by several dashed lines in Figure 13.12. Assume that the swarm members begin to update its position via this fixed communication topology at their updating time indices. In particular, swarm member $i + 1, i = 1, 2, ..., N - 1$, tries to maintain a comfortable distance d to its neighbor i so that it moves only according to the sensed position of member i.

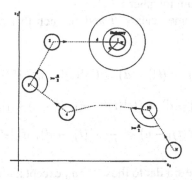

Figure 13.12. An M-dimensional ($M = 2$) N-member asynchronous swarm with a fixed communication topology (dashed line), all members moving to be adjacent to the stationary member (member 1).

Let $e^i(t) = x^i(t) - x^{i+1}(t)$, $i = 1, 2, ..., N - 1$, denote the distance vector between swarm members $i + 1$ and i. Assume the direction of $e^i(t)$ is from the position of member $i + 1$ to the position of member i and $|e^i(t)| = \sqrt{(e^i(t))^T e^i(t)}$, where $|e^i(t)|$ denotes its value. In addition, let the function $g(|e^i(t)| - d)$ denote the attractive and repelling relationship between two swarm neighbors with respect to the error between $|e^i(t)|$ and the comfortable distance d. We define two different types of g functions below, $g_a(|e^i(t)| - d)$ and $g_f(|e^i(t)| - d)$, to denote two different kinds of attractive and repelling relationships that are used to establish different swarm convergence properties. Assume that for a scalar $\beta > 1$, $g_a(|e^i(t)| - d)$ is such that

$$\frac{1}{\beta}(|e^i(t)| - d) < g_a(|e^i(t)| - d) < (|e^i(t)| - d), \text{if}(|e^i(t)| - d) > 0; \quad (9)$$

$$g_a(|e^i(t)| - d) = (|e^i(t)| - d) = 0, \text{ if } (|e^i(t)| - d) = 0; \quad (10)$$

$$(|e^i(t)| - d) < g_a(|e^i(t)| - d) < \frac{1}{\beta}(|e^i(t)| - d), \text{if}(|e^i(t)| - d) < 0. \quad (11)$$

Equation (9) indicates that if $(|e^i(t)| - d) > 0$, then swarm member position x^{i+1} is too far away from the position x^i so there is an attractive relationship between swarm members $i + 1$ and i. In addition, the low bound $\frac{1}{\beta}(|e^i(t)| - d)$ for $g_a(|e^i(t)| - d)$ guarantees that swarm member's moving step cannot be infinitely small during its movements to its desired position if $(|e^i(t)| - d)$ is not infinitely small. The constraint $g_a(|e^i(t)| - d) < (|e^i(t)| - d)$ ensures that it will not "over-correct" for the inter-swarm member distance. Equation (10) indicates that if $(|e^i(t)| - d) = 0$, then swarm member position x^{i+1} is at a comfortable distance d from the position x^i so there are no attractive or repelling relationship. Equation (11) indicates that if $(|e^i(t)| - d) < 0$, then swarm member position x^{i+1} is too close to the position x^i, so member $i + 1$ tries to move away from member i.

Assume that for some scalars β and η, such that $\beta > 1$, and $\eta > 0$, $g_f(|e^i(t)| - d)$ satisfies

$$\frac{1}{\beta}(|e^i(t)| - d) < g_f(|e^i(t)| - d) < (|e^i(t)| - d), \text{ if } (|e^i(t)| - d) > \eta; \quad (12)$$

$$g_f(|e^i(t)| - d) = (|e^i(t)| - d), \text{ if } -\eta \leq (|e^i(t)| - d) \leq \eta; \quad (13)$$

$$(|e^i(t)| - d) < g_f(|e^i(t)| - d) < \frac{1}{\beta}(|e^i(t)| - d), \text{ if } (|e^i(t)| - d) < -\eta. \quad (14)$$

These relationships are similar to those for g_a except if $-\eta < (|e^i(t)| - d) < \eta$, the two swarm members can move to be at the comfortable inter-swarm member distance within one move. It is not difficult to show that in the above M-dimensional swarm, if swarm members only update their positions according to either g function, collisions will have never happen even without proximity sensors (with appropriate initial conditions for the given topology).

A mathematical model for the above M-dimensional swarm is given by

$$x^1(t+1) = x^1(t), \forall t \in T^1$$

$$x^2(t+1) = x^2(t) + g(|x^1(\tau_1^2(t)) - x^2(t)| - d)\left[\frac{x^1(\tau_1^2(t)) - x^2(t)}{|x^1(\tau_1^2(t)) - x^2(t)|}\right],$$
$$\forall t \in T^2$$

$$\vdots = \vdots$$

$$x^{N-1}(t+1) = x^{N-1}(t) + g(|x^{N-2}(\tau_{N-2}^{N-1}(t)) - x^{N-1}(t)| - d) \cdot$$
$$\cdot \left[\frac{x^{N-2}(\tau_{N-2}^{N-1}(t)) - x^{N-1}(t)}{|x^{N-2}(\tau_{N-2}^{N-1}(t)) - x^{N-1}(t)|}\right], \quad \forall t \in T^{N-1}$$

$$x^N(t+1) = x^N(t) + g(|x^{N-1}(\tau_{N-1}^N(t)) - x^N(t)| - d) \cdot$$
$$\cdot \left[\frac{x^{N-1}(\tau_{N-1}^N(t)) - x^N(t)}{|x^{N-1}(\tau_{N-1}^N(t)) - x^N(t)|}\right], \quad \forall t \in T^N$$

$$x^i(t+1) = x^i(t), \forall t \notin T^i, \ i = 1, 2, ..., N. \tag{15}$$

Here, each item in brackets is a unit vector which represents the moving direction of each swarm member, and each g function item in front of the brackets is a scalar, which is the "step size" of each swarm member. It is simple to rewrite the "error system" for the study of inter-member dynamics.

5.1.3 Convergence Analysis of M-Dimensional Asynchronous Swarms.

Lemma 1. *For an $N = 2$ totally asynchronous swarm modeled by*

$$e^i(t+1) = e^i(t) - g(|e^i(\tau_i^{i+1}(t))| - d)\left[\frac{e^i(\tau_i^{i+1}(t))}{|e^i(\tau_i^{i+1}(t))|}\right], \forall t \in T^{i+1}$$

$$e^i(t+1) = e^i(t), \forall t \notin T^{i+1}, \tag{16}$$

where member i remains stationary, $|e^i(0)| > d$, and $g = g_a$, it is the case that for any γ, $0 < \gamma \leq |e^i(0)| - d$, there exists a time t' such that $|e^i(t')| \in [d, d+\gamma]$ and also $\lim_{t \to \infty} |e^i(t)| = d$.

Lemma 2. *For an $N = 2$ partially asynchronous swarm modeled by Equation (16) but with $g = g_f$, where member i remains stationary, $|e^i(0)| > d$, the inter-member distance of members $i + 1$ and i, $|e^i(t)|$ will converge to d in some finite time, that is bounded by $B[\frac{\beta}{\eta}(|e^i(0)| - d - \eta) + 2]$.*

We can build on the $N = 2$ case and show that the following results hold since if we have partial asynchronism and appropriate constraints on reduction of inter-member distance errors we will get finite-time convergence.

Theorem 1. (Partial Asynchronism, Finite Time Convergence): *For an N-member M-Dimensional swarm with $g = g_f$, $N \geq 2$, Assumption 2 (partial asynchronism) holds, and $|e^i(0)| > d$, the swarm members' inter-neighbor distance $|e^i(t)|$, $i = 1, 2, \ldots, N - 1$, will converge to the comfortable distance d in some finite time, that is bounded by*

$$B[\frac{\beta}{\eta}(\sum_i (|e^i(0)| - d) - \eta) + 2], \text{ for } i = 1, 2, \ldots, N - 1,$$

where $e^1(0)$ is the initial inter-neighbor distance.

The proofs are omitted due to space constraints. However, it is useful to point out that the proof of Theorem 1 entails more than a straightforward Lyapunov analysis; due the presence of asynchronism and delays typical choices for a Lyapunov function are not nonincreasing at each step (see discussion below and Figure 13.13(b)). It is the case, however, that at every time instant there exists a future time where a "Lyapunov-like" function will decrease. Our proof utilizes an induction approach over the communication topology, and relies on this essential fact. Finally, notice that for an N-member M-dimensional totally asynchronous swarm with $g = g_a$, $N \geq 2$, if Assumption 1 (total asynchronism) holds, and $|e^i(0)| > d$, we can use Lemma 1 to prove that the swarm members' inter-neighbor distance $|e^i(t)|$, $i = 1, 2, \ldots, N - 1$, will asymptotically converge to the comfortable distance d.

5.1.4   Simulation To Illustrate Swarm Dynamics.

We will simulate the swarm in Figure 13.12 converging to be adjacent to a stationary member under the assumptions of Theorem 1. In particular, we choose parameters for the attraction-repulsion function, asynchronous timing characteristics, and sensing delays for a swarm of 10 members with a communication topology that is a simple line connecting the 10 members (and member 1 is stationary). The results of the simulation are given via time "snapshots" shown in Figure 13.13(a). Figure 13.13(b) shows the inter-member distances, and illustrates that due to the delays and asynchronism they do not decrease at every step (a key feature that complicates the stability analysis).

5.1.5   Stable Mobile Swarm Formations.

Extending the results of the above sections to the mobile case (i.e., in transit swarm) where there is a single leader follows directly from our previous analysis for the one-dimensional mobile swarm case ([49, 47]). We are working now to extend the results to: (i) certain distributed dynamic reconfigurations of the communications topology, (ii) and certain "swarm formations." Notice that if we can

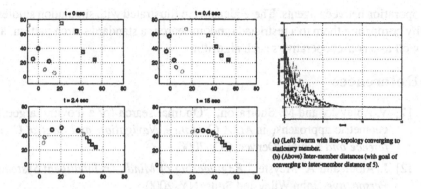

Figure 13.13. $M = 2$ dimensional asynchronous 10-member swarm converging.

extend the above results to request each member to follow the member that it is trying to converge to, not at an arbitrary following position on a fixed radius of size d, but at a fixed angle and inter-member distance following position, then we can form a "line" that sweeps out a pattern, and if we join two such lines we can get a "V" (according to which certain birds often fly during migration). Clearly, we can use basic joined line topologies to produce many swarm formations in the two dimensional case. Moreover, the results are easy to extend to the M-dimensional case so that for instance we can study stability of moving "spheres" of agents or other shapes such as those typically employed by swarming honey bees.

6.   Concluding Remarks

Advances in distributed computing and wireless communications have enabled the design of distributed agent systems. One of the key issues for a successful and wide deployment of such systems is the design of cooperative decision making and control strategies. Traditionally, feedback control methods have focused mostly on the design and analysis of centralized, inner-loop techniques. Decision and control of distributed agent systems requires a framework that is based more on cooperation between agents, and outer-loop schemes. In addition to cooperation, issues such as coordination, communication delays and robustness in the presence of losing one or more of the agents are crucial. In this chapter, we have presented a framework for cooperative search and derived stability results for multi-dimensional swarms. The proposed framework consists of two main components: learning the environment and using that knowledge to make intelligent high-level decisions on where to go (path planning) and what do to. We have presented some ideas regarding the design of a cooperative planning algorithm based on a recursive q-step ahead planning procedure and developed a real-time approach for on-line cooperation between agents. These ideas were illustrated with simulation studies by comparing them to a restricted random search, a standard search pattern, as well as a non-cooperative search algorithm.

References

[1] V. Ablavsky and M. Snorrason, "Optimal search for a moving target: a geometric approach", in *AIAA Guidance, Navigation, and Control Conference and Exhibit*, Denver, CO, 2000.

[2] J. Albus and A. Meystel, *Engineering of Mind: An Intelligent Systems Perspective*, John Wiley and Sons, NY, 2000.

[3] R. Arkin, *Behavior-Based Robotics*, MIT Press, Cambridge, MA, 1998.

[4] J. Bender and R. Fenton, "On the flow capacity of automated highways", *Transport. Sci.*, 4(1):52–63, 1970.

[5] G. Beni and J. Wang, "Swarm intelligence in cellular robotics systems", in *Proceedings of NATO Advanced Workshop on Robots and Biological System*, pages 703–712, 1989.

[6] S. Benkoski, M. Monticino and J. Weisinger, "A survey of the search theory literature", *Naval Research Logistics*, 38:469–494, 1991.

[7] D. Bertsekas and J. Tsitsiklis, *Parallel and Distributed Computation Numerical Methods*, Prentice Hall, NJ, 1989.

[8] E. Bonabeau, M. Dorigo and G. Theraulaz, *Swarm Intelligence: From Natural to Artificial Systems*, Oxford Univ. Press, NY, 1999.

[9] C. Breder, "Equations descriptive of fish schools and other animal aggregations", *Ecology*, 35:361–370, 1954.

[10] R. Brooks, "A robust layered control system for a mobile robot", *IEEE Trans. on Robotics and Automation*, 2(1), 1986.

[11] R. Brooks, editor, *Cambrian Intelligence: The Early History of the New AI*. MIT Press, Cambridge, MA, 1999.

[12] S. Cameron, "Obstacle avoidance and path planning", *Industrial Robot*, 21(5):9–14, 1994.

[13] H. Choset and P. Pignon, "Coverage path planning: the boustrophedon cellular decomposition", In *International Conference on Field and Service Robotics*, Canberra, Australia, 1997.

[14] A. Conn, K. Scheinberg and P. Toint, "Recent progress in unconstrained nonlinear optimization without derivatives", *Mathematical Programming*, 79:397–414, 1997.

[15] Z. Csahok and T. Vicsek, "Lattice-gas model for collective bilogical motion", *Physical Review E*, 52(5):5297–5303, 1995.

[16] A. Czirok, E. Ben-Jacob I. Cohen, and T. Vicsek, "Formation of complex bacterial colonies via self-generated vortices", *Physical Review E*, 54(2):1791–1801, 1996.

[17] A. Czirok and T. Vicsek, "Collective behavior of interacting self-propelled particles", *Physica A*, 281:17–29, 2000.

[18] J. Danskin, "A helicopter versus submarines search game", *Operations Research*, 16:509–517, 1968.

[19] R. Dell and J. Eagle, "Using multiple searchers in constrainted-path moving-targer search problems", *Naval Research Logistics*, 43:463–480, 1996.

[20] J. Dennis and V. Torczon, "Direct search methods on parallel machines", *SIAM Journal Optimization*, 1(4):448–474, 1991.

[21] A. Drogoul, M. Tambe and T. Fukuda, editors, *Collective Robotics*. Springer Verlag, Berlin, 1998.

[22] G. Dudek et al., "A taxonomy for swarm robots", In *IEEE/RSJ Int. Conf. on Intelligent Robots and Systems*, Yokohama, Japan, 1993.

[23] J. Eagle and J. Yee, "An optimal branch-and-bound procedure for the constrained path moving target scarch problem", *Operations Research*, 38:11–114, 1990.

[24] R. Fenton, "A headway safety policy for automated highway operation", *IEEE Trans. Veh. Technol.*, VT-28:22–28, 1979.

[25] R. Fenton and R. Mayhan, "Automated highway studies at the Ohio State University - an overview", *IEEE Trans. on Vehicular Technology*, 40(1):100–113, 1991.

[26] T. Fukuda, T. Ueyama and T. Sugiura, "Self-organization and swarm intelligence in the society of robot being", in *Proceedings of the 2nd International Symposium on Measurement and Control in Robotics*, pages 787–794, Tsukuba Science City, Japan, 1992.

[27] V. Gazi and K. M. Passino, "Stability of a one-dimensional discrete-time asynchronous swarm", in *Proc. of the IEEE Int. Symp. on Intelligent Control*, Mexico City, 2001.

[28] E. Gelenbe, N. Schmajuk, J. Staddon, and J. Reif, "Autonomous search by robots and animals: A survey", *Robotics and Autonomous Systems*, 22:23–34, 1997.

[29] D. Gillen and D. Jacques, "Cooperative behavior schemes for improving the effectiveness of autonomous wide area search munitions", in R. Murphey and P. Pardalos, editors, *Cooperative Control and Optimization*, pages 95-120, Kluwer Academic Publishers, 2002.

[30] D. Godbole, "Control and coordination in uninhabited combat air vehicles", in *Proceedings of the 1999 American Control Conference*, pages 1487–1490, 1999.

[31] S. Goldsmith and R. Robinett, "Collective search by mobile robots using alpha-beta coordination", in A. Drogoul, M. Tambe and T. Fukuda, editors, *Collective Robotics*, pages 136–146. Springer Verlag: Berlin, 1998.

[32] P. Grindrod, "Models of individual aggregation or clustering in single and multi-species communities", *Journal of Mathematical Biology*, 26:651–660, 1988.

[33] S. Hackwood and S. Beni, "Self-organization of sensors for swarm intelligence", in *IEEE Int. Conf. on Robotics and Automation*, pages 819–829, Nice, France, 1992.

[34] J. Hedrick and D. Swaroop, "Dynamic coupling in vehicles under automatic control", *Vehicle System Dynamics*, 23(SUPPL):209–220, 1994.

[35] S. Hert, S. Tiwari and V. Lumelsky, "A terrain-covering algorithm for an AUV", *Autonomous Robots*, 3:91–119, 1996.

[36] R. Hohzaki and K. Iida, "An optimal search plan for a moving target when a search path is given", *Mathematica Japonica*, 41:175–184, 1995.

[37] R. Hohzaki and K. Iida, "Path constrained search problem with reward criterion", *Journal of the Operations Research Society of Japan*, 38:254–264, 1995.

[38] R. Hohzaki and K. Iida, "A search game when a search path is given", *European Journal of Operational Reasearch*, 124:114–124, 2000.

[39] D. Hristu and K. Morgansen, "Limited communication control", *Systems & Control Letters*, 37(4):193–205, 1999.

[40] W. Jacek, *Intelligent Robotic Systems: Design, Planning, and Control*. Klwer Academic / Plenum Pub., NY, 1999.

[41] D. Jacques and R. Leblanc, "Effectiveness analysis for wide area search munitions", in *Proceedings of the AIAA Missile Sciences Conference*, Monterey, CA, 1998.

[42] K. Jin, P. Liang and G. Beni, "Stability of synchronized distributed control of discrete swarm structures", in *IEEE International Conference on Robotics and Automation*, pages 1033–1038, San Diego, California, 1994.

[43] O. Khatib, "Real-time obstacle avoidance for manipulators and mobile robots", in *Proceedings of the 1985 IEEE International Conference on Robotics and Automation*, pages 500–505, St. Louis, MO, 1985.

[44] B. S. Koontz, "A multiple vehicle mission planner to clear unexploded ordinance from a network of roadways", Master's thesis, MIT, 1997.

[45] B. Koopman, *Search and Screening: General principles with Historical Application*, Pergamon, New York, 1980.

[46] H. Levine and W.-J. Rappel, "Self-organization in systems of self-propelled particles", *Physical Review E*, 63(1):017101-1–017101-4, 2001.

[47] Y. Liu, K. Passino and M. Polycarpou, "Stability analysis of one-dimensional asynchronous mobile swarms", Submitted to the *40th IEEE Conference on Decision and Control*, 2001.

[48] Y. Liu, K. Passino and M. Polycarpou, "Stability analysis of one-dimensional asynchronous swarms", *Proceedings of the 2001 American Control Conference*, 2001.

[49] Y. Liu, K. Passino and M. Polycarpou, "Stability analysis of one-dimensional asynchronous swarms", Submitted to *IEEE Transaction on Automatic Control*, 2001.

[50] M. Lizana and V. Padron, "A specially discrete model for aggregating populations", *Journal of Mathematical Biology*, 38:79–102, 1999.

[51] S. Lucidi and M. Sciandrone, "On the global convergence of derivative free methods for unconstrained optimization", *Technical Report, Univ. di Roma "La Sapienza"*, 1997.

[52] M. Mataric, "Minimizing complexity in controlling a mobile robot population", In *IEEE Int. Conf. on Robotics and Automation*, pages 830–835, Nice, France, 1992.

[53] A. S. Mikhailov and D. H. Zanette, "Noise-induced breakdown of coherent collective motion in swarms", *Physical Review E*, 60(4):4571–4575, 1999.

[54] R. Miller and W. Stephen, "Spatial relationships in flocks of sandhill cranes (Grus canadensis)", *Ecology*, 47:323–327, 1996.

[55] M. Millonas, "Swarms, phase transitions, and collective intelligence", in *Artificial Life III*, pages 417–445, Addison-Wesley, 1994.

[56] A. Mogilner and L. Edelstein-Keshet, "A non-local model for a swarm", *Journal of Mathematical Biology*, 38:534–570, 1999.

[57] J. Nelder and R. Mead, "A simplex method for function minimization", *Computer Journal*, 7:308–313, 1965.

[58] M. Pachter and P. Chandler, "Challenges of autonomous control", *IEEE Control Systems Magazine*, pages 92–97, 1998.

[59] K. Passino, "Biomimicry of bacterial foraging for distributed optimization and control", to appear in *IEEE Control Systems Magazine*, 2001.

[60] K. Passino and K. Burgess, *Stability Analysis of Discrete Event Systems*. John Wiley and Sons Pub., New York, 1998.

[61] E. M. Rauch, M. M. Millonas and D. R. Chialvo, "Pattern formation and functionality in swarm models", *Physics Letters A*, 207:185–193, 1995.

[62] J. Reif and H. Wang, "Social potential fields: a distributed behavioral control for autonomous robots", *Robotics and Autonomous Systems*, 27:171–194, 1999.

[63] J. H. Reif and H. Wang, "Social potential fields: A distributed behavioral control for autonomous rebots", *Robotics and Autonomous Systems*, 27:171–194, 1999.

[64] C. Reynolds, "Flocks, herds, and schools: A distributed behavioral model", *Comp. Graph*, 21(4):25–34, 1987.

[65] H. Richardson, "Search theory", in D. Chudnovsky and G. Chudnovsky, editors, *Search Theory: Some Recent Developments*, pages 1–12. Marcel Dekker, New York, NY, 1987.

[66] R. Rule, "The dynamic scheduling approach to automated vehicle macroscopic control", Technical Report EES-276A-18, Transport. Contr. Lab., Ohio State Univ., Columbus, OH, 1974.

[67] E. Shaw, "The schooling of fishes", *Sci. Am.*, 206:128–138, 1962.

[68] N. Shimoyama, K. Sugawa, T. Mizuguchi, Y. Hayakawa, and M. Sano, "Collective motion in a system of motile elements", *Physical Review Letters*, 76(20):3870–3873, 1996.

[69] S. Singh, P. Chandler, C. Schumacher, S. Banda, and M. Pachter, "Adaptive feedback linearizing nonlinear close formation control of UAVs", in *Proceedings of the 2000 American Control Conference*, pages 854–858, Chicago, IL, 2000.

[70] M. Snorrason and J. Norris, "Vision based obstacle detection and path planetary rovers", in *Unmanned Ground Vehicle Technology II*, Orlanso, FL, 1999.

[71] S. Spires and S. Goldsmith, "Exhaustive geographic search with mobile robots along space-filling curves", in A. Drogoul, M. Tambe, and T. Fukuda, editors, *Collective Robotics*, pages 1–12. Springer Verlag: Berlin, 1998.

[72] T. Stewart, "Experience with a branch-and-bound algorithm for constrained searcher motion", In K. Haley and L. Stone, editors, *Search Theory and Applications*, pages 247–253. Plenum Press, New York, 1980.

[73] L. Stone, *Theory of Optimal Search*, Acadamic Press, New York, 1975.

[74] L. Stone, "The process of search planning: Current approachs and the continuing problems", *Operational Research*, 31:207–233, 1983.

[75] I. Suzuki and M. Yamashita, "Distributed anonymous mobile robots: formation of geometric patterns", *SIAM J. COMPUT.*, 28(4):1347–1363, 1997.

[76] D. Swaroop, *String Stability of Interconnected Systems: An Application to Platooning in Automated Highway Systems*. PhD thesis, Department of Mechanical Engineering, University of California, Berkeley, 1995.

[77] D. Swaroop and K. Rajagopal, "Intelligent cruise control systems and traffic flow stability", *Transportation Research Part C: Emerging Technologies*, 7(6):329–352, 1999.

[78] J. Toner and Y. Tu, "Long-range order in a two-dimensional dynamical xy model: How birds fly together", *Physical Review Letters*, 75(23):4326–4329, 1995.

[79] J. Toner and Y. Tu, "Flocks, herds, and schools: A quantitative theory of flocking", *Physical Review E*, 58(4):4828–4858, 1998.

[80] V. Torczon, "On the convergence of the multidirectional search algorithm", *SIAM Journal Optimization*, 1(1):123–145, 1991.

[81] V. Torczon, "On the convergence of pattern search algorithms", *SIAM Journal Optimization*, 7(1):1–25, 1997.

[82] T. Vicsek, A. Czirok, E. Ben-Jacob, I. Cohen, and O. Shochet, "Novel type of phase transition in a system of self-propelled particles", *Physical Review Letters*, 75(6):1226–1229, 1995.

[83] A. Washburn, "Search-evasion game in a fixed region", *Operations Research*, 28:1290–1298, 1980.

[84] S. Weaver, L. Baird and M. Polycarpou, "An analytical framework for local feedforward networks", *IEEE Transactions on Neural Networks*, 9(3):473–482, 1998.

[85] Q. Yan, G. Yang, V. Kapila, and M. Queiroz, "Nonlinear dynamics and output feedback control of multiple spacecraft in elliptical orbits", in *Proceedings of the 2000 American Control Conference*, pages 839–843, Chicago, IL, 2000.

Chapter 14

COUPLING IN UAV COOPERATIVE CONTROL

Corey J. Schumacher
Flight Control Division
Air Force Research Laboratory (AFRL/VACA)
Wright-Patterson AFB, OH 45433-7531
corey.schumacher@wpafb.af.mil

Phillip R. Chandler
Flight Control Division
Air Force Research Laboratory (AFRL/VACA)
Wright-Patterson AFB, OH 45433-7531
phillip.chandler@wpafb.af.mil

Meir Pachter
Department of Electrical and Computer Engineering
Air Force Institute of Technology (AFIT/ENG)
Wright-Patterson AFB, OH 45433-7765
mpachter@afit.af.mil

Abstract

      This chapter addresses complexity and coupling issues in cooperative decision and control of distributed autonomous UAV teams. Hierarchical decomposition is implemented where team vehicles are allocated to subteams using set partition theory. Results are presented for single assignment and multiple assignment using network flow and auction algorithms. Various methods for computing tours of multiple assignments are addressed. Simulation results are presented for wide area search munitions where complexity and coupling are incrementally addressed in the decision system, yielding improved team performance.

Keywords: Cooperation, coupling, task assignment, network flow, auction, teams

323

S. Butenko et al. (eds.), Cooperative Control: Models, Applications and Algorithms, 323-348.
© 2003 *Kluwer Academic Publishers.*

1.   Introduction

The cooperative decision and control problem can be characterized by: 1) complexity, 2) information structure, and 3) uncertainty. Uncertainty (3) for the wide area search munition, is finding objects and correctly classifying objects that are targets. Information structure (2) is the recognition that reliable full state information cannot be made available everywhere in the network, and that essential information must be relayed over noisy limited bandwidth communication links. Complexity (1), primarily caused by task coupling, is *a* major, if not *the* major issue, for cooperative control of multiple vehicles to find and attack multiple targets, with multiple time phased tasks for each target, and extensive coupling between the tasks - all to be performed at minimum cost and in minimum time.

In the operations research and weapon target assignment literature, fast and efficient static linear allocation algorithms are available for hundreds, even thousands of vehicles (n) and tasks (m). Generally $n = m$, but that is not required. These are globally optimal algorithms and require that the complete cost matrix be centrally available. The auction algorithm is a distributed form of these static linear algorithms. Problem size can be a major contributor to complexity.

However, coupling induced complexity, and not necessarily size, dominates the wide area search munition problem — this is not a static, linear, single allocation problem. The problems addressed in this chapter are of modest size, $n \leq 8$, but there are $m = 4$ tasks (search, classification, attack, verification) per target and an arbitrary number of targets (< 10 to date). The vehicles each have a default task of performing cooperative search, which introduces extensive coupling in their search trajectories. When an object is detected it needs to be classified, which can take an arbitrary number of views, performed by the same or cooperative vehicles. If classified, the target can be attacked, if a high value target, or the attack delayed for lower value targets. An important point is once a munition attacks a target, the vehicle is destroyed and no longer available for performing other tasks, including searching for high value targets. Once attacked, a target is viewed by another vehicle to ensure that it has been destroyed. This means that it takes at least 2 separate vehicles to completely service a target. Also, the tasks must be correctly ordered and sequenced in time – all this while trying to find and kill the most high value targets possible in the shortest time, due to severe fuel contraints. This problem cannot be adequately addressed by static linear allocation algorithms.

To address this complexity, a hierarchical decomposition is used [4, 6]. Figure 14.1 illustrates a general architecture for cooperative control and task apportionment among multiple vehicles. At decision level 1 are the vehicle agents that do path planning, trajectory generation, and maintain models of terrain, threats, and targets. Decision level 2 comprises the sub-team agents which coordinate the tasks of the sub-team vehicles. The team agent, at decision level 3, is responsible for meeting the mission objective, and apportioning resources (UAVs) and sub-objectives (targets) to the sub-team agents.

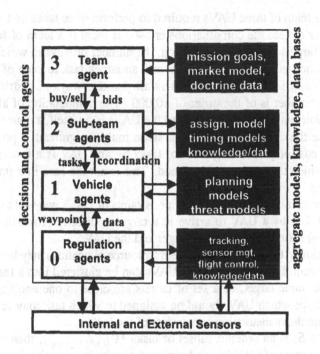

Figure 14.1. Hierarchical Decomposition

Sub-teams collapse the complexity of the overall team optimization problem so that only those vehicles that have benefit in servicing the objects are considered, greatly reducing the complexity of the optimization problem that needs to be solved. The smaller single or multiple assignment problem can then be addressed.

Multiple assignment removes much of the myopic penalties of single assignment. A longer planning horizon is essential if the performance is to be anywhere close to optimal in highly coupled systems. Iterative network flow [13, 15], binary linear programming, and auction [1, 2] are addressed in the paper as algorithms for multiple task assignment. All have their pros and cons, but some heurisitics are needed to limit the planning horizon or the complexity will explode. All require task costs to be computed from the trajectories generated at decision level 1.

The chapter is organized as follows: decomposition into sub-teams is addressed in §2; §3 addresses single assignment; §4 multiple assignment; §5 presents simulation results; and §6 the conclusions.

2.   Sub-Teams

Given a team of three UAVs required to perform three tasks on two targets, the number of feasible combinations is 240. If there is a team of four UAVs and three targets that require servicing, the number of ways in which one can assign six non-attack tasks to a UAV and an attack task to each of the other three UAVs is 1152. If one considers all possible ways of distributing the tasks, the number is of the order of 10,000. The computation of all feasible ways of distributing the tasks amongst the UAVs is, in itself, a daunting task.

To pose the current problem as a static multi-assignment is problematic, since the cost of performing a set of tasks by any UAV is a function of the order in which the tasks are performed. This is not the case in a static multi-assignment.

In order to relax the maneuvering constraints of UAVs, we will assume that the time taken by a UAV to arrive at a target is directly proportional to the (Euclidean) distance between the target and the UAV.

A solution to the simplified problem (resource allocation) may be viewed as a way to team the UAVs. A set of UAVs can be clustered into a team if they service the same target, or a set of targets are close to one another. A finer assignment of which UAV should be assigned to which task may be made by considering the maneuvering constraints of the UAVs.

Suppose S_i is an ordered subset of tasks $\{\Gamma_{i_1}, \ldots, \Gamma_{i_{p_i}}\}$, then we refer to S_i in the context of UAV V_j performing the tasks in S_i in the order in which they appear in S_i. The first and last tasks performed by V_j are Γ_{i_1} and $\Gamma_{i_{p_i}}$ respectively. One can associate a cost with V_j performing tasks in S_i. A feasible partition of tasks is an allocation of disjoint subsets of tasks for each of the UAVs to perform, so that every task is performed by some UAV and all timing (coordination) constraints on the tasks are met. The problem of resource allocation may be posed as finding the minimum cost partition of the set of tasks, where $\mathcal{P}$ is any feasible partition of tasks.

2.1.   Graph Approach

This approach combines the ideas of iterative resource allocation with those from graph theory.

The resource allocation is performed in two stages:

- In the first stage, a classical assignment is performed to allocate resources to m_1 tasks that must be serviced/performed as soon as possible. To proceed with the classical assignment, targets requiring multiple services are replicated an appropriate number of times and treated as distinct targets that are collocated. For the purpose of replication, classification followed by attack will be considered a composite task requiring only one vehicle and will be treated as a terminal task. With such a replication, one can consider each target requiring only one task to be performed on it. To avoid distinguishing between the replica and the original target, we will refer to the replica by the task required to be performed on it. The procedure is as follows:

1 Suppose there are m_1 tasks to be serviced and n UAVs with ($n \geq m$), where binary linear programming solves the classical assignment problem:

$$\min_{x_{ij} \geq 0, \ 1 \leq i \leq n, \ 1 \leq j \leq m_1} \sum_{i=1}^{n} \sum_{j=1}^{m_1} T_{ij} x_{ij}, \text{ subject to}$$

$$\sum_{i=1}^{n} x_{ij} = 1, \ \forall \ 1 \leq j \leq m_1, \ \sum_{j=1}^{m_1} x_{ij} \leq 1, \ \forall \ 1 \leq i \leq n.$$

2 There is inefficiency in this assignment, since, of the m_1 tasks, only m_2 are terminal. This implies that more than one task of the set of $m_1 - m_2$ tasks can be assigned to a UAV resulting in a faster servicing of the tasks. This leads us into considering multiple assignment.

3 We will be concerned with multiple verification assignments for a UAV that results in smaller total time required to service the tasks. Timing constraints do not appear at this stage, since UAVs that are not assigned in the first stage arrive necessarily later at a target than their counterparts in the first stage.

- In the second stage, some inefficiency in resource allocation is weeded out using graph theory.

2.2. Graph Construction

Let D_i be the distance traveled by a UAV to arrive at the i th of the $m - m_1$ targets under the classical assignment.

To specify a graph, one provides the set of nodes or vertices and the set of edges/arcs connecting the nodes. We will think of targets (equivalently verification tasks) as the nodes of a graph.

We will first construct a fully connected symmetric graph in such a way that the weight of an edge/arc connecting Γ_i and Γ_j is the Euclidean distance, d_{ij}, between the nodes. We will then construct a benefit graph as follows:

The benefit b_{ij} is the weight of an arc/edge connecting nodes Γ_i and Γ_j and is defined to be $D_j - d_{ij}$. The benefit, b_{ij} represents the saving in distance traveled in having a UAV that visits Γ_i also visit Γ_j. Clearly, it will be beneficial to have a UAV that visits Γ_i also visit Γ_j only if $b_{ij} > 0$. Clearly, the benefit graph is asymmetric, since D_i may not necessarily be equal to D_j, although $d_{ij} = d_{ji}$.

The problem of resource allocation can be thought of as partitioning the directed benefit graph into subgraphs so that

1 every node is covered by one and only one subgraph,

2 no two edges of a subgraph either start from the same node or end in the same node, and

3 the sum of the benefits of all edges in all subgraphs is maximum.

2.3.   Preliminary Results

We used a short planning horizon heuristic to solve this problem, similar to the algorithm for a minimum weight spanning tree in [12]. This algorithm decomposes the graph into isolated nodes and/or directed unary subgraphs where no two edges in the subgraphs share the same in-node or out-node. From the subgraphs, one can compose a sub-team as follows:

1 Find the targets that correspond to the nodes in a subgraph.

2 Find the vehicles that are assigned to the targets for either classification/attack or verification.

The results are shown in Figure 14.2 which indicates how the verification tasks are partitioned and allocated to different UAVs. In the plots, the numbers in black indicate the index of the UAV and its location is given by its coordinates in the plot. The numbers in blue indicate the index of the target requiring verification and its location is specified by the coordinates in the plot. There is a line starting from a UAV and joining different targets. For example, in the first plot, UAV with index 2 is connected via a poly-line to targets 1, 5 and 3 in that order. This indicates that UAV 2 is assigned to perform the verification of target 1 followed by target 5 followed further by target 3. UAVs that are not connected to any targets are released for search operations.

Figure 14.2 corresponds to the case when targets and UAVs are interspersed. In this case, UAV 2 is assigned verification tasks for targets 1, 5 and 3; UAV 3 is assigned verification task for target 2; UAV 6 is assigned the verification task for target 4; UAV 5 is assigned the verification tasks for targets 6 and 7; and all other UAVs are free to search.

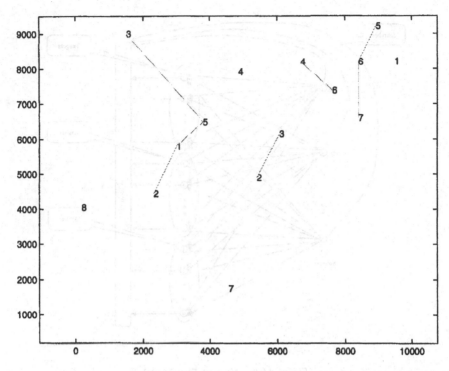

Figure 14.2. Illustration of team decomposition and verification assignments when UAVs and targets are interspersed

3. Single Assignment

A good starting point for task allocation for wide area search munitions is single assignment — assigning a single task to each vehicle at any point in time, and then assigning new tasks as old ones are completed. The results are suboptimal, since information about later tasks may not be included in the optimization, but such methods can still result in task assignments that fully service all known targets.

3.1. Network Flow

Network optimization models are typically described in terms of supplies and demands for a commodity, nodes that model transfer points, and arcs that interconnect the nodes and along which flow can take place. To model weapon system allocation, we treat the individual vehicles as discrete supplies of single units, tasks being carried out as flows on arcs through the network, and ultimate disposition of the vehicles as demands. Thus, the flows are 0 or 1. We assume that each vehicle operates independently, and makes decisions when new in-

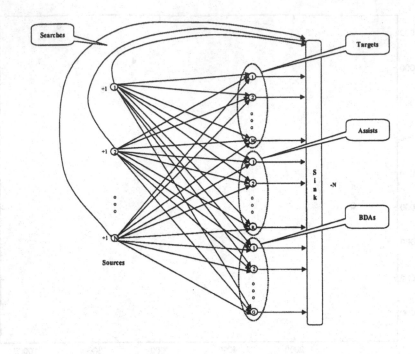

Figure 14.3. Network Flow Model

formation is received. These decisions are determined by the solution of the network optimization model. The receipt of new target information triggers the formulation and solving of a fresh optimization problem that reflects current conditions, thus achieving feedback action. At any point in time, the database onboard each vehicle contains a target set, consisting of indexes, types and locations for targets that have been classified above the probability threshold. There is also a speculative set, consisting of indexes, types and locations for potential targets that have been detected, but are classified below the probability threshold and thus require an additional look before attack.

The network flow model is shown in Figure 14.3 The model is demand driven, with N sinks, where N is the number of munitions included in the task assignment calculation, and N sources, one for each munition. There are intermediate nodes for each possible task assignment, one for each known target. An arc exists from a specific vehicle node to a target node if and only if it is a feasible vehicle/target pair. At a minimum, the feasibility requirement would mean that there is enough fuel remaining to strike the target if tasked to do so. Other feasibility conditions could also enter in, if, for example, there were differences in the onboard weapons that precluded certain vehicle/target combinations, or if the available attack angles were unsuitable. Finally, each node in the vehicle set on the left has a direct arc to the far right node labeled sink, modeling the option of continuing to search. The capacities on the arcs from the target and speculative sets are fixed at 1. Due to the integrality property, the flow values are constrained to be either 0 or 1. Each unit of flow along an arc has a "benefit" which is an expected future value. The optimal solution maximizes total value.

The network optimization model can be expressed as:

$$\max J = \sum_{i,j} c_{ij} x_{ij} \qquad (1)$$

Subject to:

$$\sum_{i,j} x_{ij} + x_{jk} = 1, \forall i = 1, \cdots, n \qquad (2)$$

$$\sum_i x_{is} + \sum_j x_{jk} = n, \forall i, j = 1, \cdots, n, \qquad (3)$$

$$x \geq 0, \qquad (4)$$

where n = number of UAVs.

This particular model is a capacitated transshipment problem (CTP), a special case of a linear programming problem. Constraint (2) enforces a condition that flow-in must equal flow-out for all nodes. Constraint (3) forces the number of assigned tasks to be equal to the number of available vehicles. Constraints (4) and (5) help enforce the binary nature of the problem. Any particular flow is either active or inactive (0 or 1). Restricting these capacities to a value of one on the arcs leading to the sink, along with the integrality property, induces binary values for the decision variables x_{ij}. Due to the special structure of the problem, there will always be an optimal solution that is all integer [1]. Solutions to this problem pose a small computational burden, making it feasible for implementation on the processors likely to be available on disposable wide area search munitions. More information on this method can be found in [13] and [15]

The goal of the optimization problem is to maximize the value of the tasks performed by the vehicles at the time the model is solved. Solving the model whenever new target information is available attempts to maximize the value of the targets destroyed over the life of the munitions. It is not possible to optimally assign multiple tasks to a vehicle using this method. Suboptimal modifications, such as the iterative approach described in the next section, are required.

4.   Multiple Assignment

Assigning only a single task to each vehicle at a time, without regard to tasks that will be required later on, results in suboptimal performance. Therefore, assignment of multiple tasks to a UAV at one time is desirable. Solving the optimization for multiple vehicles being assigned multiple tasks while servicing a number of different targets is extremely complex. Trajectory generation algorithms are used to estimate a cost for a UAV to service a sequence of targets, and to estimate the times at which those tasks will be completed. There are three criteria to which the chosen assignment should conform. First, no task should be assigned to more than one UAV. Second, no task should be assigned unless any prerequisite tasks are also assigned. That is, a UAV may be assigned to attack a target only if the target has already been classified or if another UAV is assigned to classify it. Third, the estimated time at which a task is accomplished should not be before the estimated time at which the immediately prerequisite task, if any, is assigned. Also, attacking a target is a terminal task: if a UAV is assigned to attack a target, that task must be planned for a later execution than any other task assigned to that UAV. The assignment meeting all of these criteria with the minimum cost (or maximum benefit) is desired. Small problems can be solved using an exhaustive search, and other approaches for determining a feasible, low-cost/high-benefit assignment can be used for larger problems. There is a tradeoff among computability, optimality, and fidelity.

4.1.   Iterative Methods

Nygard [10] suggests compiling tours by an iteration of the linear transshipment algorithm as a way of finding a better assignment by starting with the single assignment solution. At each stage, UAVs have a planned position and heading for the end of their assigned tour of multiple tasks. The transshipment algorithm makes a temporary assignment of these UAVs to subsequent tasks. Of these temporary assignments, the one with the earliest estimated time is locked in, and the planned position of that UAV and state of that target are updated. The process repeats until all targets are fully assigned, or all UAVs are assigned an attack task. Locking in the earliest estimated time at each stage discourages but does not entirely prevent a dependent task being planned for an earlier time than its prerequisite. One method is to associate a large cost (penalty) with any potential assignment which would create such a conflict with the prerequisite task. The other method is to adjust the cost of such a potential assignment to include loitering, so that the dependent task occurs at the earliest admissible time, and not before. This strategy is demonstrated in Figures 14.4 and 14.5. The implementation of the latter adjustment currently addresses only the order of the completion of the tasks. It could reasonably be

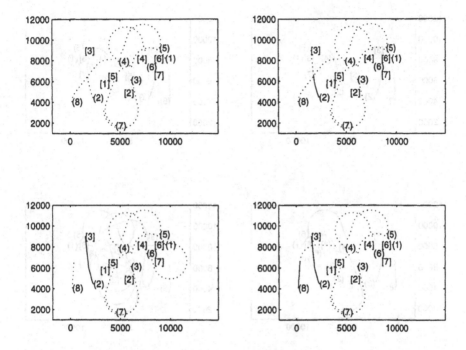

Figure 14.4. Sequence of steps in Iterated Multiple Assignment. In each figure, dotted paths represent the temporary, single assignment, and solid paths represent fixed assignments. UAVs are labeled by numbers in parentheses, and targets by numbers in square brackets. (a) Initial single assignment. (b) UAV 2 is assigned to Classify Target 3 at time 2188. (c) UAV 2 is assigned to Attack Target 3 at time 4648. (d) UAV 8 is assigned to verification Target 3 at time 4648. (e) UAV 1 is assigned to Classify Target 2 at time 4958. (f) UAV 7 is assigned to Classify Target 6 at time 7199. (g) UAV 5 is assigned to Classify Target 5 at time 7282. (h) After fifteen more steps, the final plan.

adjusted to ensure that, for example, the sensor sweep for the verification starts after the attack, and the verification is completed some time later.

4.2. Relative Benefit Methods

An intermediate approach, between single assignment and exhaustive search, is to begin with the single assignment result and look for nearby assignments with comparatively better objective values. When the objective function is simply Euclidean path distance, in [7] it is observed that the marginal benefit of one UAV assuming the duties of another UAV in addition to its own has a simple form. If Target i is separated from Target j by a distance d_{ij}, and the cost associated with servicing Target j by single assignment is D_j, the benefit of the UAV assigned to Target i subsequently servicing Target j is $D_j - d_{ij}$. A

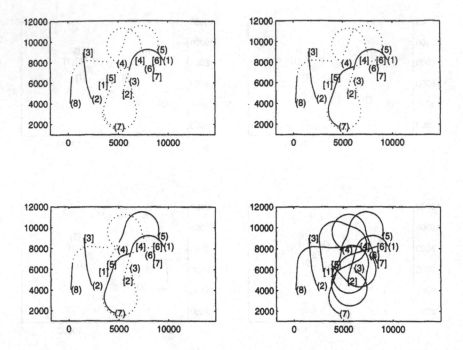

Figure 14.5. Continuation of Figure 14.4. (e) UAV 1 is assigned to Classify Target 2 at time 4958. (f) UAV 7 is assigned to Classify Target 6 at time 7199. (g) UAV 5 is assigned to Classify Target 5 at time 7282. (h) After fifteen more steps, the final plan.

matrix of benefits can be readily constructed according to

$$b_{ij} := \begin{cases} 0, & if \ i = j \\ D_j - d_{ij}, & otherwise. \end{cases}$$

Any of several algorithms can operate on this matrix, producing an assignment of targets to following targets. Such an assignment is a permutation of the targets, and represents one or more cycles. One method investigated uses an auction to make the temporary assignment resulting in D_j, uses another auction to produce the permutation of targets, evaluates the cost for each UAV to service each of the resulting loops, then uses one more auction to assign UAVs to these loops. Another method uses a greedy algorithm. At each stage, a target, Target j, is designated to be serviced by the UAV servicing another target, Target i. The ordered pair of targets chosen satisfies three conditions: Target i is the last target serviced by the UAV currently assigned to it, Target j is the first target serviced by the UAV currently assigned to it, and the UAVs currently assigned to service the two targets are distinct. The ordered pair of targets chosen at each stage maximizes b_{ij} over all ordered pairs of targets satisfying the conditions. The process is repeated until the maximum such b_{ij} is negative, or no such b_{ij} exists. The assignment by either of these algorithms could be used directly, or as a guide to partitioning targets into groups for service by sub-teams.

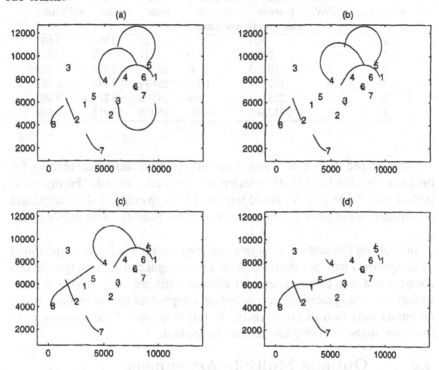

Figure 14.6. Sequence of steps in a Relative Benefit Method with a greedy algorithm and flyable paths. (a) Initial single assignment. (b) UAV 4 assumes responsibility for Target 6 from UAV 3. (c) UAV 8 assumes responsibility for Target 4 from UAV 5. (d) After two more steps, the final plan.

A Euclidean objective function is reasonable when targets and UAVs are spaced far apart in comparison with the minimum turning radius of a UAV. In more realistic conditions, the concept of relative benefit is still useful. The difference is that the benefit matrix changes over the course of the execution. Good results, shown in Figure 14.6, were produced by an algorithm similar to the one just discussed. This algorithm does not preserve the order of targets when one UAV assumes the assignment of another, but chooses the order in a greedy manner as in Section 4.1. The ordered pair of UAVs maximizing the marginal benefit is chosen at each stage until the benefit is negative or only one UAV is assigned. This heuristic allows for general objective functions of

Table 14.1. Comparison of frequency and values of possible and feasible assignments for various scenarios.

Number of Targets	Number of UAVs	Number of possible assignments	Number of feasible assignments	Best value	Mean value	Standard deviation
2	2	507	35	13 522	3 592	5 087
3	2	2 653	133	19 712	4 805	5 415
2	3	10 693	145	18 685	4 388	4 801
3	3	132 754	1 027	25 679	7 883	6 497
2	4	215 381	496	24 258	8 210	6 092
3	4	6 512 905	6 065	35 430	12 911	7 659

a UAV assigned to a set of targets. Such objective functions may involve, for instance, the cost for a UAV to return to search after its tour. Furthermore, inclusion of dummy UAVs could introduce the capability of the assignment algorithm to leave some targets unserviced, if the benefits do not outweigh the cost.

In applying this method to targets requiring multiple tasks, an initial round of assignments may be employed. UAVs not assigned to attack targets in this round would then be considered for tours of verifications in a relative benefit method. This procedure may schedule a dependent task to be planned for an earlier time than its prerequisite. In this case, one of the methods to be discussed in the following section may be applied.

4.3.   Optimal Multiple Assignment

Multiple task assignment could theoretically be formulated and solved with brute force, but the problem may be computationally intractable for large numbers of vehicles, tasks, and/or targets. This possibility is discussed in 4.3.1. Other, potentially more tractable methods are then presented.

4.3.1   Exhaustive Search.   A simplification of the problem may be solved with an exhaustive search. If the maximum number of tasks assigned to each UAV is limited, the number of feasible permutations of tasks for a UAV is small. When there is one target requiring all three tasks, and tours have up to two non-search tasks, there are six possible tours: Search, Classify, Attack, verification, Classify then Attack, or Classify then verification. With two targets, the number is 23, and with three targets, the number is 52. All combinations of tours with UAVs can be evaluated for small numbers of UAVs and small numbers of targets. Using a representative objective function based on Euclidean distances among randomly generated points, the distribution of

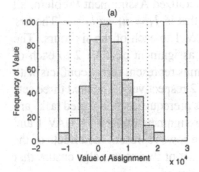
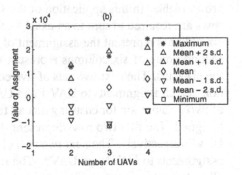

Figure 14.7. Distribution of values of feasible assignments for a particular scenario with three unclassified targets and three UAVs.

feasible assignments and their values was calculated. For each combination of two to three targets with two to four UAVs, the optimal assignment had a value about three times the mean value of feasible assignments. The distributions were qualitatively normal with a standard deviation approximately equal to the mean. As this would suggest, many feasible assignments had negative value. Table 14.1 enumerates the results. Figure 14.7 shows the distribution of values for a different three-target, three-vehicle scenario from the one shown in Table 14.1, and the statistics for similar distributions with two targets. These findings demonstrate two ways in which the problem is difficult. First, the number of feasible assignments is drastically smaller than the number of possible assignments. Second, most feasible assignments have far less value than the optimum, so a random search over feasible assignments is not likely to find a near-optimal solution with a small sample size.

4.3.2 Generalized Assignment Problem. The Generalized Assignment Problem is formulated as follows:

$$\text{Optimize} \quad \sum_i \sum_j c_{ij} x_{ij}$$
$$\text{Subject to (1)} \quad x_{ij} \in \{0,1\} \ \forall \ i,j$$
$$(2) \quad \sum_i x_{ij} = 1 \ \forall \ j$$
$$(3) \quad l_i \leq \sum_j r_{ij} x_{ij} \leq u_i \ \forall \ i.$$

In [10], Guo and Nygard used a Generalized Assignment Problem framework to assign multiple identical UAVs to targets. He represented a nonlinear objective function and relevant logical conditions by using agents and UAVs which did not directly represent physical UAVs or targets. The same concept may be applied to the assignment of tours of a given maximum length to non-identical UAVs. Figure 14.8 illustrates the Generalized Assignment Problem formulation corresponding to a simple case: tours have up to two tasks besides search; there is one target to be classified, attacked, and verified; and two UAVs are available. In this application of the Generalized Assignment Problem, all rows are required to sum to an exact value; that is, $l_i = u_i =: b_i \, \forall \, i$. The first six columns represent the assignment of UAV 1 to each of the six tours. The second group of six columns represent the assignment of UAV 2 to each of the six tours. The next two sets of five columns represent direct conflicts with non-search assignments to UAV 1 and UAV 2 respectively. The next three sets of two columns are for ensuring that all tasks prerequisite to assigned tasks are assigned. The first two rows represent the assignment chosen for UAV 1 and UAV2 respectively. The next two sets of five rows represent the effects of the assignments to UAV1 and UAV2. The next set of two rows helps ensure that all tasks prerequisite to assigned tasks are assigned.

		Tours Available to UAV 1						Tours Available to UAV 2						Tours Disable Arbiter for UAV 1					Tours Disable Arbiter for UAV 2					Tours Enabler		Tours Enabled		Tours Enable Arbiter	
		S	C	A	B	CA	CB	S	C	A	B	CA	CB	C	A	B	CA	CB	C	A	B	CA	CB	A	B	A	B	A	B
	UAV1	a	a	a	a	a	a																							
	UAV2							a	a	a	a	a	a																	
Tour Choice Messages for UAV 1	C		β											γ	γ	γ	γ	γ	γ	γ	γ	γ	γ	δ	δ			θ	θ
	A			β										γ	γ	γ	γ	γ	γ	γ	γ	γ	γ	δ	δ			θ	θ
	B				β									γ	γ	γ	γ	γ	γ	γ	γ	γ	γ	δ	δ			θ	θ
	CA					β								γ	γ	γ	γ	γ	γ	γ	γ	γ	γ	δ	δ			θ	θ
	CB						β							γ	γ	γ	γ	γ	γ	γ	γ	γ	γ	δ	δ			θ	θ
Tour Choice Messages for UAV 2	C								β					γ	γ	γ	γ	γ	γ	γ	γ	γ	γ	δ	δ			θ	θ
	A									β				γ	γ	γ	γ	γ	γ	γ	γ	γ	γ	δ	δ			θ	θ
	B										β			γ	γ	γ	γ	γ	γ	γ	γ	γ	γ	δ	δ			θ	θ
	CA											β		γ	γ	γ	γ	γ	γ	γ	γ	γ	γ	δ	δ			θ	θ
	CB												β	γ	γ	γ	γ	γ	γ	γ	γ	γ	γ	δ	δ			θ	θ
Tour Enable Messages	A																								ε		ζ		η	
	B																									ε		ζ		η

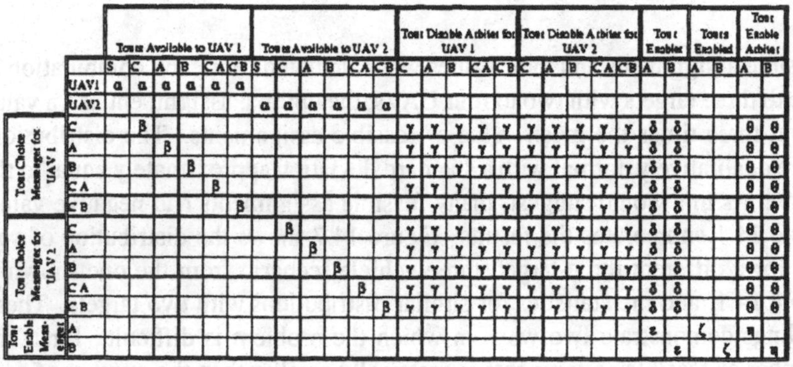

Figure 14.8. Map of a small implementation of the Generalized Assignment Problem formulation.

Each cell in the rows and columns described represents an x_{ij} to which a binary value will be assigned. The effect of condition 2 in the Generalized Assignment Problem is that each column has exactly one of these values equal to 1, and the rest 0. The only cells with corresponding objective function values c_{ij} are those labeled with "α" in Figure 14.8. These cells represent the actual assignment. These cells also correspond to $r_{ij} = 1$, and for these rows $b_i = 1$, so that each UAV is assigned to exactly one tour. The assignments chosen are communicated to the rest of the tableau by the cells labeled "β." For example, if UAV 1 is assigned to verify the target, x_{14} is assigned a value of 1. This means that all other x_{i4} are assigned a value of 0. Most importantly, x_{54}, labeled with "β," has a value of 0. The r_{5j} labeled with "γ," "δ," and "θ," in that row, and b_5, are chosen so that each of the corresponding x_{5j} is assigned a value of 1 if and only if x_{54} is assigned a value of 0. This allows for logical tests of the criteria for a feasible assignment. For example, suppose that, if UAV 1 can Classify and Attack the target, completing the attack at time 10, and UAV 2 can verify the target at time 5, that combination of assignments should be disallowed because the target would be verified before being Classified. The structure of the Generalized Assignment Problem prevents this combination by using nonzero values for four r_{ij}. The relevant values of i correspond to the C A Tour Choice Messenger for UAV 1 and the B Tour Choice Messenger for UAV 2. The relevant values of j correspond to the C A Disable Arbiter for UAV 1 and the B Disable Arbiter for UAV 2. If the x_{ij} for the cells labeled "α" corresponding to this assignment are given a value of 1, the respective x_{ij} for cells labeled "β" must be given a value of 0. The row sum condition requires that all other x_{ij} for cells labeled "γ" must have a value of 1, specifically the four described above. However, this creates an illegal state because only one x_{ij} in any column can be nonzero, and two Disable Arbiter columns each have two nonzero x_{ij}. The same logic can prevent any pair of assignments deemed mutually exclusive. Specifically, the cells labeled "γ" are chosen to prevent more than one UAV from choosing the same task: if UAV 1 is assigned to Classify, then UAV 2 cannot be assigned the C, C A, or C B tour.

Similar logic ensures that no task is assigned unless its prerequisite is assigned. The Tour Choice Messenger for a tour with one or more prerequisite tasks has nonzero x_{ij} in the Tour Enabler columns for the respective dependent tasks, labeled "δ." For example, if UAV 1 is assigned the C A tour, $x_{15} = 1$, the corresponding cell labeled "β" has $x_{ij} = 0$, and any corresponding cells labeled "ε" with nonzero r_{ij} have $x_{ij} = 1$. This means that the cell labeled "ε" in that Tour Enabler has $x_{ij} = 0$. This information is relayed to the relevant Tour Enable Arbiter via the relevant Tour Enable Messenger. Cells labeled "ε" or "η" have $r_{ij} = 1$, cells labeled "ζ" have $r_{ij} = 2$, and Tour Enable Messengers have $b_i = 2$. This means that, within a Tour Enable Messenger, the x_{ij} labeled "ε" has the same value as the x_{ij} labeled "η." The cell linking the A Tour Enable Messenger with the A Tour Enable Arbiter will have $x_{ij} = 0$ if and only if some UAV is assigned a Classify task. For any tour, the cells labeled "θ" have nonzero values for the dependent tasks in the tour. This causes a conflict with the cell labeled "η" if the corresponding prerequisite task has not been assigned.

Finally, there are cases in which no labeled cell in a column has $x_{ij} = 1$. In this case, any unlabeled cell in that column may have $x_{ij} = 1$, since for unlabeled cells have $r_{ij} = 0$ and therefore condition 2 is unaffected.

The Generalized Assignment Problem formulation encodes all of the criteria for a feasible assignment. The optimal solution to the Generalized Assignment Problem will optimize the objective function among all feasible assignments in which no UAV is assigned more than the predetermined number of tasks.

There are solvers available for the Generalized Assignment Problem, but there are indications that a specialized solver would be needed. Due to the complex constraints, the problem is expected to be poorly conditioned with respect to general solvers. Also, the problem size scales quickly. The number of rows and the number of columns each scale with the product of the number of UAVs and the number of possible tours, and the number of tours scales with the number of available tasks raised to the power of the tour length. Some possible tours could be heuristically eliminated in the formulation stage, but the problem remains large. The number of variables is equal to the number of cells, and the number of constraints is equal to the number of rows plus the number of columns. For the simple case discussed, this is 42 constraints on 392 variables. For eight UAVs, three targets each ready to be classified, and tours of two tasks, the problem becomes one of 1,264 constraints on 355,324 variables. It is unlikely that such a problem would be solved in a few seconds, particularly with a general solver.

4.3.3 Binary Linear Programming.

Concepts similar to those used in structuring the Generalized Assignment Problem formulation can be used with a more efficient method. A Binary Linear Programming framework can encode all of the same information in a much smaller problem which is faster to set up and to solve. The general form of a Binary Linear Programming problem is as follows:

$$
\begin{aligned}
\text{Optimize} \quad & A \cdot x \\
\text{Subject to (1)} \quad & x_j \in \{0, 1\} \; \forall \, j \\
\text{(2)} \quad & (F \cdot x)_i \leq d_i \; \forall \, i \\
\text{(3)} \quad & (F_{eq} \cdot x)_i = d_{eq_i} \; \forall \, i.
\end{aligned}
$$

Figure 14.9 is a map of the Binary Linear Programming problem for the same parameters as used for Figure 14.8. The first six variables (and, correspondingly, the first six columns of C_o, F, and F_{eq}) represent the assignment of UAV 1 to each of the six tours considered. The next six represent the same for UAV 2. The last two help ensure that timing conflicts are handled properly. C_o is the linear objective function. The first three rows of F ensure that no task is assigned if it has a prerequisite which is not assigned, and that no task is assigned to more than one UAV. The other two rows of F ensure that no task occurs before its prerequisite. The first two rows of F_{eq} count the number of tours assigned to each UAV. The other two rows of F_{eq} help ensure that timing conflicts are handled properly.

Elements labeled "α" are determined by the objective function. The other elements in C_o are 0 because the objective function is independent of the other variables. The elements labeled "β" have an integer value of -1 to 2. If a tour includes a Classify task, the element in the C Tasks Assigned constraint is 1, and -1 is contributed to the A Tasks Assigned constraint. If a tour includes an Attack task, -1 is contributed to the B Tasks Assigned constraint. If a tour includes an Attack task or a verification task, 2 is contributed to the respective Tasks Assigned constraint. This means that the C A tour has 1, 1, -1 as the first three elements of that column of F, while the B tour has 0, 0, 2 as the first three elements of that column of F. Since the B Tasks Assigned row multiplied by the variable must be no more than 1, a verification task may be assigned only if the corresponding Attack task is assigned. The last two columns in F are 0 for the same reason as the last two elements in C_o : they do not represent tours and therefore do not affect the accounting governed by these constraints. A tour with a Classify task has the element in the C-A row, one labeled "γ," equal to the estimated time at which the target will be classified. A tour with an Attack task has the element in the C-A row equal to the negative of the estimated time at which the target will be attacked. The constraint is that, if a Classify and an Attack task are chosen, the attack must come after the classification, i.e., its estimated time of achievement must be larger than that of the classification. If a Classify task is assigned but no Attack task is assigned, this constraint should be satisfied, and the elements labeled "δ" allow for this case. The A No Dependent Task variable is 1 if and only if no UAV is assigned to attack the target. The corresponding element labeled "δ" is a sufficiently negative number to ensure that the constraint is satisfied no matter when the classification is accomplished. The last two rows of F_{eq} constitute a messenger similar to the last two rows in the Generalized Assignment Problem formulation. The elements labeled "ε" are 1 if the tour includes the dependent task, and 0 otherwise. The identity matrix in the last two columns of these equality constraints means that the last two variables are 0 if the dependent task is assigned, and 1 otherwise. This induces the proper conditions for the last two rows of F.

The Binary Linear Programming formulation thus also encodes all of the criteria for a feasible assignment. The optimal solution to the Binary Linear Programming problem will optimize the objective function among all feasible assignments in which no UAV is assigned more than the predetermined number of tasks. There are solvers available for Binary Linear Programming, but there are indications that a specialized solver would be needed. Due to the complex constraints, the problem is expected to be poorly conditioned with respect to general solvers. The problem size does not scale as quickly as the Generalized Assignment Problem. The number of columns each scale with the product of the number of UAVs and the number of possible tours, though with a smaller coefficient. The number of constraints scales only with the number of tasks or the number of UAVs. A formulation with tours of three tasks would have more variables than a formulation with tours of two tasks, but the same number of constraints. As before, some possible tours could be heuristically eliminated in the formulation stage. For the simple case discussed, this is 9 constraints on 14

		Tours Available to UAV 1						Tours Available to UAV 2						No Dep. Task		
		S	C	A	B	CA	CB	S	C	A	B	CA	CB	A	B	
	C_o															
Value		α	α	α	α	α	α	α	α	α	α	α	α	0	0	
	F															d
Tasks	C	β	β	β	β	β	β	β	β	β	β	β	β	0	0	1
Assigned	A	β	β	β	β	β	β	β	β	β	β	β	β	0	0	1
	B	β	β	β	β	β	β	β	β	β	β	β	β	0	0	1
ETAs	C-A	γ	γ	γ	γ	γ	γ	γ	γ	γ	γ	γ	γ	δ	0	0
	A-B	γ	γ	γ	γ	γ	γ	γ	γ	γ	γ	γ	γ	0	δ	0
	F_{eq}															d_{eq}
UAV	UAV 1	1	1	1	1	1	1	0	0	0	0	0	0	0	0	1
Ass't	UAV 2	0	0	0	0	0	0	1	1	1	1	1	1	0	0	1
Dep't	A	ε	ε	ε	ε	ε	ε	ε	ε	ε	ε	ε	ε	1	0	1
Task	B	ε	ε	ε	ε	ε	ε	ε	ε	ε	ε	ε	ε	0	1	1

Figure 14.9.   Map of a small implementation of the Binary Linear Programming formulation.

variables. For eight UAVs, three targets each ready to be classified, and tours of two tasks, the problem becomes one of 23 constraints on 422 variables. It is reasonable that a specialized solver might be written which would solve the Binary Linear Programming problem in a reasonably short time.

5.   Simulation Results

The authors have developed a MatLab Simulink based multi-UAV simulation of a wide area search munition scenario with a hierarchical distributed decision and control system as depicted in Figure 14.1. The vehicles cooperatively search for and destroy high value targets. The vehicles travel at $120m/s$, $225m$ altitude, and have a sensor footprint of $600m$ by $250m$ which is $1000m$ in front of the vehicle. The vehicles have a maximum turn rate of $2g's$, during which time the sensor is inoperative, and have an endurance of 30min.

The scenario, shown in Figure 14.10, has 8 vehicles following a serpentine search strategy. The scenario starts with the vehicles in an echelon formation with the sensor footprints edge to edge to provide complete coverage of the designated area. Figure 14.10 is a snapshot 98sec into the scenario where the vehicles path and footprint are color-coded. There are 3 high value rectangular targets in the search space that have an arbitrary orientation on the ground. When these potential targets, or objects, pass completely through the footprint (not in a turn), the object is declared detected. The object cannot be attacked until it has been classified, with sufficiently high confidence, as 1 of 5 target classes. The probability of classification is a function of the aspect angle at which it is viewed and is encoded in templates carried onboard. Oftentimes the object is not viewed at an aspect angle that allows it to be classified as a target with sufficient confidence. Therefore, additional passes or views by the same, or other vehicles, must be made over the object at more favorable aspect angles. These views are then combined statistically. More details of the cooperative classification are covered in [5].

Once classified, a target can then be attacked. When attacked, the vehicle is destroyed, but the target may not be. Hence, there is an optional (for low value targets) verification task where another vehicle views the attacked target to confirm the kill. If not destroyed, the verification vehicle could also attack the target, although this is not implemented in this scenario. The implications of this scenario is that it takes, at minimum, 2 vehicles to completely service a high value target. It can also take anywhere from 1 to 3 separate views to classify the target.

Figure 14.10 uses the decision and control structure of Figure 14.1 where there is no team agent to decide which resources are assigned to sub-teams. Alternatively, this can be viewed as all of the vehicles and all of the objects are always assigned to one sub-team. The sub-team agent here solves a static binary assignment problem using a capacitated transshipment or network flow analogy, as discussed earlier and in more detail in [13, 15]. The n vehicle by m task matrix includes the costs for every surviving vehicle to perform a task that transitions each object to the next state (detect, classify, attack, verify). The vehicle agent calculates these costs by generating optimal (minimum time) trajectories that satisfy kinematic (though not dynamic) constraints with specified terminal (position, heading) conditions.

Figure 14.10 shows the typical consequences of this static, single step look ahead, readily computable decision and control system. Following their pre-specified waypoints, vehicles 1,2 detect targets 1-3. Vehicles 5-7 are assigned to classify targets 1-3 respectively. These vehicles have the lowest cost (time) to view the objects at aspect angles that have the highest probability of classifying the objects. Once classified, which in this case takes only 2 views, the classifying vehicles are assigned, as they generally are, to attack the target (rarely is another vehicle closer). The 3 "racetracks" are a consequence of 1) the single assignment binary optimization where the 3 remaining verification tasks are assigned to 3 different vehicles, and 2) the default policy of returning the vehicles to the point of search departure when reassigned back to search. The optimization is triggered only when an object changes state or a task is completed. An interesting artifact is the near circular trajectory of vehicle 1. This is an example of "churning" – vehicles being continuously reassigned and never completing a task. This is inefficient. Vehicle 1 was originally assigned to verify target 3 and vehicle 2 to verify target 2. But vehicle 2, on the way to target 2, serendipitously encounters target 3. Once target 3 is verified, vehicle 2 is then assigned to target 2 and vehicle 1 is reassigned back to search. The major consequences of this "myopic" optimization is that it takes 6 vehicles

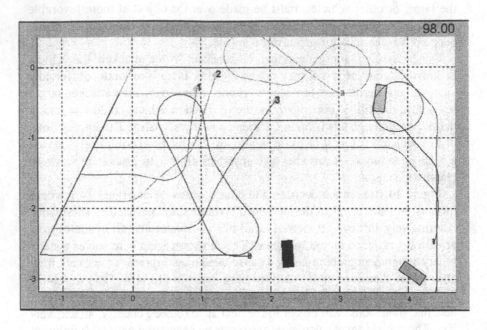

Figure 14.10.   Decision Hierarchy with no Sub-Teams

to service 3 targets, taking crucial time away from searching for high value targets. The next "iteration" in the design of the cooperative control system addresses this shortcoming.

Iteration 2 of the design process is shown in Figure 14.11 where the full 3 level decision hierarchy in Figure 14.1 is used. The global optimization does not scale well and explodes computationally with size. The sub-team concept is a significant attack on complexity. Basically, not every vehicle is considered for every task. Here, the team agent allocates resources to an attack sub-team or a search sub-team. Initially, all the vehicles are assigned to the search sub-team. As more objects are detected, more vehicles are allocated to the attack sub-team for a maximum of 4 vehicles, which is the minimum needed to attack and verify 3 high value targets. Which vehicles are assigned is based on time to target. Initially, the graph based set partition with Euclidian distance, as discussed earlier, was used, yielding vehicles 1-4 for attack. Then the more realistic flyable trajectory distances were used in the graph and gave much more satisfactory results, where vehicles 4-8 are "closest". At the sub-team level, the binary network flow assignment algorithm is sequentially applied as resources, both vehicles and objects, are allocated. Only now, a maximun of 4 vehicles are considered for each task, yielding a more tractable problem. This is a single

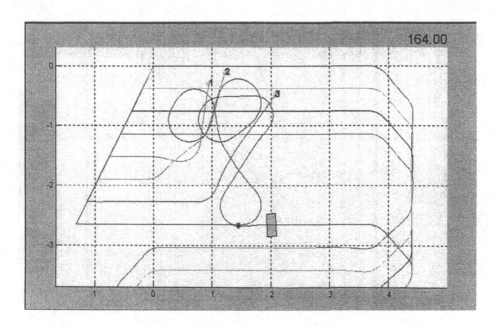

Figure 14.11. Decision Hierarchy with Sub-Teams

assignment algorithm and is still myopic, but not as much as before. Vehicle 8 does the verification for all 3 targets, but this is not enforced in the algorithm. Basically, vehicle 8 does the verification of target 1, when completed it is assigned target 2, and then target 3. The convoluted and inefficient trajectories are a consequence of the myopic assignment, which is addressed in iteration 3 of the design using multiple assignments.

Multiple assignments remove the planning caused myopic inefficiency, but at an appreciable increase in complexity, both at the decision level and at the trajectory generation level, because tours are involved. In the general case, this would be equivalent to solving an n traveling salesmen problem. This is intractable even for moderate sized problems. The sub-team concept reduces the scope to a possibly feasible level, but even then, heuristics are required. Here, the generalized assignment, graph based, and the iterative network, all with tours of maximum length 3, have been used. The iterative network assignment is by far the most computationally efficient, but though not guaranteed to yield the optimal, the results have been very good. Figure 14.12 shows the less myopic 3 task planning horizon for the same 4 vehicle attack sub-team. Now, vehicle 8 has been assigned a 3 verification task tour. The vehicle 8 agent only has to generate 6 sequence combinations, where 3-2-1 is optimal, compared

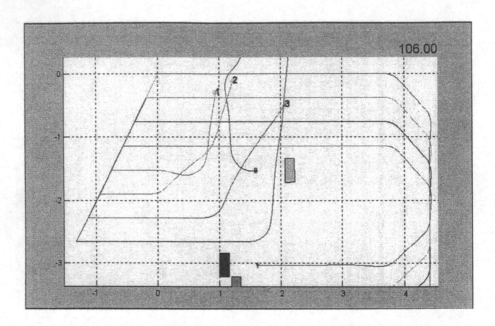

Figure 14.12.   Decision Hierarchy with Sub-Teams and Multiple Assignment

to 1-2-3 before. As can be seen, the complicated inefficient trajectory seen earlier is eliminated. Not only is the sequence important, but the coupling between the trajectory segments is critical. The headings into the targets have to be determined based on the sequence. Not only that, but the width of the sensor footprint has to be exploited when the targets are close together, to avoid convoluted trajectories. The coupling of target tasks with search tasks is also seen in this scenario. Here, vehicle 8 is reassigned back to the closest "lane" that has not been completely searched. Ideally, the search pattern should be reoptimized each time a vehicle is lost, or even when a vehicle departs from the search pattern to service a target. A dynamic search strategy is addressed in [14].

6.   Conclusions

We have shown that task coupling induces such complexity into the cooperative wide area search munition scenario that straightforward decision and control algorithms or naive heuristics yield poor performance. A general performance metric for the wide area search scenario is to maximize the probability of finding and killing high value targets with the available fuel. However, the measure of performance pursued here is to service the objects while minimizing the total time taken away from searching for targets in unsearched areas. The complexity is such that it is generally not possible to determine the globally optimal solution in a reasonable time, even offline. It is difficult, therefore, to evaluate how good a result may be. The problem is radically simplified if the coupling is significantly reduced. For example, the vehicles and objects are widely dispersed so that, realistically, only one vehicle can service an object; or, the verification task is eliminated.

The primary attack upon coupling-induced complexity is hierarchical decomposition. This yields a suboptimal solution, but the optimal solution can only be found by direct enumeration, which is computationally prohibitive. The graph partition technique with flyable trajectories appears to be the most promising for determining sub-team composition. The planning horizon is still necessarily short, but does capture what is meant by "near" or "close".

The solution of the binary linear program for the single assignment problem is the starting point for much of what has been done. Network flow solvers, auction implementations, and binary linear programming algorithms all yield the optimal solution for the binary linear single assignment problem. All are fast, but the auction mechanism is more readily implemented in a distributed fashion. However, when the auction is distributed, it can require much more communication than the network flow solver.

Multiple assignment extends the planning horizon and yields significant improvements in performance and robustness. The network flow optimization algorithm is not readily extendable to multiple assignment because constraints cannot be added to prevent duplicate assignments. However, the iterative network flow optimization algorithm is a short planning horizon heuristic that has been found to work extremely well in practice. Robustness is an issue, as it is in all heuristic algorithms, and the algorithm could occasionally give poor performance in some situations. Task order and timing can be enforced in the generalized assignment problem, the graph partition, binary linear programming, and presumably also an auction algorithm could be devised that does not violate these constraints. All these algorithms are much more computationally intensive, however.

Finally, trajectory or cost generation is a fundamental part of the decision system. All of the techniques discussed require a database/sequence of costs/bids to be generated. The higher the performance desired, the more alternatives are considered, which generally requires a trajectory to be planned for each one.

References

[1] D. Bertsekas, "Auction algorithms for network flow problems: A tutorial introduction", *Computational Optimization and Applications*, 1: 7–66, 1992.

[2] D. Bertsekas, D. Castanon, and H. Tsaknakis, "Reverse auction and the solution of inequality constrained assignment problems", *SIAM J. on Optimization*, 3: 268–299, 1993.

[3] S. Bortoff, "Path planning for UAVs", *AIAA GNC 2000*, Denver, CO, August 2000.

[4] P. Chandler and M. Pachter, "Hierarchical Control for Autonomous Teams", *AIAA GNC 2001*, Montreal, Quebec, Canada, August 2001.

[5] P. Chandler, M. Pachter, K. Nygard, and D. Swaroop, "Cooperative control for target classification", in R. Murphey and P. Pardalos, editors, *Cooperative Control and Optimization*, pages 1-19, Kluwer Academic Publishers, 2002.

[6] P. Chandler, M. Pachter, and S. Rasmussen, "Cooperative control for UAVs", *ACC 2001*, Arlington, VA, June 2001.

[7] P. Chandler, M. Pachter, D. Swaroop, J. Fowler, J. Howlett, S. Rasmussen, C. Schumacher, and K. Nygard, "Complexity in UAV cooperative control", to appear in the *Proceedings of the 2002 ACC Conference*.

[8] L. Dubins, "On curves of minimal length with a constraint on average curvature and with prescribed initial and terminal positions and tangents", *American J. of Math*, 79: 497–516, 1957.

[9] P. Chandler, S. Rasmussen, and M. Pachter, "UAV cooperative path planning", *AIAA GNC 2000*, Denver, CO, August 2000.

[10] W. Guo and K. Nygard, "Combinatorial trading mechanism for task allocation", *13th International Conference on Computer Applications in Industry and Engineering*, June 2001.

[11] K. Judd and T. McLain, "Spline based path planning for unmanned air vehicles", *AIAA GNC 2001*, Montreal, Quebec, Canada, August 2001.

[12] G. Nemhauser and B. Wolsey, *Integer and Combinatorial Optimization*, Wiley, 1999.

[13] K. Nygard, P. Chandler, and M. Pachter, "Dynamic network optimization models for air vehicle resource allocation", *ACC 2001*, Arlington, VA, June 2001.

[14] K. Passino, M. Polycarpu, D. Jacques, and M. Pachter, "Distributed cooperation and control for autonomous air vehicles", in R. Murphey and P. Pardalos, editors, *Cooperative Control and Optimization*, pages 233-271, Kluwer Academic Publishers, 2002.

[15] C. Schumacher and P. Chandler, "Task allocation for wide area search munitions via network flow optimization", *AIAA GNC 2001*, Montreal, Quebec, Canada, August 2001.

Chapter 15

OPTIMAL ADAPTIVE TARGET SHOOTING WITH IMPERFECT FEEDBACK SYSTEMS

Samuel S. Wu
Department of Statistics,
University of Florida, Gainesville, FL 32611.
samwu@biostat.ufl.edu

Zuo-Jun Max Shen
Department of Industrial & Systems Engineering,
University of Florida, Gainesville, FL 32611.
shen@ise.ufl.edu

Mark C. K. Yang
Department of Statistics,
University of Florida, Gainesville, FL 32611.
yang@stat.ufl.edu

Abstract   In military combats, a common task is to destroy multiple targets with limited number of missiles. The weapons are not perfect and each can eliminate its target with a certain probability. Suppose after each shot we may receive some information on the state of the target. Realistically these feedbacks are subject to error, thus we will not know for sure whether the target is really destroyed. This research focuses on finding the optimal allocation of missiles so that maximum number of targets can be destroyed, or the probability of destroying all targets is maximized.

Formally, we consider the following sequential target shooting problem. Given a fixed number of targets and missiles, the objective is to find the optimal strategy, where the missiles are fired sequentially at the targets in the attempt to destroy as many targets as possible. The probability of destroying a target at each shot is known and after each shot, a report becomes available on the state of the target: either destroyed or intact. However, the reports are subject to two types

S. Butenko et al. (eds.), *Cooperative Control: Models, Applications and Algorithms*, 349-364.
© 2003 *Kluwer Academic Publishers.*

of errors and the probabilities of making these errors are also known. We call the report system imperfect if the probability of getting wrong reports is positive.

Note that after each shot, the probability that a target is intact can be evaluated based on the imperfect reports. We have shown that the myopic decision strategy, which always shoots the target with the highest intact probability, is optimal when all the missiles have the same hitting probability and the targets are homogeneous. In addition, techniques for comparing imperfect feedback systems are developed. A partial ordering of the systems is provided when the optimal strategy is carried out.

Keywords: Imperfect Feedback Systems; Optimal Sequential Decision; Target Shooting; Adaptive Decision Rule

1.   Introduction

We consider a class of optimization problems that deal with sequential allocations of efforts among a number of competing projects. The efforts and the projects may take a variety of forms. For instance, allocating investment to different mutual funds over time based on observed performance; firing at a given number of missiles sequentially to destroy a fixed number of targets with imperfect feedback information; adaptive allocation of patients to two different treatment options with information on treatment outcomes available. In each case, decisions are made based on the feedbacks from previous trials, taking into account historical information.

The multi-armed bandit is a prototype of this class of problems [1-6]. It can be stated in the following general, non-Markovian formulation: there are k independent "arms", only one of which may be pulled at any given time, while the others remain "frozen". By pulling a particular arm one receives a certain random reward, depending on time and on the history of the arms that have been pulled. The objective here is to develop an optimal strategy in which the arms are pulled at the right time so as to maximize the total expected discounted awards over an infinite time horizon.

Questions of this sort are so difficult that they were once recognized as unsolvable. In the 1970s, Gittins and his collaborators made pioneering contributions that amounted to a real breakthrough. Gittins demonstrated that it is possible to assign to each project an "index function" of its state, computable (in principle) in terms of that project's dynamics only and such that the optimal policy takes the following form: at any time, compute the indices of different projects and engage the project with maximal index.

Another type of sequential decision in the statistical literature tries to find the solution to some stochastic equations, e.g., the search for the correct dose level for a required efficacy in Robbins-Monro procedure and Kiefer-Wolfowitz procedure. Both procedures try to optimize allocation of experimental resources [7].

In this work, we consider the following sequential target shooting problem. Suppose a shooter wishes to destroy k targets with N missiles. After each shot, the target aimed at is either completely destroyed or intact. All targets are initially intact, but a target once destroyed remains destroyed. The probability, p, to destroy a target by each shot is known. Right after each shot, a report is generated to inform the shooter of the state of the target just fired at and the shooter has this information before using the next missile. The report may be in error, however. There are two usual types of errors and their probabilities are known:

$$\alpha = \text{Pr}\{\text{An intact target is reported as destroyed}\}$$
$$\beta = \text{Pr}\{\text{A destroyed target is reported as intact}\}.$$

All shots and reports are stochastically independent. The goal is to find the optimal strategy that minimizes the expected number of intact targets when all the N missiles have been fired.

Suppose t missiles have been fired and t reports have been collected. We can then calculate the posterior probability for the state of each target. A reasonable strategy is to fire the next missile at the target that has the highest probability of being intact. We call this strategy the best myopic strategy (BMS). We have shown that this strategy is indeed optimal if the targets are homegeneous [8]. The BMS strategy is also optimal for allocating the missiles under several other criteria.

If we have more than one imperfect report systems to choose from, we need to choose one so that the objective of the problem is best served. Techniques for comparing imperfect feedback systems have been developed in [9].

In our problem, the rewards are not identically distributed random variables for each target when an additional shot is fired. Thus, our problem differs from the multi-armed bandit problems. Dirickx and Jennergren [10] stated a general necessary and sufficient condition for a myopic optimal decision rule to be globally optimal. Unfortunately, their condition requires the final expected values at each stage, which are problem specific and usually difficult to calculate. Many special cases such as [11, 12] and the current problem have to find their own rules based on the special structures of the problem.

This chapter is organized as follows. In Section 2 we introduce the algorithm for calculating posterior intact probabilities. Section 3 deals with best myopic strategy. In Section 4 results for comparing two imperfect feedback systems are provided. We conclude the chapter with a discussion in Section 5.

2.   Posterior Intact Probability

N missiles will be fired sequentially in an attempt to destroy as many targets as possible from k targets. The probability of destroying a target at each shot, p, is known. After each shot, there is a report on the state of the target; destroyed or intact. The reports are subject to the usual two types of errors and the probabilities of making these errors, α and β, are also known.

Let $\bar{\alpha} = 1 - \alpha$, $\bar{\beta} = 1 - \beta$, $q = 1 - p$, and let state 0 and 1 denote destroyed and intact, respectively. We consider time and the number of missiles used to be synonyms, i.e., at time t means that we have already fired the tth missile and got the tth report back, $t = 1, 2, ..., N$. The next shot is the missile $(t + 1)$. Since there are two actions at each time step, we may use "just after (before) shooting" or "just after (before) reporting" to clarify some particular situations. If the tth missile is fired at target j, then we let $Y_j(t)$ be the report and $Z_j(t)$ be the true state of target j from the consequence of the shooting and define

$$\pi_j^{(i)}(t) = \Pr\{Y_j(t) = i\}, i = 0, 1;$$

$$u_j^{(i)}(t) = \Pr\{Z_j(t) = 1 | Y_j(t) = i\}, i = 0, 1;$$

$$u_j(t) = \begin{cases} u_j^{(0)}(t) & \text{if } Y_j(t) = 0, \\ u_j^{(1)}(t) & \text{if } Y_j(t) = 1. \end{cases} \tag{1}$$

And for $j' \neq j$, we have $u_{j'}(t) = u_{j'}(t - 1)$. Initially, $u_j(0) = 1$ for all j. For example, if the first shot is at target j, then

$$\pi_j^{(0)}(1) = q\alpha + p\bar{\beta}, \quad \pi_j^{(1)}(1) = q\bar{\alpha} + p\beta;$$

$$u_j^{(0)}(1) = \frac{q\alpha}{q\alpha + p\bar{\beta}}, \quad u_j^{(1)}(1) = \frac{q\bar{\alpha}}{q\bar{\alpha} + p\beta},$$

and $u_{j'}(1) = 1$ for all $j' \neq j$.

Note that $u_j(t)$ is the posterior probability of target j being intact after the tth report and it is also the prior probability of target j being intact just before time $t + 1$. Let the vector $u(t) = (u_1(t), u_2(t), ..., u_k(t))$ with the initial conditions $u_j(0) = 1$ for all $j = 1, ..., k$. Based on Bayes theorem and the fact that given all the previous reports, $u_j(t)$ is a Markov chain, we showed the following properties in [8]:

Theorem 2.1. *Assume that target j has intact probability $u_j(t - 1)$ just before time t, based on the past history of shots and reports. If it is fired at time $t \geq 1$, then we have*

$$\pi_j^{(1)}(t) = u_j(t - 1)q\bar{\alpha} + (1 - u_j(t - 1)q)\beta,$$
$$\pi_j^{(0)}(t) = u_j(t - 1)q\alpha + (1 - u_j(t - 1)q)\bar{\beta}, \tag{2}$$

before the report is received, and

$$u_j^{(1)}(t) = \frac{u_j(t-1)q\bar{\alpha}}{u_j(t-1)q\bar{\alpha}+(1-u_j(t-1)q)\beta},$$
$$u_j^{(0)}(t) = \frac{u_j(t-1)q\alpha}{u_j(t-1)q\alpha+(1-u_j(t-1)q)\bar{\beta}}, \tag{3}$$

after the report is received.

The initial condition $u_j(0) = 1$ for all j can be changed to any initial prior $u_j(0)$. If the targets' initial conditions are known with certainty, then $u_j(0)$ is either 1 or 0. The properties still hold. Based on the above results we can derive the following recursive algorithm for calculating the posterior probabilities.

Corollary 1. *Supposed target j has been fired at ℓ times and reports $y_1, y_2, ..., y_\ell$ are received in the order they are generated. Then: $Pr\{Z_j(t) = 1|y_1, y_2, ..., y_\ell\} = [1 + \xi(\ell)]^{-1}$, where $\xi(\ell)$ can be computed recursively by*

$$\xi(\ell) = [p + \xi(\ell-1)]\frac{1}{q}(\frac{\beta}{1-\alpha})^{y_\ell}(\frac{1-\beta}{\alpha})^{1-y_\ell}, \text{ with } \xi(0) = 0.$$

Note that the order of the y's is important. The intact probability cannot be determined by the number of 0s and 1s in $y_1, y_2, ..., y_\ell$ alone.

3.   Best Myopic Strategy

3.1.   Optimality of the Best Myopic Strategy

We define **Best Myopic Strategy (BMS)** as the one where each missile fires at the target with the greatest expected immediate gain, e.g. at each time t allocate the missile as if it is the last one left. When all the missiles have same hitting probability and the targets are homogeneous, it is easy to see that the BMS is the one where each missile fires at the target with the highest intact prior probability.

The following result was shown in [8].

Theorem 3.1. *If $\alpha + \beta < 1$, then the BMS is optimal for minimizing the expected number of intact targets when all the bullets are fired.*

This result can be proven by backward induction. The idea is to first show that the last missile should be fired at the target with the highest prior before $t = N$, regardless of the strategy used previously. Then just before time $t = N - 1$, we need only to compare two strategies: BMS and another 'optimal' strategy that fires at a different target. However, the non-BMS 'optimal' strategy has to use also the BMS to allocate the last shot already shown to be optimal. In general, we need to compare the two strategies that differ only at the immediate next shooting, because both of them have to follow the BMS afterwards by induction.

Since we need only to compare the next shooting, we may assume without loss of generality that the target with the largest prior is target 1, and the target to be fired at by the other 'optimal' strategy (S2) is target 2. In other words, at time t, the BMS fires the missile $(t + 1)$ at target 1, and the S2 fires at target 2. Note that both of them will follow BMS afterwards according to backward induction. To reverse the time frame for backward induction, we let $n = N - t$ be the number of missiles left at time t. We define $\mathcal{E}_B^{(n)}(u(t))$ be the expected number of intact targets at the end of the shooting, if BMS is followed for the missile $(t + 1)$ and afterwards. Similarly, we let $\mathcal{E}_2^{(n)}(u(t))$ be the expected numbers when S2 are followed. Our proof of the optimal property essentially relies on quantity

$$\mathcal{M}^{(n)}(u(t)) = \frac{1}{u_1(t)u_2(t)} \left[\mathcal{E}_2^{(n)}(u(t)) - \mathcal{E}_B^{(n)}(u(t)) \right],$$

and the following fact:
Suppose

$$u_1(t) \geq u_1'(t) \geq u_2'(t) \geq max(u_2(t), u_3(t), \cdots, u_k(t)) \text{ and } \alpha + \beta < 1,$$

then

$$\mathcal{M}^{(n)}(u_1(t), u_2(t), u_3(t), \cdots, u_k(t)) \geq \mathcal{M}^{(n)}(u_1'(t), u_2'(t), u_3(t), \cdots, u_k(t)).$$

The assumption that $\alpha + \beta < 1$ can actually be relaxed. Because if $\alpha + \beta > 1$, we can reverse the reported states and let $\bar{\alpha}$ and $\bar{\beta}$ be the report errors. After this reversing, Theorem 3.1 holds. If $\alpha + \beta = 1$, e.g. there is no feedback information, Theorem 3.1 is still true. Its BMS is to allocate the missiles as evenly as possible so that the difference in shots between any two targets is at most 1.

We are able to show that the BMS is also optimal in two other important situations. One situation is to change the goal from destroying the maximum expected number of targets to maximizing the probability of destroying all the targets. In many situations, even one intact target can cause severe consequences. Thus, the concern is not the number of targets left, but whether all targets are destroyed. This leads to the criterion of probability maximization. BMS is still the optimal solution in this case. The other situation is when the targets are weighted by their importance, i.e., we wish to minimize $G = E\left(\sum_{j=1}^{k} w_j Z_j(N)\right)$ for a set of positive weights $(w_1, w_2, \cdots, w_k)$. In this case the BMS is still optimal. However, for each missile, the BMS fires at the target with the highest $w_j u_j(t)$ instead of the highest intact prior probability.

3.2. Trade-Off Study

Also of interest are the trade-offs between report accuracy, hitting chance and number of missiles. For example, should a limited budget goes to improvement of hitting probability by manufacturing better missiles or weapons, or to improvement of the report accuracy instead? Another example would be to spend money to purchase more missiles or to improve hitting probability?

Our next three figures illustrate the improvement of sequential allocation strategy comparing to the fixed strategy, and the possible trade-off. Figure 15.1 plotted the probability of destroying all targets vs number of targets k. $p=68.2\%$ is chosen so that there is 95% chance of destroying all 5 targets when $N = 20$. It clearly shows that sequential strategy enhanced a lot over the fixed strategy. For example, for the following three strategies: fixed strategy, BMS with $\alpha = 0.10$ and $\beta = 0.15$, and BMS with perfect feedback, the probabilities of destroying all 10 targets with 20 missiles are 34%, 69%, and 97% , respectively. From another point of view, BMS can destroy all 7 (or 10) targets with more than 95% probability when we have feedback system with $\alpha = 0.10$ and $\beta = 0.15$ (or perfect feedback).

Figure 15.2 plots the probability of destroying all targets vs individual hitting probability p. The dotted lines are for the sequential strategy while the solid lines are for fixed strategy. At $p = 0.4$, the probabilities of destroying all 5 targets are 0.01 and 0.01 for $N = 5$, 0.11 and 0.28 for $N = 10$, 0.50 and 0.86 for $N = 20$. Also it shows that when $N = 10$, the imperfect feedback is approximately equivalent to 0.10 improvement in individual hitting probability, that is, the probability of destroying all 5 targets using 10 missiles with feedback $\alpha = 0.10$ and $\beta = 0.15$ and hitting probability p is the same as when there is no feedback and hitting probability $p + 0.1$.

Figure 15.3 plots the probability of destroying all targets vs feedback report errors for different hitting probability p. It shows that the accuracy of the report have a big impact on the power of destroying all targets. Also from this graph, it is observed that the imperfect feedback system has the biggest effect for intermediate hitting probability.

One of the future work is to investigate the trade-offs among report accuracy, hitting chance and number of missiles for the general setup, in which the probabilities of destroying different targets depend on the types of missiles and the time of shooting.

4. Comparison of Two Imperfect Feedback Systems

When designing target shooting systems, we have to consider the trade-off between the two types of report errors. In other words, if we have more than one imperfect report systems to choose from, how can we select one so that

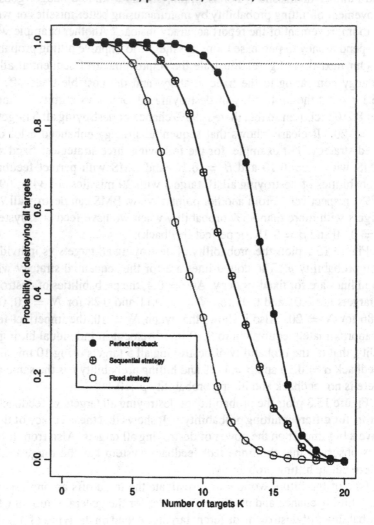

Figure 15.1. Illustration of advantage of sequential strategy vs fixed strategy. For example, for the following three strategies: fixed strategy, BMS with $\alpha = 0.10$ and $\beta = 0.15$, and BMS with perfect feedback, the probabilities of destroying all 10 targets with 20 missiles are 34%, 69%, and 97% , respectively.

K=5, a=10%, b=15%, curves parametrized by N

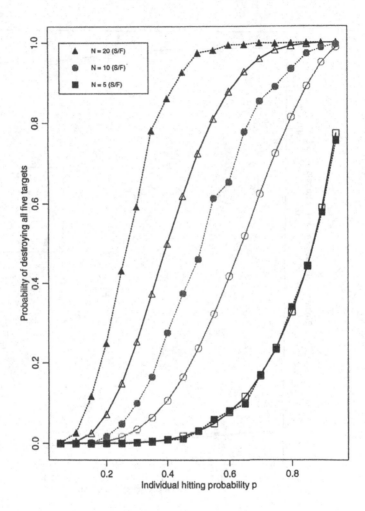

Figure 15.2. Illustration of advantage of sequential strategy at different number of missiles N and hitting probability p. The top two curves compare the sequential strategy with the fixed strategy when $N = 20$. The middle two curves are for $N = 10$. The figure shows that when $N = 10$, the imperfect feedback system $\alpha = 0.10$ and $\beta = 0.15$ is approximately equivalent to 0.10 improvement in individual hitting probability.

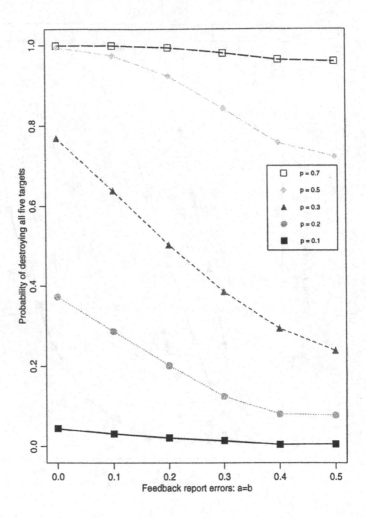

Figure 15.3. Illustration of the impact of accuracies of the report on the power of destroying all targets. Imperfect feedback system has the biggest effect on cases with intermediate hitting probabilities.

the objective of the problem is best served? In this section we present results for comparing imperfect feedback systems. A partial ordering of the systems is provided when the optimal strategy is carried out.

In [9], we demonstrated that for any given report system (α, β) with $\alpha + \beta \leq 1$, we can classify all other imperfect feedback systems (α', β') into four categories:

- *Superior*: At least as good for all (N, k, p), and strictly better for at least one (N, k, p);

- *Inferior*: There is no case such that (α', β') is strictly better;

- *Not deterministic*: (α', β') is better for some (N, k, p) and worse for some other (N, k, p);

- *Unknown*: Not the above.

The results are illustrated in Figure 15.4. Given any feedback system (α, β) with $\alpha + \beta < 1$, we draw two lines through point $G = (\alpha, \beta)$: the first line AD passes $(0, 1)$ and the second line CF connects $(1, 0)$. We can show that any system with report errors (α', β') falls in region $OAGF$ (below the two lines including the boundary) are superior than (α, β). In particular, if $\alpha' \leq \alpha$ and $\beta' \leq \beta$, then the one with the reporting error rates α' and β' has at least as good chance to destroy more targets as one with α and β when the BMS are followed by both. Conversely any system with report errors (α', β') in the region CDG (above the two lines including the boundary) are inferior. Furthermore if we draw a line BE through G with slope -1, then points in the regions ABG and GDE are not deterministic. For the remaining regions, BCG and GEF, we conjecture that it is also not deterministic. But it is still unknown except that boundary BC is not deterministic when $\beta > 0$. Our simulation study found that certain points in the other two regions are also not deterministic.

4.1.   Deterministic Cases

As stated above, given any report system (α, β) with $\alpha + \beta < 1$, we may define two deterministic regions $OAGF$ and CDG such that system in $OAGF$ is better than (α, β) for any (N, k, p), while the other is always worse than (α, β). Notice that the two lines AD and CF defined above can be expressed as $\alpha'/(1 - \beta') = \alpha/(1 - \beta)$, and $\beta'/(1 - \alpha') = \beta/(1 - \alpha)$. Therefore a point (α', β') below the two lines satisfies conditions:

$$(1 - \beta')(\alpha' - \alpha) + \alpha'(\beta' - \beta) \leq 0; \quad (1 - \alpha')(\beta' - \beta) + \beta'(\alpha' - \alpha) \leq 0.$$
$$(4)$$

These results are summarized in the following.

Theorem 4.1. *A feedback system with the reporting error rates (α', β') is at least as good as one with (α, β) under the two conditions in (4). Furthermore,*

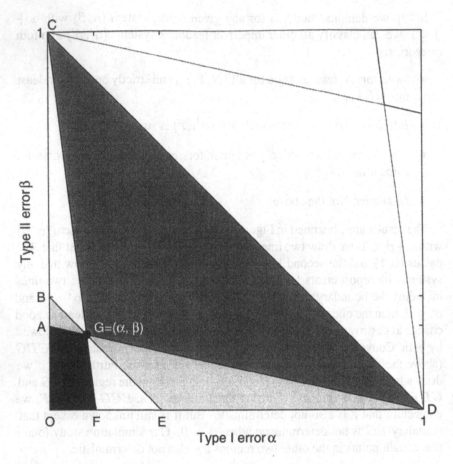

Figure 15.4. Illustration of four categories of report systems comparing to $G = (\alpha, \beta)$. A system with report errors (α', β') falls in region $OAGF$ (including the boundary) are superior than (α, β), while the systems in region CDG (also including the boundary) are inferior. Furthermore points in the regions ABG and GDE are not deterministic. It is also known that boundary BC is not deterministic when $\beta > 0$.

if at least one of the conditions holds with strict inequality, then report system (α', β') *is superior.*

Let R denote the report under feedback system (α', β'), we define the new report $\tilde{R}$ as following:

$$\tilde{R} = \begin{cases} 1 \text{ with probability } \theta; 0 \text{ with probability } 1 - \theta; & \text{if } R = 1; \\ 1 \text{ with probability } 1 - \phi; 0 \text{ with probability } \phi; & \text{if } R = 0, \end{cases} \quad (5)$$

where

$$\theta = \left[(1 - \beta')(1 - \alpha) - \alpha'\beta \right] / (1 - \alpha' - \beta'),$$

$$\phi = \left[(1 - \alpha')(1 - \beta) - \beta'\alpha \right] / (1 - \alpha' - \beta').$$

Under the conditions of Theorem 4.1, the auxiliary report system $\tilde{R}$ has errors (α, β). Therefore, there is a (non-optimal) strategy based on system (α', β') that is at least as good as the optimal strategy using report system with errors (α, β) for any case (N, k, p).

Furthermore, using the fact that the expected number of intact target is $(q^2 + q) + p^2 q(\alpha + \beta - 1)$ for a target shooting problem with $N = 3, k = 2$, we know that there are cases such that (α', β') in region $OAGF$ is strictly better than (α, β) when $a' + \beta' < \alpha + \beta$. This completes the proof of the superiority of points in the $OAGF$ area. Inferiority of points in area GCD can be derived similarly.

A simple geometric proof of the results can be derived based on the boundary points. For any point X on line GC, $X = wC + (1 - w)G$ for some $0 \le w \le 1$, so the report system X can be obtained using (α, β) along with $(0, 1)$, thus X is inferior. Same is true for points on line GD. Similarly for a point X on line AG, $G = wX + (1 - w)D$ for some $0 \le w \le 1$, so the report system $G = (\alpha, \beta)$ can be obtained using X along with $(1, 0)$ which implies that X is superior than G. These boundary results imply that the inferiority (or superiority) of interior points.

4.2.   Non-Deterministic Cases

The comparison results for points in the regions ABG and GDE can be restated as the following.

Theorem 4.2. *A feedback system with the reporting error rates (α', β') is not deterministic comparing to (α, β) if it falls between two lines given by*

$$a' + \beta' = \alpha + \beta; \qquad \beta'/(1 - \alpha') = \beta/(1 - \alpha). \tag{6}$$

Recall that being not deterministic means that it could be better than (α, β) for some (N, k, p) and worse for some other (N, k, p). For (α', β') in the left region ABG, we have $a' + \beta' < \alpha + \beta$ so it is better than (α, β) for cases when $N = 3, k = 2$. On the other hand, we can show that for cases $(N = k+1, k, p)$ with large k, (α', β') in the left half region is worse than (α, β). This is because that for a report system with errors (α, β), it can be shown that the expected number of intact target after $N = k + 1$ shots are

$$E_{k+1} = -pq \left(\bar{\alpha} \frac{1 - \pi_0^k}{1 - \pi_0} + \alpha \pi_0^{k-1} \right) + kq,$$

where $\pi_0 = q\alpha + p\bar{\beta}$. Thus as $k \to \infty$, $(E_{k+1} - kq)/pq$ converges to

$$-\frac{\bar{\alpha}}{1 - \pi_0} = -\frac{\bar{\alpha}}{q\bar{\alpha} + p\beta} = -\frac{1}{q + p\frac{\beta}{1-\alpha}}.$$

Since system (α', β') in the left half of the region satisfies $\frac{\beta'}{1-\alpha'} > \frac{\beta}{1-\alpha}$, it is worse than (α, β) for this case.

A system (α', β') in the right half region is the opposite: it is better than (α, β) for the large k cases while it is worse in the $k = 2$ cases.

4.3.   Remaining Cases

For (α', β') in the remaining two regions, BCG and GEF, we know by simulation that certain points are not deterministic. However, we are not able to prove the result for the whole regions. Nevertheless, for the case when $N = 3$ and $k = 2$ points in region BCG are worse than (α, β), while those in region GEF are better.

In addition, we do know that boundary BC is not deterministic when $\alpha > 0, \beta > 0$. This is established based on the cases when $N = nk$ for large enough p. Let the "fixed" strategy be the one that hits the k targets in order acting like that there is no feedback. If $N = nk$, this will end up hit each target n times. Our next theorem shows that BMS based on (α, β) is only as good as the "fixed" strategy when p is large enough and N equals nk.

Theorem 4.3. *Suppose $N = nk$, $\alpha > 0, \beta > 0$, then BMS based on feedback system (α, β) performs same as the "fixed" strategy when p is large enough.*

Thus for the case when $N = 4$, $k = 2$ and very large p, the BMS has expected number of intact targets $2q^2$ with system (α, β), which is worse than the expectation of $2q^2 - 2p^2q^2(1 - \beta')$ for feedback system $(0, \beta')$.

5.   Discussions

5.1.   Optimal Strategy for General Case

The BMS is optimal when missiles and targets are homogeneous. However, for more realistic cases where the probabilities of hitting different targets depend on the types of missiles and the time of shooting, BMS is no longer optimal. This is true even when there is no feedback information.

A special case is where the N missiles have hitting probabilities $p_1, p_2, \ldots, p_N$, and we do not have any feedback at all. The difference of the hitting probabilities may be due to either the quality of missiles or the time of shooting. Let $w_i = -ln(1 - p_i)$, $i = 1, 2, \ldots, N$ and $A_j = \sum_{i \in I_j} w_i$, where I_j is the index set for the missiles allocated to the jth target, $j = 1, 2, \ldots, k$. Then the expected number of intact targets after all N missiles have been shot are

$$E = \sum_{j=1}^{k} \left(\prod_{i \in I_j} (1 - p_i) \right) = \sum_{j=1}^{k} \exp(-A_j). \qquad (7)$$

Thus an optimal strategy is equivalent to dividing N objects with weights $w_1, w_2, \ldots, w_N$ into k groups such that $\sum_{j=1}^{k} exp(-A_j)$ is minimized, which could be different from BMS.

If there exists a partition of the N objectives into k sets so that

$$A_1 = A_2 = \ldots = A_k. \qquad (8)$$

then it corresponds to the optimal solution to (7). However, to check if such a partition exists or not is NP-Complete even for $k = 2$ [13, 14].

We proposed several algorithms to solve (7) [15]. For large N, we have an $O(N log N)$ algorithm which is asymptotically optimal.

5.2.   More on Comparison Results

First we notice that the procedure for constructing auxiliary reports in section 4.1 does not depend on the setup of our sequential target shooting problem. For all scenarios that have the usual two types of errors, we may use those deterministic results to compare two "systems". For example, if two statistical tests have errors (α, β) and (α', β') respectively, we may compare them use our results. For the same reason, our results might be used in other sequential allocation problems with imperfect feedbacks.

Secondly, we conjecture that all systems in the regions BCG and GEF are not deterministic comparing to G. It can be shown that points in a region near the BC boundary are not deterministic. This region can be analytically specified but for simplicity we omit it here. We also know that the results for the two regions BCG and GEF are equivalent, and that is if BCG is not deterministic for any (α, β) then same thing is true for GEF. Furthermore, we know that a sufficient condition to prove the above conjecture is that there always exist a case such that (α', β') is better than $(0, \alpha)$ whenever $\bar{\beta}'/\alpha' > 1/\alpha$.

References

[1] J. C. Gittins, "Bandit processes and dynamic allocation indices", *J. Roy. Stat. Soc. B.*, 41: 148-164, 1979.

[2] J. C. Gittins, *Multi-armed Bandit Allocation indices*, John Wiley, New York, 1989.

[3] N. E. Karoui and I. Karatzas, "General Gittins index processes in discrete time", *Proc. Natl. Acad. Sci. USA*, 90: 1232-1236, 1993.

[4] A. N. Burnetas and M. N. Kathehakis, "On large deviation properties of sequential allocation problems", *Stochastic Analysis and Applications*, 14: 23-31, 1996.

[5] A. N. Burnetas and M. N. Kathehakis, "Optimal adaptive policies for sequential allocation problem", *Advances in Applied Mathematics*, 17: 122-142, 1996.

[6] D. A. Berry and B. Fristedt, *Bandit Problems*, Chapman and Hall, New York, 1985.

[7] Y. M. Dirickx and L. P. Jennergren, "On the optimality of myopic policy in Sequential decision problems", *Management Science*, 21: 550-556, 1975.

[8] M. Duflo, *Random Iterative Models*, Springer, Berlin, 1990.

[9] C. Y. C. Mo, S. S. Wu, R. Chen, and M. C. K. Yang, "Optimal sequential allocation with imperfect feedback information", *J. Appl. Prob.*, 38: 248-254, 2001.

[10] S. S. Wu, C. X. Ma, and M. C. K. Yang, "On imperfect feedback systems in sequential allocation problem", submitted to *Stat. & Prob. Letters*.

[11] L. Benkherouf and J. A. Bather, "Oil Exploration: sequential decision in the face of uncertainty", *J. Appl. Prob.*, 21: 529-543, 1988.

[12] G. Schroeter, "Distribution of number of point targets killed and higher moments of coverage of area targets", *Naval Research Logistics Quarterly*, 31: 373-385, 1984.

[13] M. R. Garey and D. S. Johnson, *Computers and Intractability*, W. H. Freeman and Company, New York, 1979.

[14] T. Ibaraki and N. Katoh, *Resource Allocation Problems*, MIT Press, 1988.

[15] Z.-J. Shen, S. S. Wu, and Y. Zhu, "K-partitioning problem: applications and solutions algorithms", *Technical Note, Dept. of Industrial and Systems Engineering, University of Florida*, 2001.